Lecture Notes in Statistics

Edited by J. Berger, S. Fienberg, J. Gani, K. Krickeberg, I. Olkin, and B. Singer

61

Jens Breckling

The Analysis of Directional Time Series: Applications to Wind Speed and Direction

Springer-Verlag

Berlin Heidelberg New York London Paris Tokyo Hong Kong

Author

Jens Breckling
Australian Bureau of Agricultural and Resource Economics
GPO Box 1563, Canberra City, ACT 2601, Australia

Mathematical Subject Classification: 62-02, 62-07, 62H20, 62M10, 62H05, 62F10, 60G35, 62P99

ISBN 0-387-97182-3 Springer-Verlag New York Berlin Heidelberg
ISBN 3-540-97182-3 Springer-Verlag Berlin Heidelberg New York

Printing and binding: Druckhaus Beltz, Hemsbach/Bergstr.
2847/3140-543210 – Printed on acid-free paper

ACKNOWLEDGEMENTS

This monograph is based on my PhD thesis written at the University of Western Australia in Perth. My greatest thanks therefore go to Prof. Terry Speed for his extensive contributions during my years as a postgraduate student. In particular, I would like to thank him for introducing the topic and closely supervising my research. Further, I am indebted to Dr Robin Milne and Prof. Tony Pakes whose numerous suggestions helped to substantially improve the presentation of this work.

The wind data were kindly provided by R.K. Steedman & Associates in Subiaco, and permission to use the data for this study is very much appreciated. It is also a pleasure to acknowledge the assistance of the Australian Bureau of Meteorology in Perth. Finally I wish to thank Ann Milligan for her excellent typing of the manuscript, and Sabine Bockholt and Jan Dlugosz for their patience and careful preparation of the diagrams and figures.

SUMMARY

The subject of this study is the analysis of directional time series. Corresponding to empirical and theoretical aspects, the monograph has been divided into two parts, both of which are self-contained. In Part I we present a full and comprehensive analysis of a series of hourly records of wind speed and direction, which was taken at the port of Fremantle in Western Australia. In Part II we take a more general approach and develop the theoretical framework for the analysis of directional time series, using the methods of mathematical statistics.

The objective in Part I is to find regularities in the record of wind speed and direction, to detect general weather patterns, and to describe their seasonal characteristics and interdependencies. It is shown that the complete record of observations can be divided into a number of components which are amenable to physical interpretation. Initially the series is divided into a geostrophic and a daily component so as to separate the influence of the daily circulation and other short-term disturbances from the prevailing wind.

It is shown that the geostrophic component agrees well with the synoptic-scale wind as depicted on weather charts. By classifying the geostrophic component according to specified wind configurations, it is possible to establish a general flow of low and high pressure systems. An investigation of the daily component reveals that the land and sea breeze circulation has a dominant influence on the local weather throughout the year, even though it cannot be observed directly in winter. Furthermore, it is shown how the strength and feature of this circulation depend on the time of the year and on the geostrophic wind.

We then remove the sea breeze effect and study the resultant series in order to detect and describe other short-term events such as calms, storms and oscillating winds. These events are subsequently characterized by their distribution over the year and their time of onset. By relating the events to both the geostrophic wind and the sea breeze circulation, it is found that calms are confined to the winter months and are evidence of a high pressure system extending to the local region, while storms, the way they are defined in this study, are associated with cold fronts approaching from the south-west in winter and a depression over the north-west of the continent in summer. After removing the storms we obtain a residual series of short-term fluctuations, which in general cannot be related to any physical phenomenon.

Part II is motivated by the actual data analysis and focusses on two autoregressive models for directional time series. These models are related to the von Mises and wrapped normal distribution, and will be called von Mises and wrapped autoregressive process, respectively. Fundamental in this context is a concept of angular dependence. It is shown that the von Mises process can be associated with a new measure of angular correlation. When this measure is compared with other measures in the literature, in particular in the context of a wrapped autoregressive process, it clearly demonstrates the best performance. The autocovariance function of a directional process is therefore defined in terms of this measure of association. Given an estimator of this function, various methods of fitting the two models to time series of directional data are developed and compared. When applying these techniques to the directional component of the residual series of short-term fluctuations, it is established that successive wind direction changes are virtually independent.

CONTENTS

PART I : WIND DATA ANALYSIS

PART II : TIME SERIES OF DIRECTIONAL DATA

PART I

WIND DATA ANALYSIS

CHAPTER 1

INTRODUCTION

The objective in Part I of this monograph is to analyse a particular series of wind speeds and directions, which was recorded at the port of Fremantle in Western Australia. In order to detect seasonal characteristics and to establish general weather patterns, it is supposed that the complete record of wind speed and direction can be divided into the following four components:

(i) a prevailing wind, as determined by the configuration of high and low pressure systems;
(ii) a land and sea breeze cycle, which dominates the weather pattern for most of the year, even though it cannot be observed directly in winter;
(iii) a storm component, which is associated with extratropical cyclones approaching from the south-west in winter and a depression over the north-western part of the continent in summer; and
(iv) a residual component of short-term fluctuations.

It is shown that the components, which are constructed from the raw data series, are in fact amenable to the interpretation given above. Each of these components will subsequently be discussed in detail.

A brief description of the data source is given in section 1.1, while the general weather pattern for the Fremantle area is summarised in section 1.2. We will concentrate on those aspects that are relevant to the local wind, and briefly relate them to the given record of wind speed and direction. An outline of the analysis will be provided in section 1.3.

1.1 Surface Wind Observation

Speed and direction of the surface wind at the port of Fremantle (latitude: 32°03′ south; longitude: 115°44′ east) have been recorded on an hourly basis since January 1971. The elevation of the anemometer is $z_0 = 60$ m, but to compare the records with those taken at other stations, wind speeds are adjusted to 10 m average mean sea level (AMSL) according to the power law

$$v = v_0 \, [z/z_0]^{1/7} \quad \text{with} \quad z = 10 \text{ m} ,$$

where v is the adjusted speed and v_0 the recorded speed at 60 m AMSL. The wind direction, on the other hand, is assumed to be constant at all elevations between sea level and the anemometer at 60 m. This time series of hourly records of wind speed and direction will be denoted by $(\mathbf{w}_t)$, with the index t referring to time, and will provide the basis of the analysis.

An investigation of the wind field across the area by Steedman & Craig (1979) led to the following conclusions:

(i) The horizontal wind field is reasonably uniform under moderate conditions, that is when wind speed is between 6 and 30 knots.
(ii) In the case of light winds with speeds below 6 knots the directional records are not necessarily representative of the local wind field.
(iii) The passage of a dissipating tropical cyclone is reflected in a systematic change of wind direction, whilst the change in wind speed is not nearly as pronounced.
(iv) Large variations in wind speed are usually associated with cold fronts passing through the area. As a result of substantial veering, the data may not be representative during these periods.

The aim of this study is to detect seasonal characteristics in the time series $(\mathbf{w}_t)$ and to determine general weather patterns. Questions concerning the interaction of atmospheric forces are usually studied in the framework of dynamic meteorology. The primary goal in this field is to interpret the observed structure of atmospheric circulation systems in terms of physical equations. On a short-term basis the laws of momentum, mass and energy conservation do indeed describe the large-scale atmospheric disturbances to a good approximation. However, questions concerning seasonal occurrence and the interaction of general weather patterns have an implicit long-term horizon. Hence, in order to characterize types of atmospheric circulation systems in terms of wind speed and direction, and to study their frequencies and interdependencies, the methods of mathematical statistics are required.

The wind series $(\mathbf{w}_t)$ will be broken up into seasonal components as follows:

Winter from 1 June to 31 August,
Spring from 1 September to 30 November,
Summer from 1 December to 28/29 February, and
Autumn from 1 March to 31 May.

To illustrate the principles and results of the analysis we will generally refer to the winter season of 1971 and to the summer season of 1971/72. The outcome has been compared with other years and found to provide a representative description of the seasonal patterns.

1.2 General Weather Pattern

The wind regime during the year is largely controlled by the north and south movement of the anticyclonic belt. From late April to early October it extends in an east-west direction right across the Australian continent, when westerly winds along its southern edge produce cool cloudy weather and rain over the Fremantle area. For the remainder of the year the anticyclonic belt lies just south of the continent, giving rise to hot and dry weather conditions as a result of a predominantly easterly air-stream (Bureau of Meteorology 1973).

From June to August the northern fringe of the *roaring forties* extends to the southern parts of the continent giving rise to frequent westerly gales along coastal districts. These winds are maintained by a series of low pressure systems moving in an easterly direction south of the continent. The extent to which they affect the Fremantle area depends largely on the intensity and location of these systems. In particular, if the depressions move in a south-eastward direction the region may come under the influence of the high pressure system situated over central or southern Australia. In this case moderate easterlies dominate the weather pattern for an extended period of time.

To illustrate this pattern let us refer to the period from 20 to 27 June 1971. Synoptic charts showing the atmospheric pressure distribution during this period are presented in Figure 1.1. Wind speed is depicted in Figure 1.2 (b), while Figure 1.2 (a) shows hourly records of wind direction with 0°, 90°, 180° and 270° corresponding to northerly, easterly, southerly and westerly winds, respectively. Of course, 360° refers again to a northerly direction. The synoptic charts in Figure 1.1 demonstrate that the inland division of the Australian continent is under the influence of a large high pressure system. Note, however, that initially it does not extend to the lower south-west and that this area is under the influence of a depression situated south of the Bight. A cold front passing through the region on 20 June is reflected in relatively high wind speeds for that day as can be seen in Figure 1.2 (b). Once the depression has moved further east, the lower south-west comes under the influence of a high pressure system centered over the western part of the continent. During this period moderate easterlies prevail as evident in the record of wind speed and direction in Figure 1.2. On 25 June the high starts to move slowly eastwards, and a situation similar to that depicted on the first weather chart begins to emerge.

Towards the end of winter the anticyclonic belt begins to move southward and a heat low over the inland division begins to develop. As summer approaches the westerlies are confined more and more to the lower southwest and south coastal areas. By November, December, easterlies prevail over most of the state as the anticyclonic belt has migrated so far south that its axis lies off the south coast of the continent. At the same time, the heat low over the tropical region is fully developed producing

Figure 1.1. Synoptic charts for 20 to 27 June 1971 (0700 h). Source: Bureau of Meteorology.

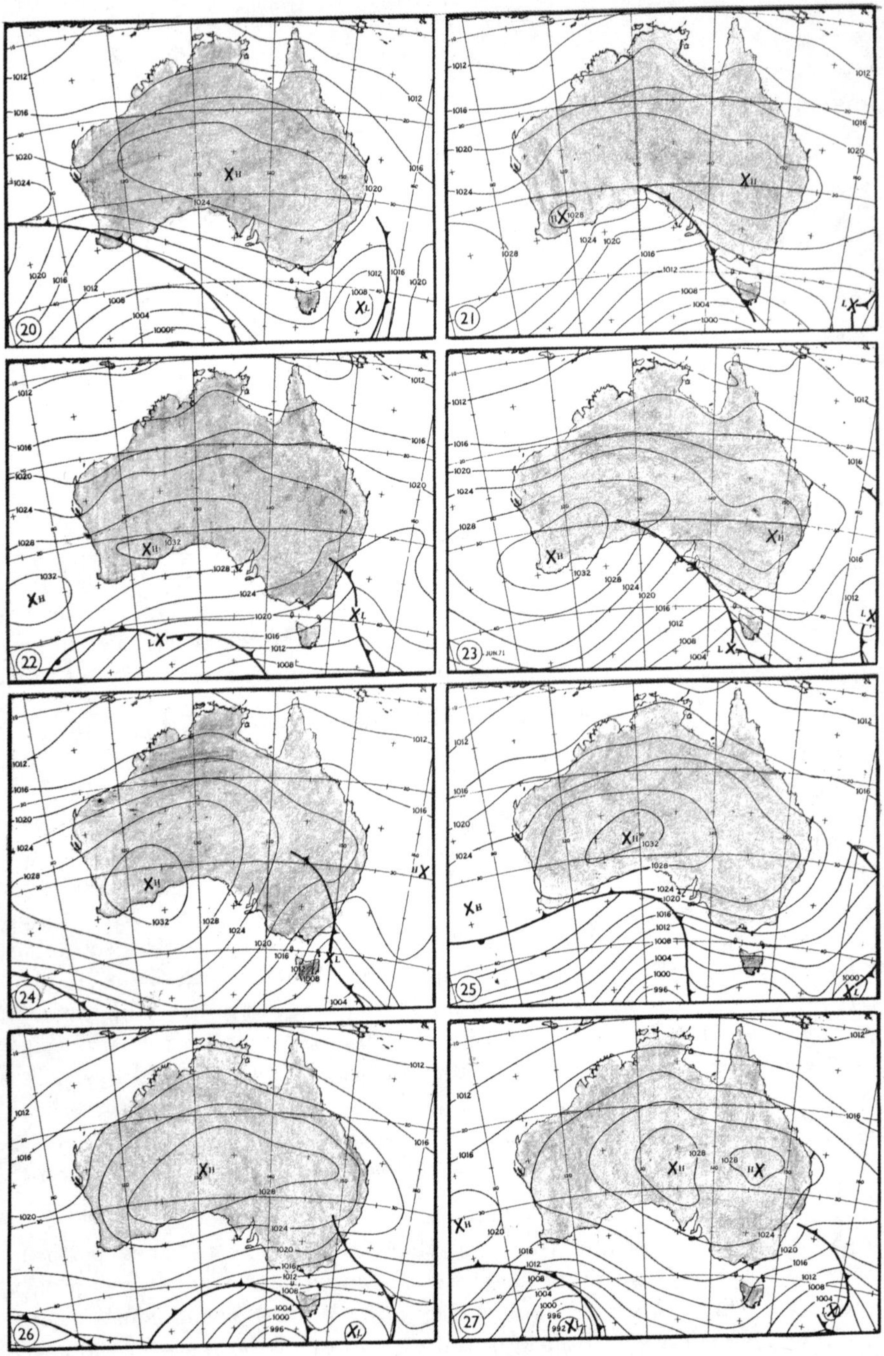

Figure 1.2 (a). Wind direction in $(\mathbf{w}_t)$ *from 20 to 27 June 1971.*

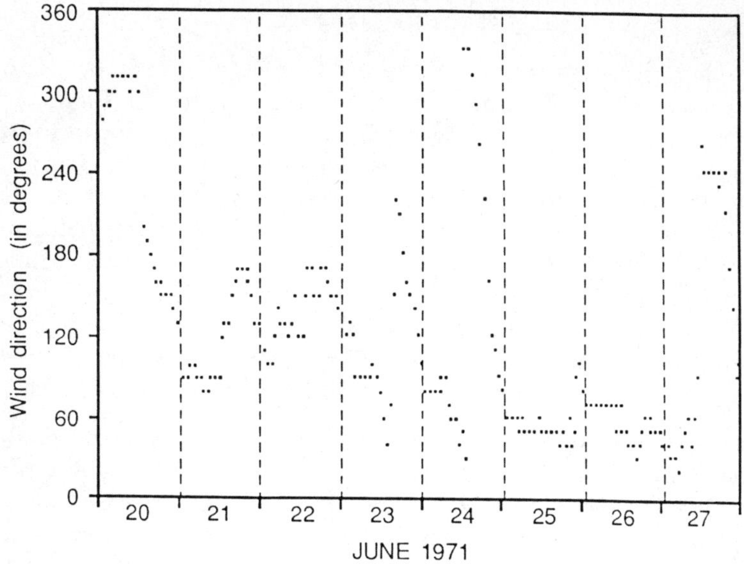

Figure 1.2 (b). Wind speed in $(\mathbf{w}_t)$ *from 20 to 27 June 1971.*

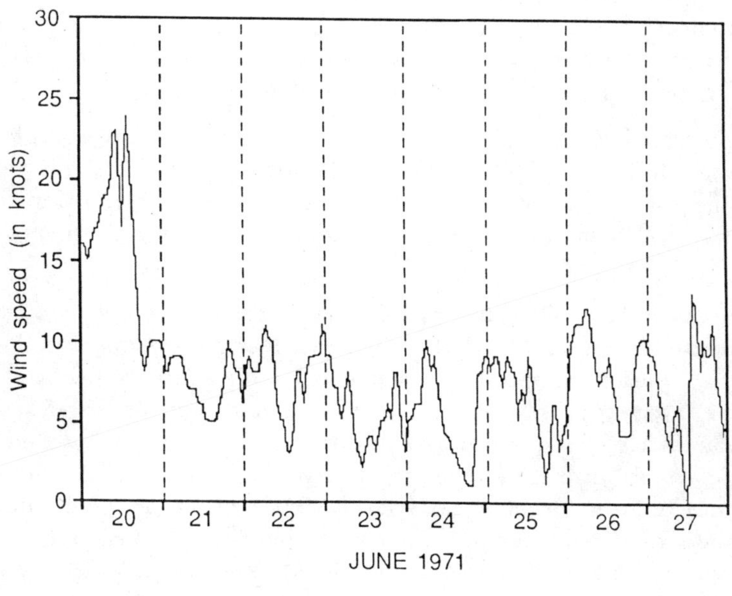

easterlies along its southern edge and thus amplifying the effect of the anticyclonic belt. The winds are further influenced by a persistent high pressure system over the Indian Ocean which contributes mainly to the southerly sector.

To exemplify the decomposition of the wind series $(\mathbf{w}_t)$ we will refer to the record of 1 to 8 December 1971. From the synoptic charts presented in Figure 1.3 it can be seen that for the whole period a low depression is centered over the north-west division of the continent. Further note that between 1 and 6 December a high pressure ridge extends from the eastern Indian Ocean right across the Bight into Victoria. Even though two weak fronts pass through the area on 1 and 5 December the region remains under the influence of this high pressure system. On 7 December a trough begins to develop from the heat low over the north-west in a southerly direction and to break the high pressure ridge into a western cell over the Indian Ocean and an eastern cell centered over Tasmania. As a result the south-west division of the continent comes under the influence of a depression with westerlies extending to the lower south-west.

The high temperatures during daytime generate a daily land and sea breeze cycle which dominates the summer pattern and often extends well into autumn and spring. Its distinctive feature is also reflected in the wind data of 1 to 8 December presented in Figure 1.4. Except for the 8 December the records exhibit a clear pattern. In the morning the wind blows from a south-easterly direction, then swings around to a south-westerly direction in the afternoon and gradually moves back to the southerly sector. A similar cycle can be detected in the record of wind speed plotted in Figure 1.4 (b), with wind strengthening in the afternoon and weakening in the morning. The fact that this pattern is not observed on 8 December is related to the cold front passing through the region that day and preventing the development of a sea breeze. Also note that southerlies dominate in the record of wind direction shown in Figure 1.4 (a), although the synoptic charts indicate that an easterly air flow prevails.

The wind records presented in Figure 1.2 do not suggest that a similar daily cycle exists in winter. It is generally believed that the day temperatures are too low during this time of the year to generate a distinct land and sea breeze cycle. However, it will be shown that this cycle can also be detected in winter even though it cannot be observed directly.

The general patterns described in this section are also reflected in the 'bivariate' histograms plotted in Figure 1.5. Each ring and each angular segment in these diagrams correspond to a particular speed and a particular direction, respectively. Each cell defined as the intersection of such a ring and angular segment therefore corresponds to a particular pair of wind speed and direction, with different hatchings indicating the number of recordings. Note that the marginal distribution of wind direction in winter is almost uniform although easterlies occur slightly more often. In summer, on the other hand, the wind is confined almost completely to the easterly and southerly sectors. However, during both seasons south-westerlies seem to be associated with stronger winds.

Figure 1.6 shows the frequency distributions of wind speed and direction for each day of the year using hourly records from January 1971 to December 1977. The speed is divided into intervals of 2 m/s $\approx$ 3.9 knots while the direction is divided into northerly,

Figure 1.3. Synoptic charts for 1 to 8 December 1971 (0700 h).
Source: Bureau of Meteorology.

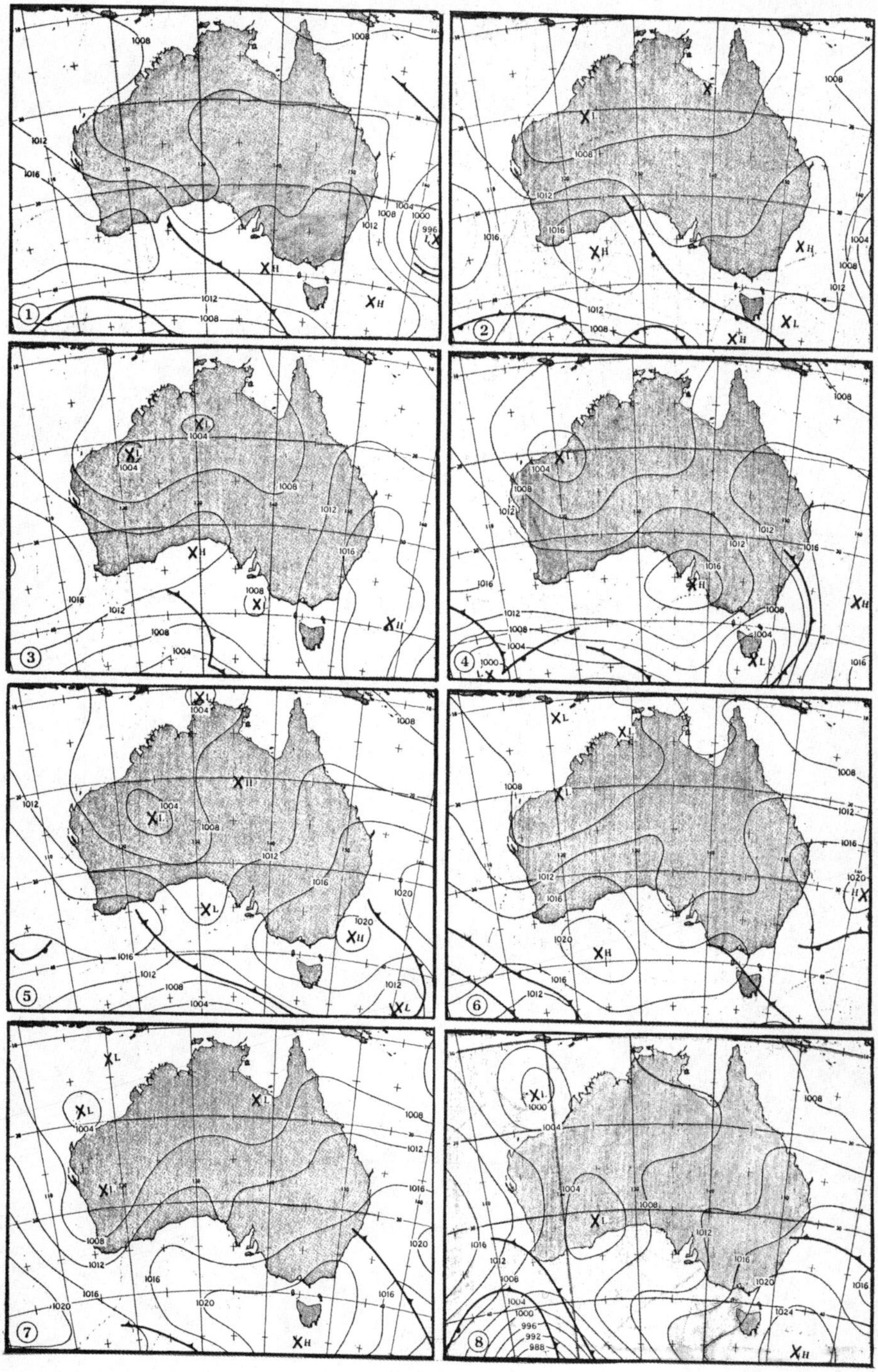

Figure 1.4 (a). Wind direction in $(\mathbf{w}_t)$ *from 1 to 8 December 1971.*

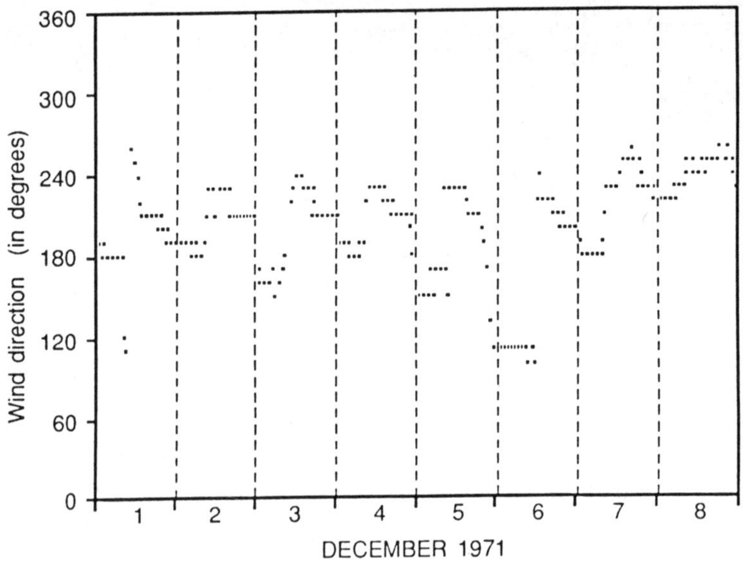

Figure 1.4 (b). Wind speed in $(\mathbf{w}_t)$ *from 1 to 8 December 1971.*

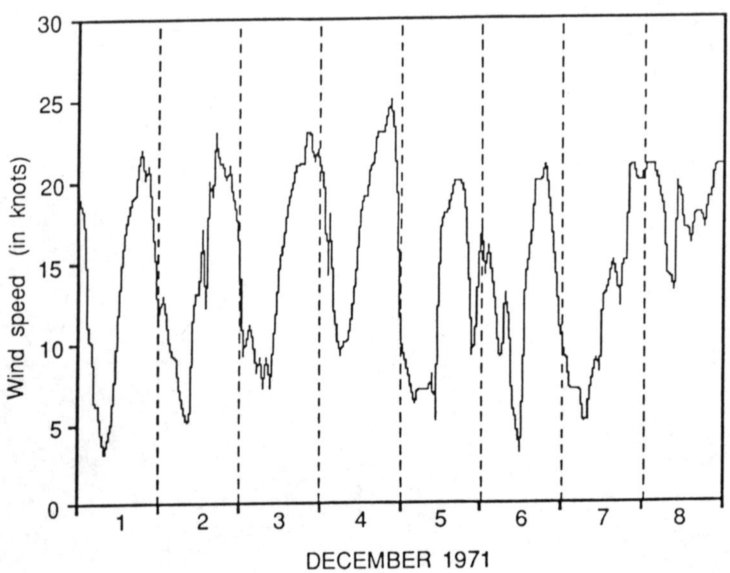

Figure 1.5. Frequency of wind speed and direction in $(\mathbf{w}_t)$.

(a) Winter 1971.

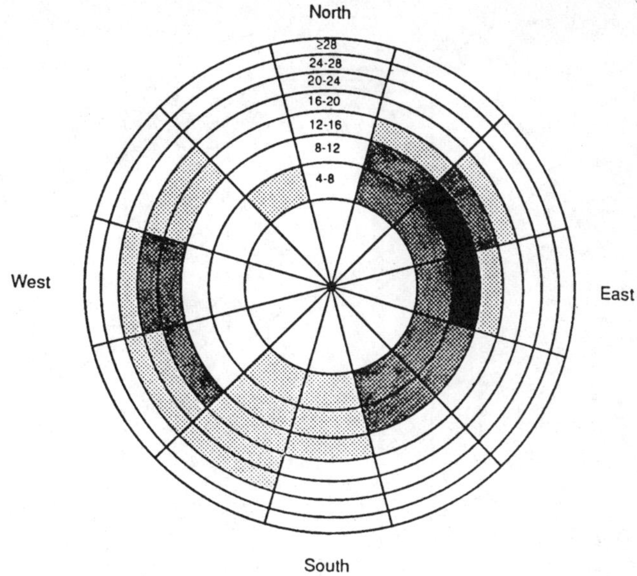

(b) Summer 1971/72.

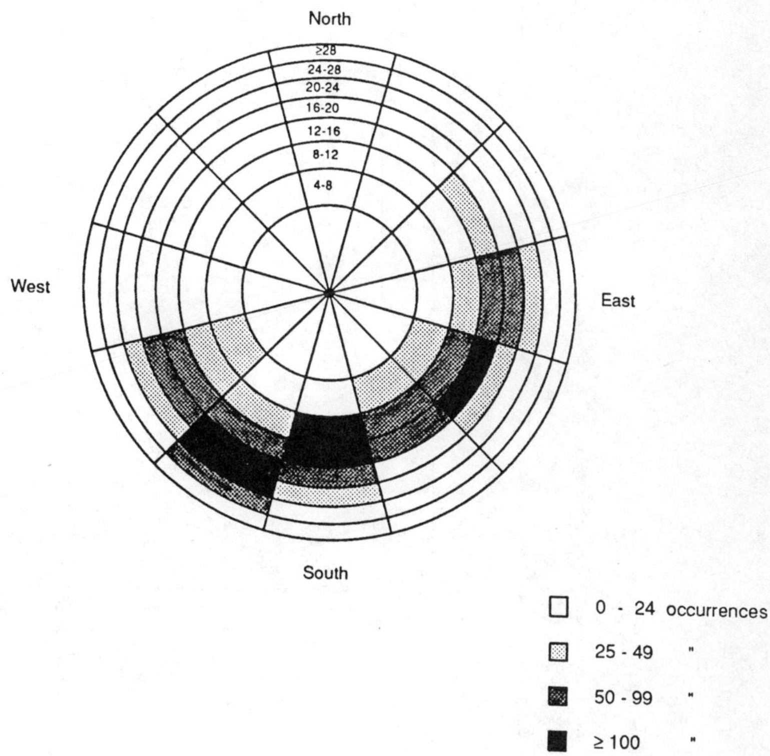

Figure 1.6. Distribution of wind speed and direction for each day of the year (based on hourly records from 1971 to 1977). Source: Steedman & Craig (1979).

(a) Wind direction.

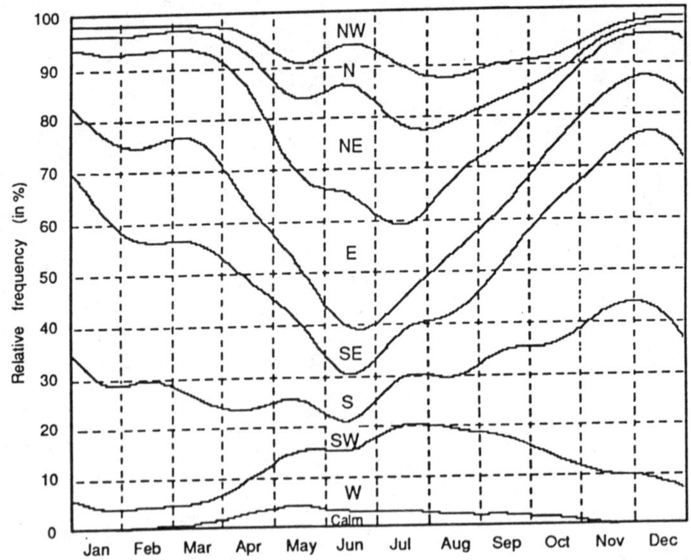

(b) Wind Speed (1 m/s ≈ 1.94 knots).

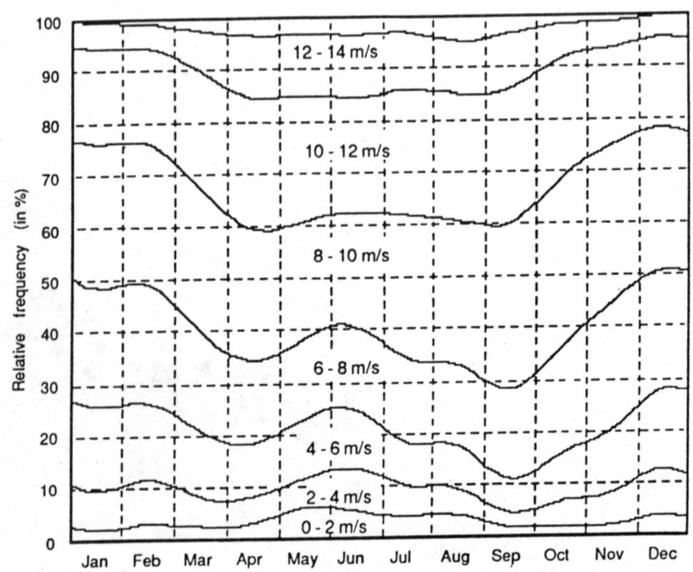

north-easterly, easterly *etc.* sectors. Throughout the year winds with a speed between 12 and 24 knots tend to dominate. However, whereas there is a significant proportion of winds with speeds higher than 24 knots in winter there are only few corresponding records in summer. Moderate winds with speeds below 12 knots, on the other hand, occur more often in summer than in winter. From the distribution of wind direction in Figure 1.6 (a) one can see that in summer easterlies to south-westerlies occur almost exclusively, whereas in winter winds from the westerly and north-easterly sectors are also important.

1.3 Outline of this Monograph

The subject in this study is the analysis of directional time series. Corresponding to an empirical and a theoretical aspect the monograph has been divided into two parts. Part I contains a complete analysis of the wind series $(\mathbf{w}_t)$ while the theoretical background necessary for the analysis of directional time series is presented in Part II. Each part is self-contained and more comprehensive than is required for the other part. Part I provides the motivation for the theoretical development in Part II, and at the same time, serves to illustrate the approach presented in that part.

The objective in Part I is to detect seasonal characteristics in the record of wind speed and direction, and to establish general weather patterns. First the prevailing wind as determined by the configuration of low and high pressure systems is analysed, then an empirical land and sea breeze cycle is defined and finally the characteristics of certain short-term events such as calms, storms and oscillating winds are investigated. Accordingly, it is supposed that the wind series $(\mathbf{w}_t)$ can be regarded as a superposition of the following components:

(i) a geostrophic component $(\mathbf{g}_t)$,
(ii) a land and sea breeze cycle $(\mathbf{b}_t)$,
(iii) a storm component $(\mathbf{e}_t)$, and
(iv) a residual component $(\mathbf{f}_t)$ of short-term fluctuations.

The decomposition into these components is the subject of Part I and is best explained by reference to the diagram in Figure 1.7.

In chapter 2 it is first established that there is a pronounced 24-hour pattern in $(\mathbf{w}_t)$. The series is then divided into a geostrophic component $(\mathbf{g}_t)$ and a daily component $(\mathbf{d}_t)$ so as to separate the influence of the daily circulation from the prevailing wind. In order to reduce the effect of gusty winds and other short-term disturbances, commonly used robust techniques are adapted to these data by introducing a 2-stage estimation procedure and extending the concept of M-estimation to the multivariate situation. In particular, spatial versions of a median and Hampel's 3-part descending estimator are introduced and are compared numerically, as moving estimators of location, with other filters in a number of situations.

In chapter 3 a simple model is introduced to predict the geostrophic wind from the atmospheric pressure field and to establish the extent of association between this series, which describes a meteorological phenomenon and is denoted by $(\mathbf{h}_t)$ in Figure 1.7, and the geostrophic component $(\mathbf{g}_t)$ constructed from $(\mathbf{w}_t)$. Although the atmospheric data, represented by the series $(\mathbf{p}_t)$, consist only of the location and central pressure of all pressure systems relevant to the weather in Fremantle, the series $(\mathbf{h}_t)$ of estimated geostrophic winds agrees well with $(\mathbf{g}_t)$. It can therefore be assumed that the geostrophic component $(\mathbf{g}_t)$ captures the main characteristics of the geostrophic wind and that the decomposition is physically meaningful. Finally, the series $(\mathbf{g}_t)$ is partitioned into sequences of synoptic patterns, which are defined in terms of wind speed and direction, and roughly reflect the general flow of high and low pressure systems as outlined in section 1.2.

The daily component $(\mathbf{d}_t) = (\mathbf{b}_t) + (\mathbf{e}_t) + (\mathbf{f}_t)$ is the subject of chapter 4. Based on those days when the wind swings around anticlockwise in a full circle the land and sea breeze cycle will be quantified, and an empirical circulation pattern $(\mathbf{b}_t)$ be derived. In contrast with the approach in dynamic meteorology here we are interested in the *types* of circulation that occur depending on the time of the year and on the prevailing wind. By focussing on $(\mathbf{d}_t)$ rather than $(\mathbf{w}_t)$ it is possible to detect this circulation also in winter when it is generally believed not to exist. However, when studying the influence of the geostrophic component $(\mathbf{g}_t)$ on the daily circulation it is found that the sea breeze is unlikely to occur when westerlies prevail. This means, given the increased frequency of these winds in winter, that the sea breeze is registered on fewer days than in summer. After the removal of the sea breeze component $(\mathbf{b}_t)$ from $(\mathbf{d}_t)$ the residual series $(\mathbf{r}_t)$ is obtaind.

The aim of chapter 5 is to detect and study particular short-term events in $(\mathbf{r}_t)$ such as calms, storms and oscillating winds. Following a brief description of the meteorological phenomena suitable definitions of these events are presented in terms of wind speed and direction. On the basis of $(\mathbf{r}_t)$ the events are characterized by their day-time of onset and their distribution over the year. By relating the events to the geostrophic and sea breeze components it is found that calms tend to occur more often in winter and are evidence of the high pressure belt extending to the Fremantle area, whereas storms, the way they are defined in this study, are associated with extratropical cyclones in winter and troughs developing from the low over the north-west in summer. Removing the storm effect $(\mathbf{e}_t)$ from $(\mathbf{r}_t)$ produces a residual series $(\mathbf{f}_t)$ of short-term fluctuations.

The theoretical background for the analysis of directional time series is presented in Part II. Motivated by the actual data analysis, this part focusses on the discussion of two parametric models for directional time series. Corresponding to the distributions associated with these models they will be called *von Mises* and *wrapped autoregressive* process, respectively. Whereas the latter is obtained by simply wrapping an ordinary autoregressive process in $\mathcal{R}$ around the circle, an explanation of the former requires more development than can be provided in this summary. Most of the discussion in Part II is devoted to fitting a wrapped autoregressive model to a time series of angular data. Given a directional time series, we first determine the *autocovariance function* of the underlying circular process, then estimate the autocovariance function of the unwrapped process and finally derive the parameters of this process using standard techniques. It is noted that standard theory developed for time series in $\mathcal{R}$ cannot readily be applied to angular data because of their 2π-*periodic* nature. An application to the directional component (θ_t) of the residual series $(\mathbf{f}_t)$ will serve to illustrate the approach presented in this monograph.

In chapter 6 the two models for directional time series are introduced. In view of the dilemma that there is no single distribution which plays a similarly central role for directional variables as the normal distribution for random variables in $\mathcal{R}$, two different autoregressive models are proposed which are related to the von Mises and wrapped normal distribution, respectively. In the same way these two distributions approximate each other, it is shown that the two time series models can be approximated by one another. Since each of the models has some but not all of the desirable properties, one may choose whatever model is more appropriate for the intended purposes. However, whereas it will be assumed that the wrapped autoregressive process is stationary, the von Mises process will generally exhibit a certain kind of non-stationarity.

In order to define a circular autocovariance function, a concept of angular dependence needs to be developed. In chapter 7 we explore various methods of measuring the association between two circular variables by a scalar and propose a new correlation coefficient for bivariate angular data. Following the specification of desirable properties for measures of association the reader is reminded that these measures are generally related to bivariate density functions through the maximum entropy characterization. It is shown that the new proposal is related to the von Mises process, and further that one is confronted with the *same* dilemma as above in that each particular measure of association has some but not all of the desirable properties. Finally, the properties of the proposed measure are examined and compared with those of other measures in the literature.

In chapter 8 the different measures of angular association are compared numerically, first in the case of independent and identically distributed variables and then in the context of a wrapped autoregressive process. Amongst the correlation coefficients considered the new proposal demonstrates the best behaviour. Based on the results of the simulation study performed in this chapter it is decided to define the circular autocovariance function in terms of this measure.

Given the circular autocovariance function of a wrapped autoregressive process, chapter 9 focusses on the estimation of the autocovariance function of the underlying unwrapped process. Five different techniques are described and then compared in a number of Monte-Carlo studies. The approach EQ which is based on equating empirical and theoretical covariances is found to be almost as good as maximum likelihood estimation if the time series are sufficiently long. Given that the directional component (θ_t) of the residual series $(\mathbf{f}_t)$ is rather long, it is decided to use EQ in the estimation process.

The concepts presented in chapters 6 to 9 are applied in chapter 10 in the analysis of the wind series $(\mathbf{w}_t)$. Both, a first order wrapped autoregressive and a first order von Mises model are fitted to the series (θ_t) of directional fluctuations. It is further shown that the two processes can be well approximated by one another. However, on the basis of this approximation it is also found that the von Mises model is slightly better suited to describe a time series in which periods with a prevailing wind direction alternate with periods when the direction changes more rapidly. Consecutive wind direction changes in $(\mathbf{f}_t)$ are almost independent which implies that the wind direction can be regarded as the result of a random fluctuation around the previous record.

The chapters in Part I are generally structured according to the following pattern. First the objective is described in a meteorological context, then suitable definitions of the physical structures are given in terms of wind speed and direction. The results of applying these definitions to the actual data are presented and discussed again in the meteorological framework. Mathematical details are generally relegated to the appendices. Whilst all chapters are preceded by short summaries, the main results will again be drawn together and presented with the conclusions in chapter 11. A glossary describing the symbols used in this study is provided at the back.

Figure 1.7. Structural diagram of data analysis.

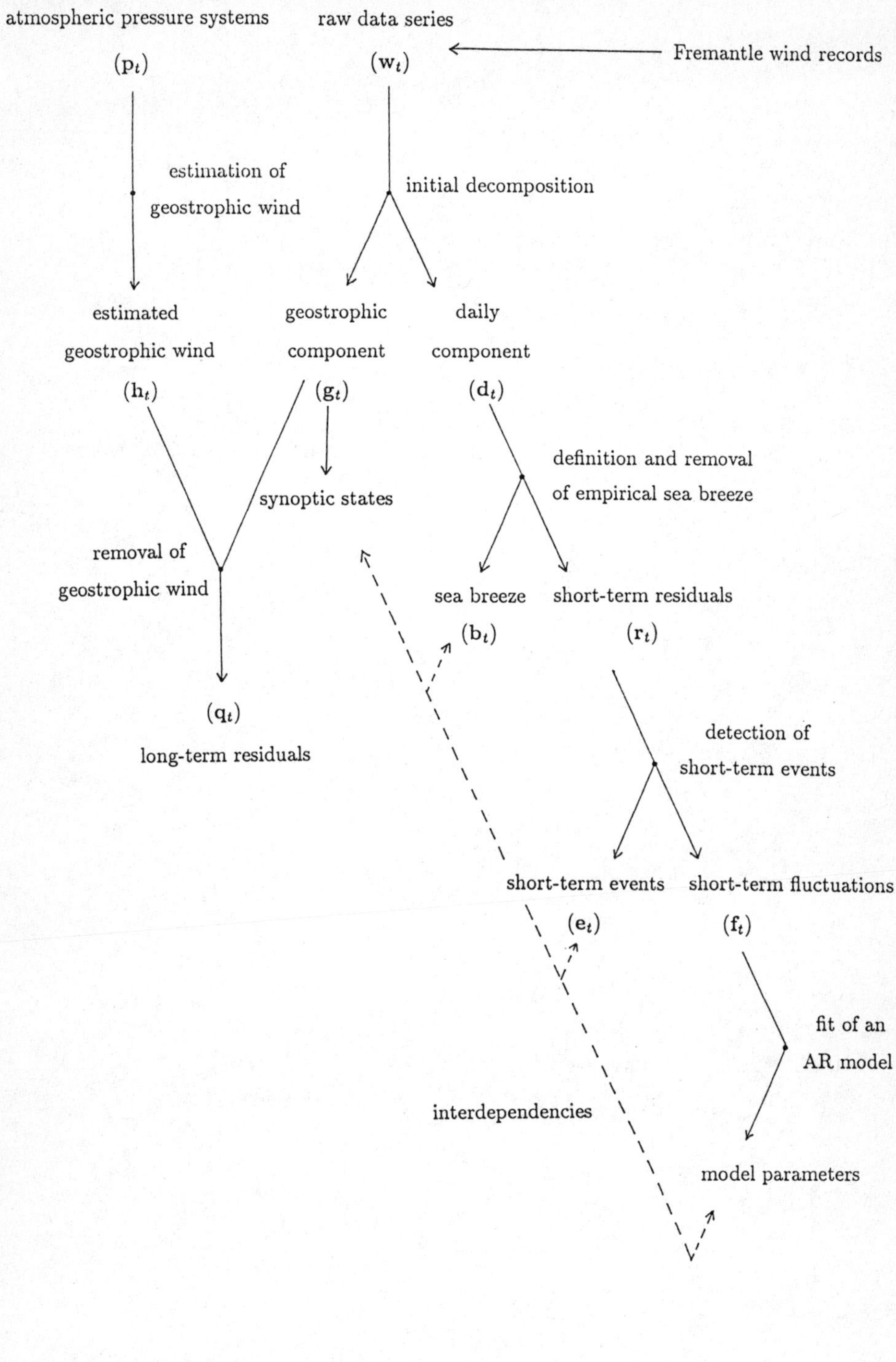

CHAPTER 2

THE INITIAL DECOMPOSITION

In this chapter it is first established that there is a significant seasonal component in $(\mathbf{w}_t)$ *corresponding to a 24-hour period. In order to separate this daily cycle from the prevailing wind, the series is divided into a geostrophic component* $(\mathbf{g}_t)$ *and a daily component* $(\mathbf{d}_t)$*. Commonly used robust techniques are adapted to data of the given kind, so as to reduce the effect of gusty winds and other short-term disturbances.*

In section 2.1 the decomposition model is explained in a meteorological context. In section 2.2 we briefly review robust estimation techniques developed for independent and univariate data, and discuss their application in dependent situations. The modifications of these techniques required for the long record of wind speed and direction are presented in sections 2.3 to 2.5. First, an iterated 2-stage estimation procedure is devised, which is less expensive to implement, then the concept of M*-estimation is extended to multivariate situations. More specifically, spatial versions of a median and Hampel's 3-part descending estimator are introduced. The corresponding filters, which are defined as moving estimators of location, are compared numerically with other filters in a number of situations that are typical of the given wind series. Mathematical details concerning the data representation and the use of a spatial median are placed in the appendix 2.6.*

2.1 General Background

The daily land and sea breeze circulation is one of the main weather phenomena in the Fremantle area especially in summer. In Part II of this monograph we will introduce a *univariate* spectrum for vector valued time series (cf. formula (7.23)). Based on a measure of vector association, it can be interpreted in the same way as the spectra for univariate time series. In this study the spectrum is used to confirm the presence of a daily circulation in the series of wind speed and direction.

Indeed, Figure 2.1 showing the spectra of the time series $(\mathbf{w_t})$ provides ample evidence that there is a 24-hour periodicity inherent in the data. This feature can mainly be attributed to the daily circulation. The 12-hour peaks in the spectra further support this finding, since the absolute value of the underlying measure of vector association is greatest whenever the wind comes from either the same or the opposite direction. Now, if the wind moves around in a full circle, then it will come from the same direction every 24 hours and from the opposite direction every 12 hours. Comparing the amplitudes of the 24-hour cycle, it is obvious that the circulation is much stronger in summer than in winter. We will return to these issues in chapter 4 when the daily land and sea breeze cycle will be examined in detail.

Figure 2.1. Raw wind spectra based on speeds and directions.

(a) Winter 1971. *(b) Summer 1971/72.*

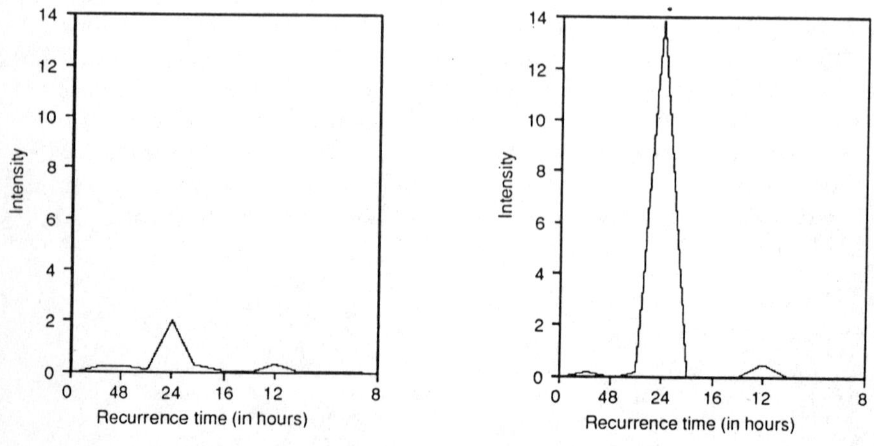

A *moving statistic* or *filter* based on a statistic Φ is defined as the operator which converts an observed input series (x_t) into an output series (y_t) according to

$$y_t = \Phi(x_{t+s}, \ldots, x_t, \ldots, x_{t-s})$$

where $s \in \mathcal{N}_0$ is a parameter of the filter. The seasonal component can be removed from the wind series $(\mathbf{w_t})$ by applying a filter which assigns constant weights to consecutive periods comprising 24 observations. The output and residual series will be called *geostrophic component* ($\mathbf{g_t}$) and *daily component* ($\mathbf{d_t}$), respectively. The former is supposed to capture the prevailing wind, whilst the latter should reflect the land and sea breeze circulation. To illustrate the decomposition techniques we will refer to the record of wind speed and direction of 3 to 5 December 1971 presented in Figure 1.2.

In order to find the most suitable filter for the separation of the geostrophic component from the daily cycle, a simulation study was performed which is described in detail in sections 2.3 and 2.4 below. Take $T \in \mathcal{Z}$ and let $(w_t) = (w_t)_{t=0,\ldots,T}$ be a univariate time series in $\mathcal{R}$ involving dependent data as well as clustered outliers which will generally be interpreted as storms. The performance of various filters is investigated using a simulated series (w_t) that is generated as the sum of the following four components:

(i) a geostrophic component (g_t°),
(ii) a sea breeze component (b_t°) with a period of 24 hours and with the sum over any full period equal to 0,
(iii) a storm component (e_t°) where

$$e_t^\circ = \begin{cases} c_e & \text{if } t_1 \leq t \leq t_2, \text{ with } c_e \in \mathcal{R} \text{ and } 0 \leq t_1,\, t_2 \leq T \\ 0 & \text{otherwise} \end{cases}, \text{ and}$$

(iv) a residual series (f_t°) of short-term fluctuations.

Given the objective to remove the geostrophic component from the given wind series, this study is designed to find a filter which can recapture (g_t°) and $(d_t^\circ) = (b_t^\circ) + (e_t^\circ) + (f_t^\circ)$ from the simulated series (w_t) as well as possible.

One might expect the mean filter to more or less retain its optimal properties if the underlying distribution of the data is only slightly different from a normal distribution. It is well known, however, that the mean is very sensitive to outliers, so that even mild deviations from the normal distribution may have a strong influence on the quality of the estimate (Tukey 1960). To illustrate this effect consider the following example.

Let $T = 72$ and construct (w_t) using the values and components below:

(i) a constant geostrophic component with $g_t^\circ = 4$,
(ii) a sinusoidal land and sea breeze cycle with $b_t^\circ = 2\sin(2\pi t/24)$,
(iii) a storm of 6 hours duration with $t_1 = 25$, $t_2 = 30$ and $c_e = 24$, and
(iv) negligible residuals, that is $f_t^\circ = 0$.

The series (w_t) and (d_t°) are plotted in Figure 2.2 (a). Note that the geostrophic component (g_t°) amounts to a uniform shift of (d_t°) by a constant of 4. Even though the physical unit of wind speed is redundant in this example and the simulation study below, we will generally refer to *knots*.

The result of applying a moving average, which is defined as a simple mean over 24-hour periods, to (w_t) is shown in Figure 2.2 (b). Since the mean blurs the storm (e_t°) $(t_1 \leq t \leq t_2)$ into the geostrophic component, the shape of the series (g_t) and (d_t) obtained by filtering becomes highly distorted. This suggests that we should use a technique that is more robust against storms and patchy outliers. Various robust filters will subsequently be introduced and gradually adapted to the situation of a multivariate time series.

Figure 2.2.

(a) Simulated series (w_t) and (d_t°), 6 hours storm.

(b) Decomposition of (w_t) using a moving average.

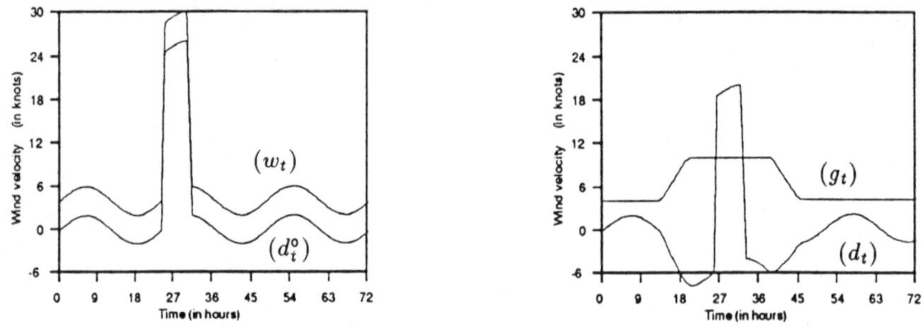

2.2 Robust Filtering

Any robust estimator should

(i) be little affected by rounding, grouping and other local inaccuracies, that is *reject smoothly*,

(ii) have a *breakdown point* as high as possible, that is remain reliable even in the presence of large contamination,

(iii) keep a low bound on the *gross-error-sensitivity*; that is, change as little as possible if one moves away from the parametric model in any direction, and

(iv) have a *variance* as small as possible under the ideal parametric model.

Furthermore, it is often desirable that the estimator has a *rejection point* beyond which the data are ignored. Since conditions (iii) and (iv) conflict with each other, one usually has to compromise between efficiency and gross-error-sensitivity. However, with only a few percentage points loss in efficiency the gross-error-sensitivity can be decreased substantially. For a detailed discussion of robust statistics and its principles the reader is referred to Barnett & Lewis (1978), Serfling (1980), Huber (1981) and Hampel *et al.* (1985).

Given a random sample $\{x_i, \ldots, x_n\}$ of identically distributed random variables, let us assume that the underlying distribution has a symmetric density and that we want to estimate its centre of symmetry. One of the most important concepts of robust statistics is the so-called *M-estimator* which is defined as the value $\hat{\tau}$ that minimises the function

$$L(\tau) = \sum_{j=1}^{n} \rho\left(\frac{x_j - \tau}{s}\right)$$

where s is a scale parameter and the function $\rho : \mathcal{R} \to \mathcal{R}_0^+$ is symmetric. Hence, putting $\psi(x) = d\rho(x)/dx$, $\hat{\tau}$ is obtained as the solution of the following equation

$$\sum_{j=1}^{n} \psi\left(\frac{x_j - \hat{\tau}}{s}\right) = 0 \, .$$

The function $\psi : \mathcal{R} \to \mathcal{R}$ is, apart from a scaling constant, identical with the *influence curve* which was introduced by Hampel (1968) and is characteristic of the estimator. Asymptotic normality of the M-estimator was initially proved by Huber (1964) for independent samples and later, in a more general form, by Portnoy (1977) who considers a particular type of dependence. As opposed to order statistics where the influence of the data depend on their relative position in the ordered sample, in an M-estimator the influence is determined by their distance to the centre of symmetry. All the filters which will subsequently be compared numerically are now introduced. Stylized influence curves of the corresponding estimators are presented in Figure 2.3.

In an extensive Monte-Carlo study conducted for independent samples by Andrews *et al.* (1972) at Princeton, the M-estimators generally gave better results than linear functions of order statistics and estimators based on rank tests. The following eight points emerged from this comparative study:

(i) The mean M is highly non-robust and very sensitive to extreme outliers. This behaviour corresponds to an unbounded influence curve as shown in Figure 2.3 (a).

(ii) The trimmed mean $T(\alpha)$ is reasonably reliable if the proportion α of data points removed from each end is not too small. In the centre of the data it behaves like the mean as illustrated by the mid-part of the curve in Figure 2.3 (c). The adjacent horizontal lines are supposed to indicate that the influence of outliers is restricted by trimming the sample.

(iii) The median N is the most robust, though not very efficient estimator. As can be seen in Figure 2.3 (d) the information contained in each observation is reduced to ± 1 depending on whether it is larger or smaller than the centre of symmetry.

(iv) The clearly most successful group were the 3-part descending M-estimators $H(a,\, b,\, c)$ which are based on

$$\psi_H(x) = \begin{cases} x & \text{if} \quad |x| \le a \\ a \,\text{sign}(x) & \text{if } a \le |x| \le b \\ a\,[\,x - c\,\text{sign}(x)\,]\,/\,[\,b - c\,] & \text{if } b \le |x| \le c \\ 0 & \text{if } c \le |x| \end{cases} .$$

These estimators were first introduced by Hampel (1968) and possess all the robustness properties listed above including a rejection point as shown in Figure 2.3 (e). The modified estimator depicted in Figure 2.3 (f) has been proposed to ensure greater numerical stability. As for a robust estimator of scale Hampel suggests the median deviation

$$s = \text{med}_j \; |x_j - \text{med}_k \; x_k| \; , \tag{1}$$

that is the median of all the absolute deviations from the median.

Figure 2.3. Stylized influence curves.
Sources: Andrews et al. (1972) and Huber (1981).

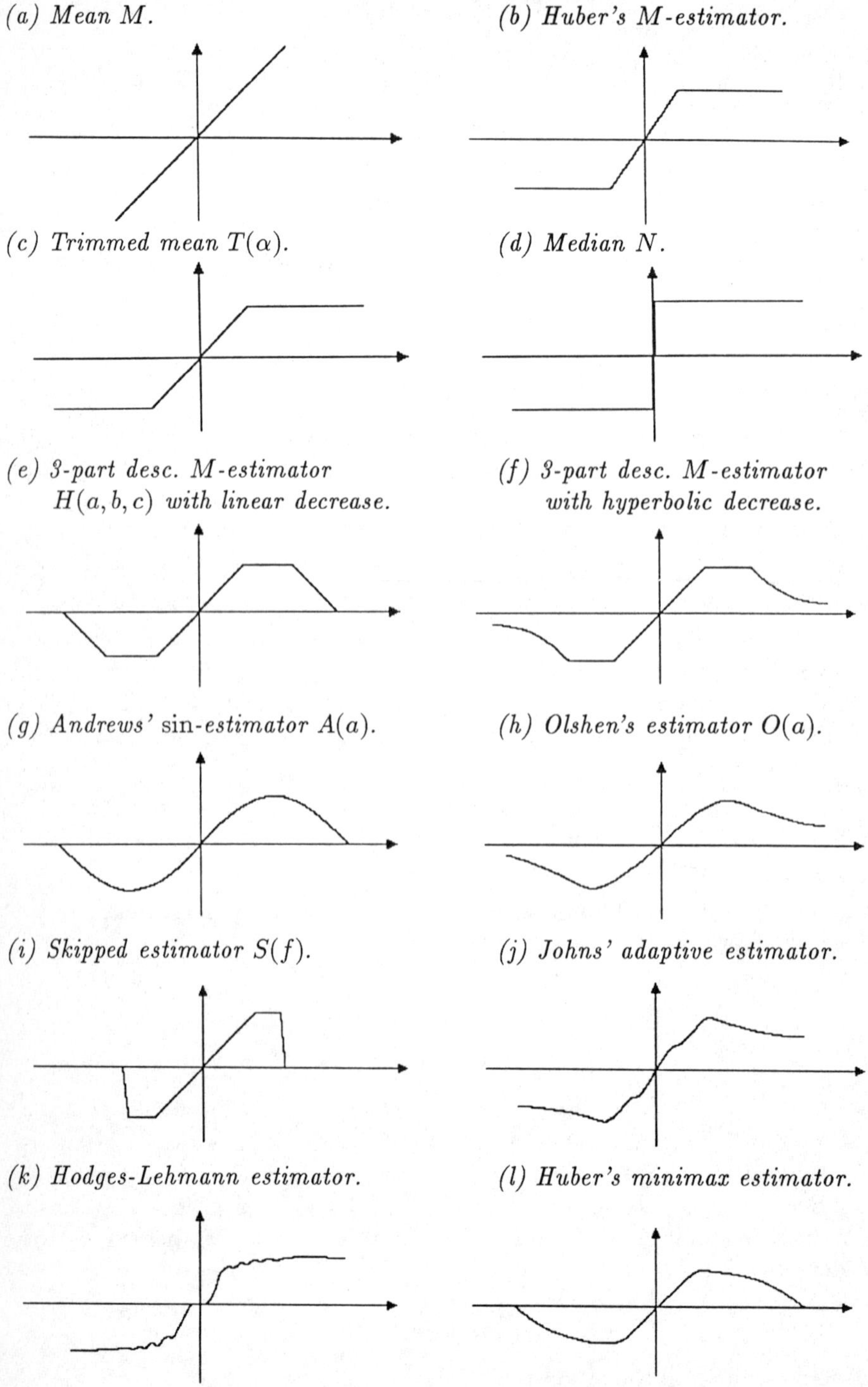

(v) Closely related are Andrews' M-estimator $A(a)$ and Olshen's $O(a)$ which are defined by

$$\psi_A(x) = \begin{cases} \sin(x/a) & \text{if } |x| \leq a\tau \\ 0 & \text{otherwise} \end{cases}$$

and

$$\psi_O(x) = x \,/\, [\, a + x^2 \,] \,,$$

respectively. As can be guessed from Figures 2.3 (g) and (h) their behaviour is similar to that of Hampel's 3-part descending estimator.

(vi) The skipped estimators $S(f)$ with $f : \mathcal{N}_0 \rightarrow \mathcal{N}_0$, which combine some rejection procedure with moderately adaptive trimming, also showed a good general performance. With q_l and q_u denoting the lower and upper quartile, respectively, first all observations outside the interval $[\, q_l - 1.5(q_u - q_l),\ q_u + 1.5(q_u - q_l)]$ are deleted. If a total of k points is rejected in this way, another $f(k)$ points are deleted from each end before the mean of the remainder is calculated. As for the trimmed mean the influence curve in Figure 2.3 (i) is supposed to be indicative only.

(vii) Good results were also obtained when using adaptive estimators such as that proposed by M.V. Johns which is defined as a linear combination of two trimmed means. Although this particular estimator loses efficiency for intermediate contamination it gains at the extremes, that is near the normal and the long-tailed distributions, and thus attains a more uniform overall-behaviour. For its stylized influence curve see Figure 2.3 (j).

(viii) Other estimators which are commonly used but did not perform as well include the Winsorized mean, the Hodges-Lehmann estimator sketched in Figure 2.3 (k), and the average of two symmetric percentiles.

Furthermore, Huber (1981) suggests minimizing the maximum asymptotic variance over a specified set of distribution models, yielding an influence function of the type shown in Figure 2.3 (l). Although these estimators have good robustness properties this approach may not be appropriate if the amount of contamination is unknown or the sample size is too small to identify the nature of this contamination to any sufficient degree of accuracy. Since both the amount and the nature of contamination varies widely over the given wind series $(\mathbf{w_t})$, we will confine our study to the more rigid non-adaptive estimators.

Our final comment in this section is concerned with robustness in situations involving serial dependence and clustering outliers. It is known that the presence of relatively small correlation can drastically inflate the variance of an estimator, and many estimators which appear to be reasonably good for independent samples can be extremely poor for correlated data (Portnoy 1977). However, comparatively little is known about robustness in cases of serial dependence.

Gastwirth & Rubin (1975) studied the effect of serial dependence on the efficiency of the Hodges-Lehmann estimator and various order statistics. Their results were slightly generalized by Justusson (1979) who studied pairs of moving order statistics in the context of signal and image processing, and in particular, investigated the effect of applying a moving median to a strictly stationary process. These filters had been suggested earlier by Tukey (1977) as a means of smoothing time series. They are very robust towards single outliers, and preserve the sharpness of jumps in cases where the underlying process suddenly changes the mean level. Numerical studies based on a first order autoregressive process show that the median loses only a few percent efficiency relative to the mean if the process has high positive correlation (Justusson 1979).

In a paper by Portnoy (1979) approximately optimal estimators are presented, in the sense that they minimize the maximum asymptotic variance over a class of contaminated normals. Although these estimators showed the best behaviour in a numerical study conducted by Portnoy (1979), they have the disadvantage that the correlation and the amount of contamination have to be known. Amongst the more rigid non-adaptive estimators the best choice seemed to be $O(2)$ which had an only marginally greater variance than the optimal adaptive estimators. The best Hampel-type estimators were those in which the slope did not become too negative, for example $H(1.7, 3.5, 8.5)$.

In recent years a number of contributions have been made towards robustifying parametric ARMA models with built-in correlation structure. When adapting robustness concepts to time series analysis Denby & Martin (1979) distinguished between single and patchy outliers. In the context of the given wind series ($\mathbf{w_t}$) the former would correspond to some sort of recording error while the latter could be attributed to storms. In order to center an observed series, which in the context of the given wind series ($\mathbf{w}_t$) is equivalent to removing the geostrophic component ($\mathbf{g}_t$), Martin (1978) suggests the use of an M-estimator even though in practice they are expensive to implement.

In summarising, it can be said that different filters may be appropriate under different circumstances. We will therefore simulate a number of time series which are similar to the given wind series, so as to assess the adequacy of the various filters for the initial decomposition.

2.3 Univariate Filter Study

In the following two sections we discuss the performance of the estimators introduced above using simulated time series. In section 2.3 we investigate the case of univariate and in section 2.4 the case of multivariate data. Various factors such as computing costs, the required precision of the analysis, the proportion of outliers, or the quality of the data in general demand a variety of filtering techniques. Besides, a number of problems concerning dependent situations have not previously been thoroughly explored. We therefore conduct a numerical study in order to determine the most appropriate filter for the decomposition of the wind series ($\mathbf{w_t}$).

Each of the simulated series is defined as the sum of four different components as explained in section 2.1. By varying these components we generated a number of series which were then used to examine the properties of the various filters. The following components were considered for $t = 0, ..., 72$:

Figure 2.4. Decomposition into a geostrophic and a daily component, $\delta = 1$.

(a) Simulated series (w_t) and (d_t°).

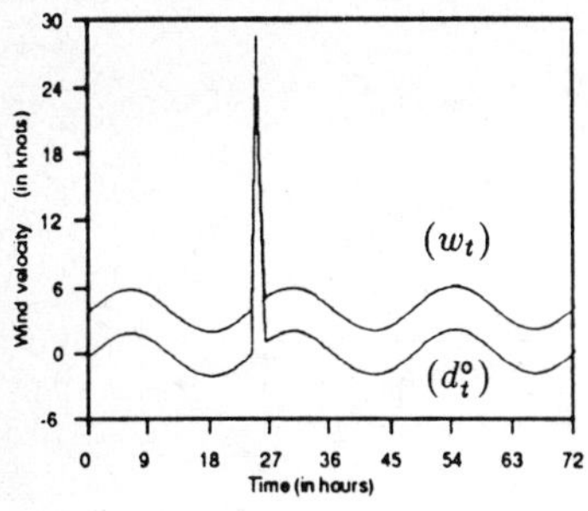

(b) Mean filter M.

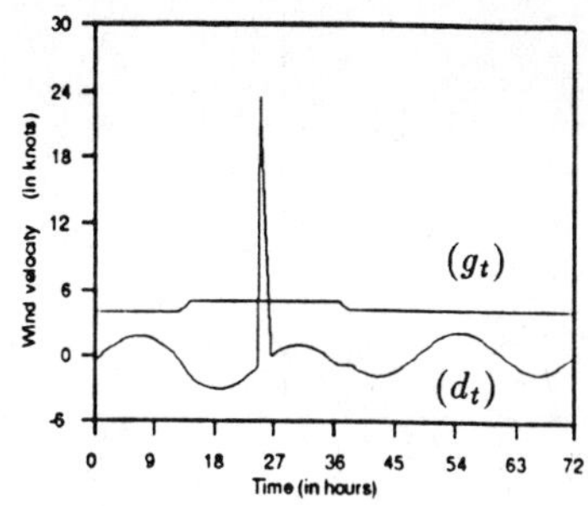

(i) the geostrophic component

$$g_1^\circ(t) = 4$$

$$g_2^\circ(t) = 6 - \frac{4}{72}\, t \;;$$

(ii) the land and sea breeze cycle

$$b_c^\circ(t) = c \cdot \sin \frac{2\pi}{24}\, t \quad \text{where} \quad c = 2, 4 \;;$$

(iii) the storm component

$$e_{\delta,a}^\circ(t) = \begin{cases} a & \text{if } t \in (24,\, 24 + \delta\,] \\ 0 & \text{otherwise} \end{cases}$$

where $\delta = 1, 3, 6, 10, 12$ and $a = 6, 12, 24$; and

(iv) the residual component $f_{\sigma^2}^\circ(t)$ where the data were independently generated from $N(0, \sigma^2)$ with $\sigma^2 = 0, 1, 2$.

The study indicated that the best estimators were those which had also shown the best performance in the Monte-Carlo study at Princeton (Andrews *et al.* 1972). It turned out that the filters differed mainly in their sensitivity towards the duration parameter δ. Hence, for the type of series under investigation the breakdown point became the most important concept for assessing the robustness properties, with a high breakdown point guaranteeing a diminishing influence of even longer lasting storms. It also determined our simulation strategy. The results are now discussed in detail.

When a seasonal component is to be removed from an observed time series one commonly uses a filter which gives the same weight to all observations representing the same period. This means that we require a filter that assigns an equal weight to 24 consecutive observations in order to separate the geostrophic component from the daily fluctuation. However, a moving statistic that has to be evaluated, at each point in time and possibly by iteration, from 24 observations is costly to implement and thus impractical. We therefore propose a *composite* procedure connecting two filters Φ and Ψ in series, both of which assign the same weight to sets of only 5 observations.

Figure 2.5. Decomposition into a geostrophic and a daily component, $\delta = 3$.

(a) Simulated series (w_t) and (d_t°).

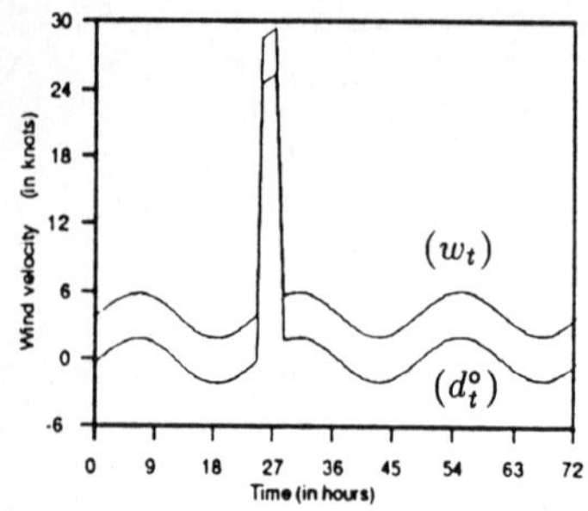

(b) Mean filter M.

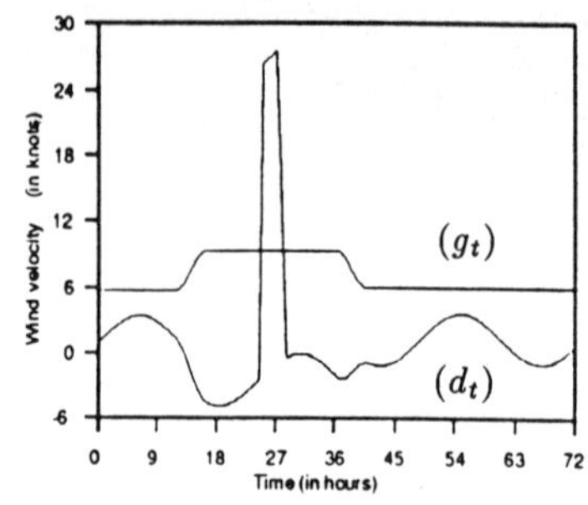

(c) Median filter N.

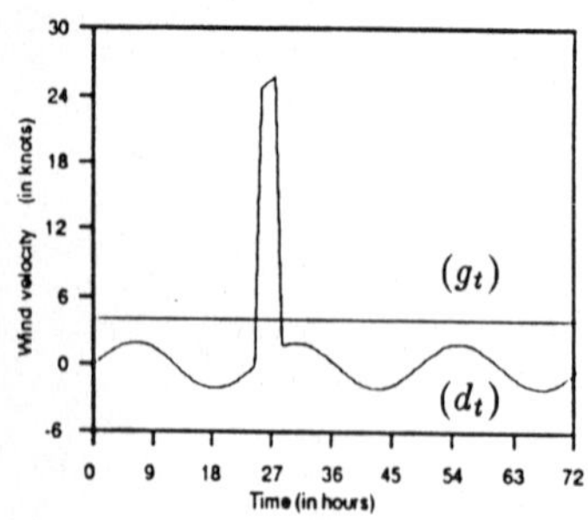

(d) Composite $(5H)^2$.

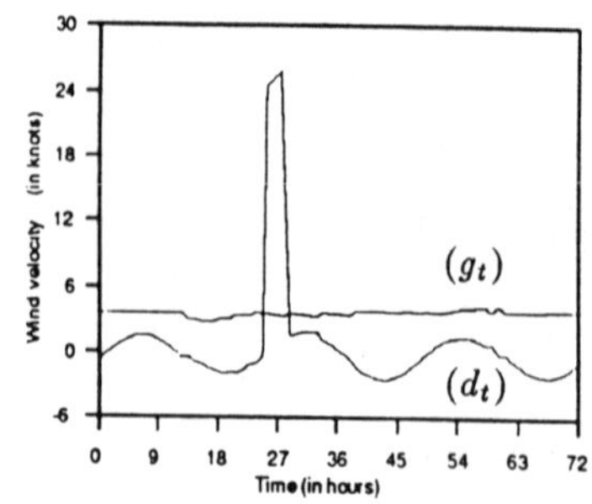

First, the filter Φ is applied to (w_t) yielding a series (v_t) with

(2a) $$v_t = \Phi(w_{t+2},\ w_{t+1},\ w_t,\ w_{t-1},\ w_{t-2})\ .$$

Then in order to estimate the geostrophic component the second filter Ψ is applied to (v_t) as follows

(2b) $$g_t = \Psi(v_{t+10},\ v_{t+5},\ v_t,\ v_{t-5},\ v_{t-10})\ .$$

The composite filter assigns equal weight to 25 consecutive observations and will be denoted by $5\Phi\,5\Psi$ or $(5\Phi)^2$ in cases where $\Phi = \Psi$.

The effect of various filters on the decomposition is shown in Figures 2.4 to 2.8 for the following series $(t = 0, ..., 72)$:

$$d_\delta^\circ(t) = b_2^\circ(t) + e_{\delta,24}^\circ(t) + f_0^\circ(t)$$
$$w_\delta(t) = g_1^\circ(t) + d_\delta^\circ(t)\ .$$

If the value of δ is not ambiguous we will also write (d_t°) and (w_t), respectively. All the filters depicted in Figures 2.4 to 2.8, except for the composite ones, assign the same weight to 24 consecutive observations. Both H and the composite filter $(5H)^2$ are based on the Hampel-type estimator $H(1.7,\ 3.4,\ 8.5)$ where the scale parameter was estimated

Figure 2.6. Decomposition into a geostrophic and a daily component, $\delta = 6$.

(a) Simulated series (w_t) *and* (d_t°).

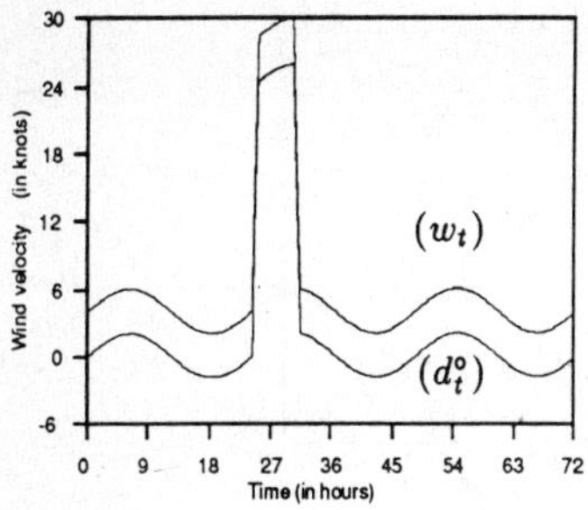

(b) Mean filter M.

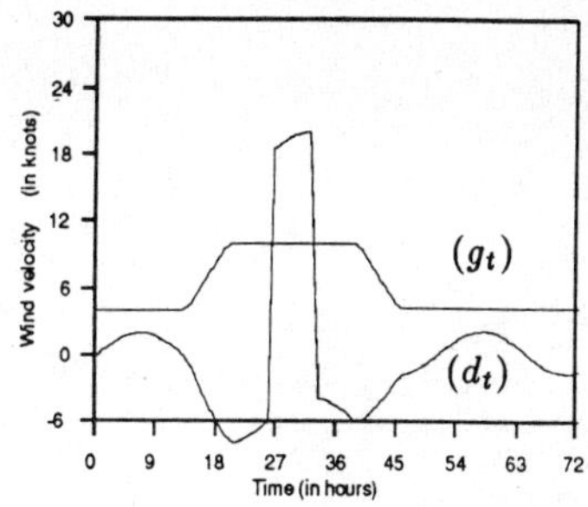

(c) Median filter N.

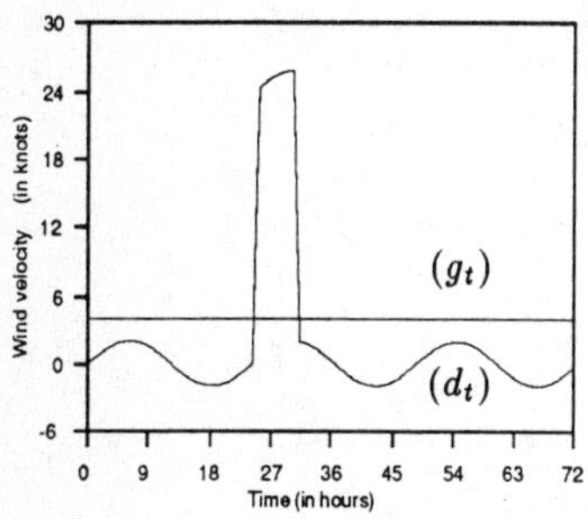

(d) Composite $(5H)^2$.

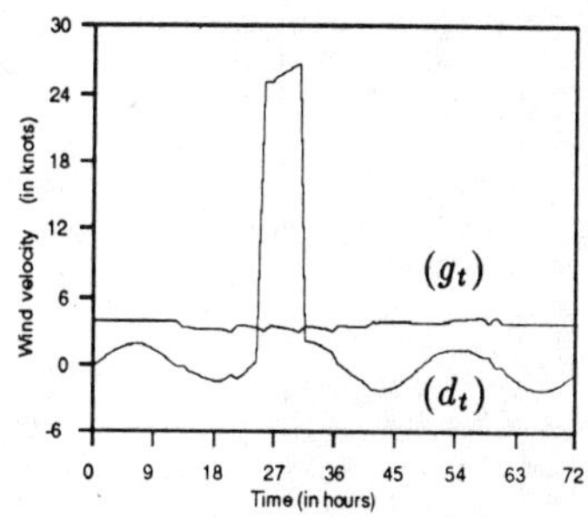

(e) Composite $(5N)^2$.

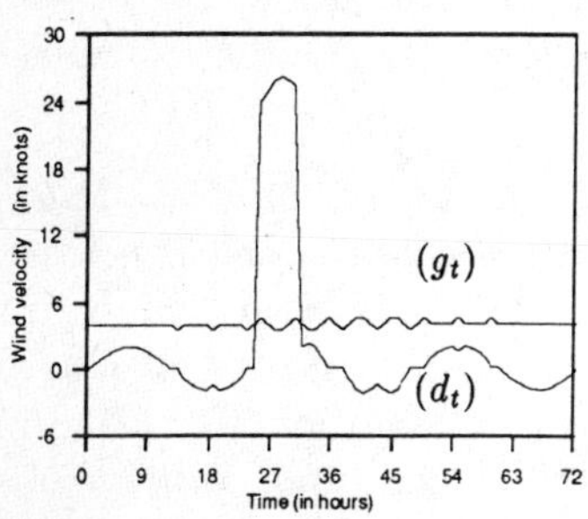

(f) Hampel's $H = H(1.7, 3.4, 8.5)$.

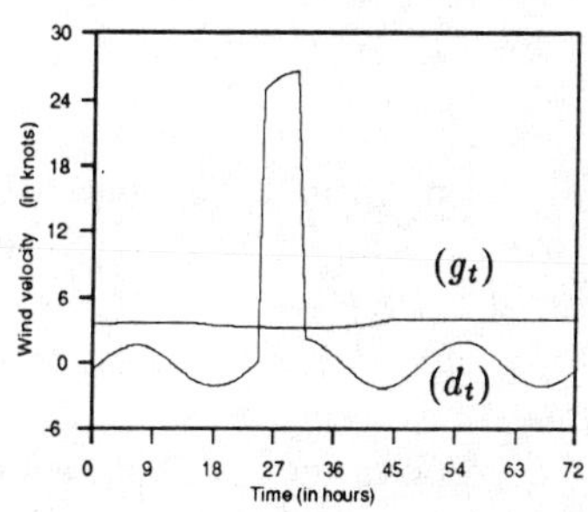

(g) Trimmed mean $T(0.25)$.

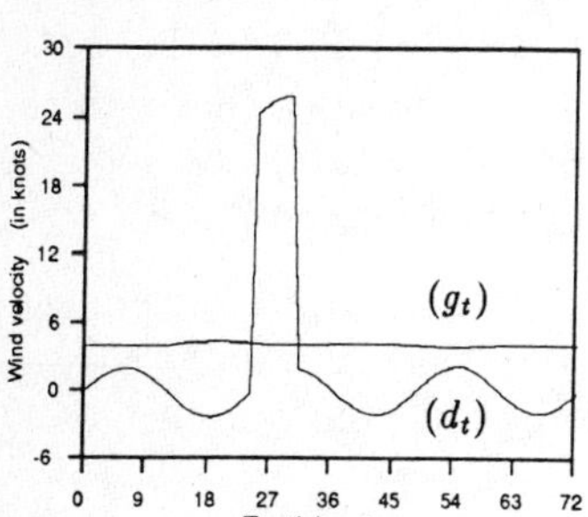

(h) $S(\min\{\max\{2k, 1\}, 0.6n - k\})$.

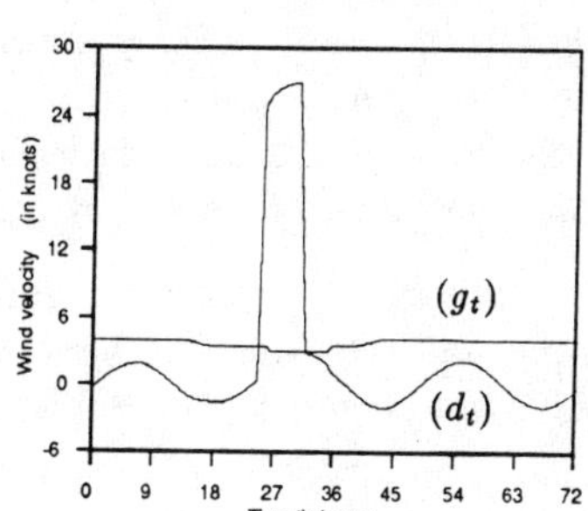

Figure 2.7. Decomposition into a geostrophic and a daily component, $\delta = 10$.

(a) Simulated series (w_t) and (d_t°).

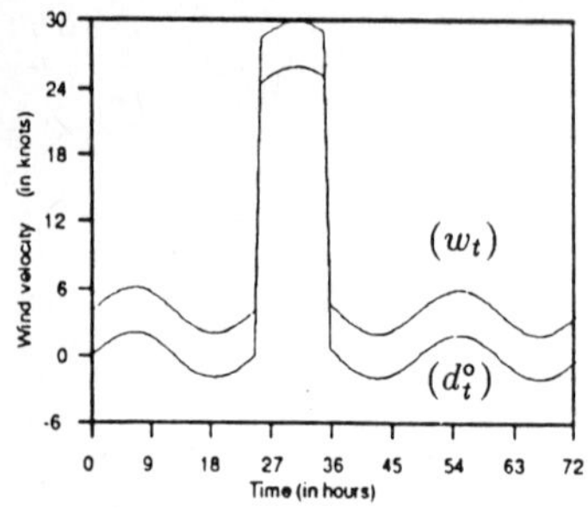

(b) Mean filter M.

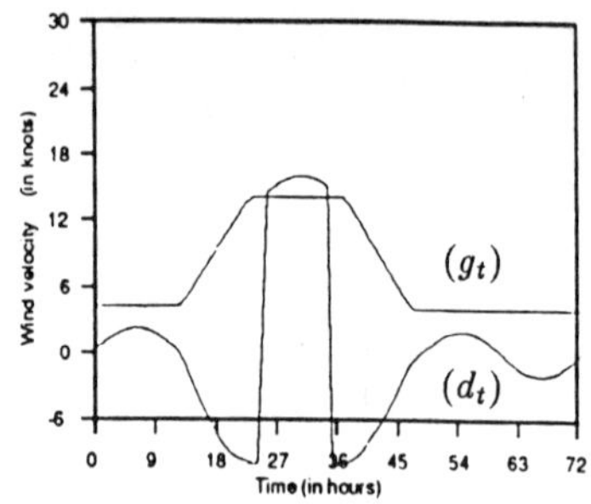

(c) Median filter N.

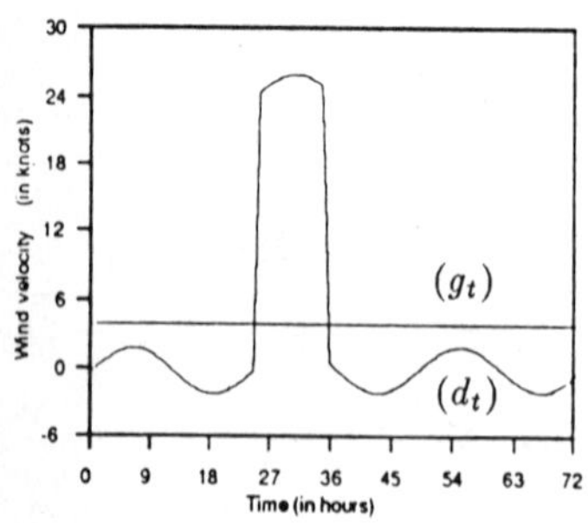

(d) Composite $(5H)^2$.

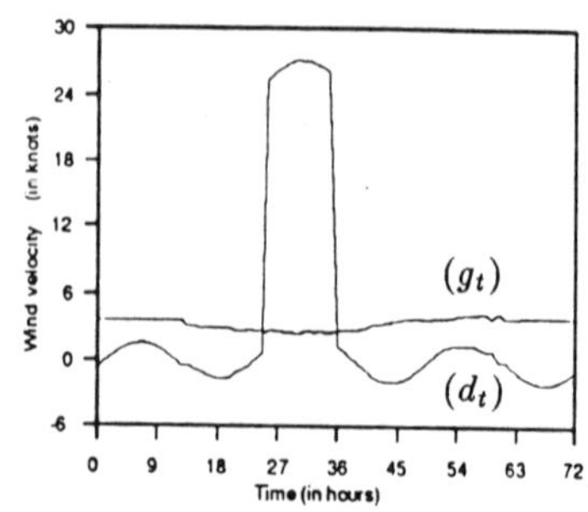

using the median deviation as defined in (1). As for the skipped estimator $S(f)$ we chose the function $f(k) = \min\{\max\{2k, 1\}, 0.6\,n - k\}$ with $n = 24$ for the number of sample points under consideration.

The simulated series (w_t) is shown in diagram (a) of Figures 2.4 to 2.8 for a storm duration of 1, 3, 6, 10 and 12 hours, respectively. Also presented is the generated daily component (d_t°) which we attempt to reproduce from (w_t) by filtering. Since the geostrophic component (g_t°) is identical to 4 knots in all cases it has not been plotted in these diagrams. From Figure 2.4 (b) it is seen that a single outlier, like a storm which lasts for only one hour, hardly affects the results of the mean filter. However, as the storm duration increases to 3 hours the change in the geostrophic component (g_t) becomes noticeable (cf. Figure 2.5 (b)). As a result the values in (d_t) are 3 knots below those in (d_t°) for a period of about 20 hours. The median filter, on the other hand, exactly reproduces the geostrophic and daily components as shown in Figure 2.5 (c). The same is true for all the filters which were introduced in the last section and which assign the same weight to 24 consecutive observations. The composite filter $(5H)^2$ depicted in Figure 2.5 (d) produces a few ripples but otherwise gives satisfactory results.

In Part II of this monograph a measure of vector correlation will be introduced, which is used here to determine the agreement between the generated daily component (d_t°) and the residual series $(d_t) = (w_t) - (g_t)$ obtained by filtering. For both these

Figure 2.8. Decomposition into a geostrophic and a daily component, $\delta = 12$.

(a) Simulated series (w_t) *and* (d_t°).

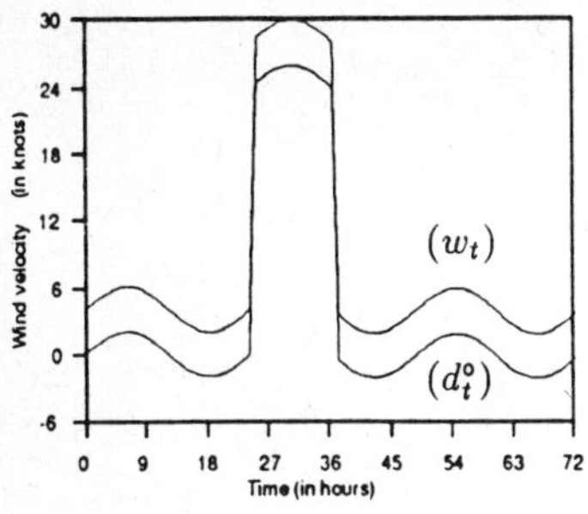

(b) Mean filter M.

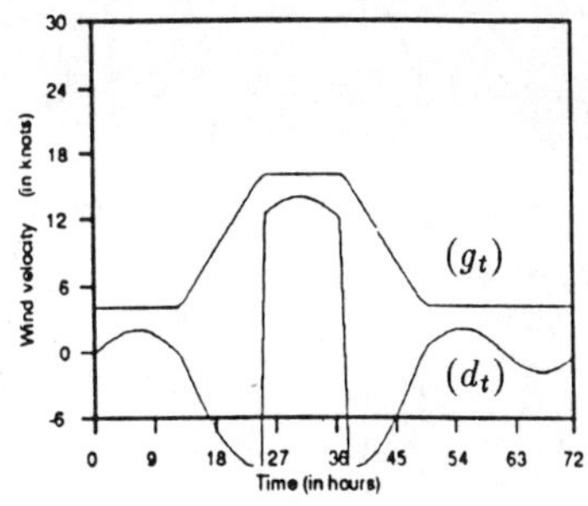

(c) Median filter N.

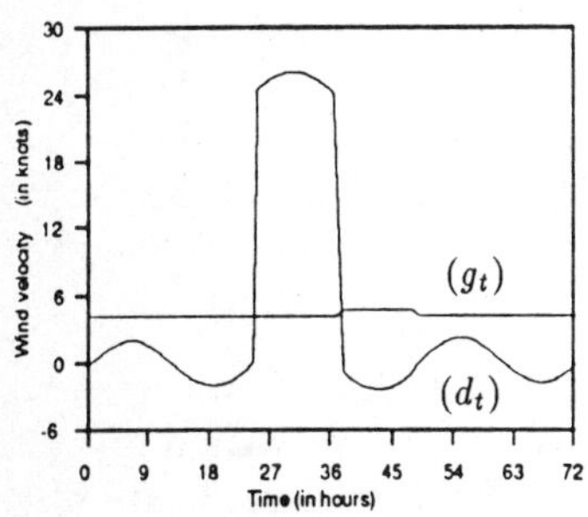

(d) Composite $(5H)^3$.

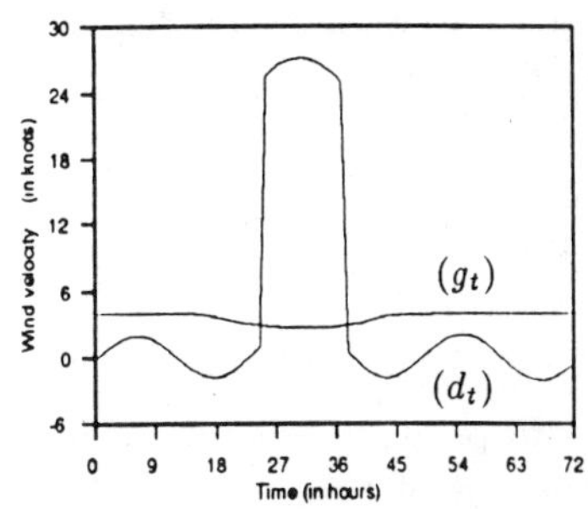

series the 24-hour periods just before and just after the storm are regarded as components of one 48-dimensional vector. Values of this measure of association are presented in Table 2.1 for all the filters shown in Figures 2.4 to 2.8 and for various storm durations. A value of 1 corresponds to perfect correlation, whilst a value of 0 suggests that there is no association at all. Table 2.1 thus gives an indication of how well the various filters recapture the daily component (d_t°).

From the second row in Table 2.1 it follows that in case of a 3-hour storm duration all filters, except for the mean but including the composites, achieve a very good result. The situation is basically the same if the storm lasts for 6 hours as is evident from the third row in Table 2.1 as well as the plots in Figure 2.6. However, the ripples produced by the composite filters are more pronounced than before. This is the case in particular for the median filter $(5N)^2$ presented in Figure 2.6 (e).

Of the 7 filters depicted in Figure 2.6 the clearly best results are obtained for the median, Hampel's 3-part descending estimator and the trimmed mean. Note, however, that the latter would start producing distorted results if the storm lasted only one hour longer, due to a breakdown point of only 0.25. This is also reflected in the relatively small degree of association for the trimmed mean in case of a 10-hour storm duration. In this situation the M-estimators such as the median N or Hampel's H are clearly superior to the mean and other filters based on order statistics, for example the trimmed

Table 2.1. Degree of association between the generated component (d_t°) and the reconstructed component (d_t).

Storm duration δ	Mean M	Trimed mean $T(.25)$	Skipped $S(f)$	Median N	Hampel H	Comp. $(5N)^2$	Comp. $(5H)^2$	Comp. $(5H)^3$
1	0.90	1.00	1.00	1.00	1.00	0.98	0.99	
3	0.62	1.00	0.99	1.00	1.00	0.98	0.98	
6	0.54	1.00	0.96	1.00	0.98	0.98	0.96	
10	0.65	0.60	0.53	1.00	0.95	0.98	0.92	
12	0.71	0.61	0.60	0.99	0.96	0.96	0.93	0.96

mean and skipped estimator. The distorting effect of the mean is demonstrated by the curve in Figure 2.7 (b). There is hardly an indication of the three cycles imposed on the daily component by (b_t°). It is also evident from Figure 2.7 (d) that the ripples produced by the composite filter $(5H)^2$ do not further deteriorate. Indeed, its performance is comparable to that of the one-step filter H which assigns the same weight to 24 consecutive observations.

The situation is virtually the same for a storm duration of 12 hours. Whereas the M-estimators still produce reliable results, the other filters give no indication of what the simulated daily and geostrophic component looked like. Even the composite filter $(5H)^2$ essentially recaptures the shape of (d_t°) although some additional smoothing may be required. By applying another $5H$ filter to the series produced by $(5H)^2$ we get the filter $(5H)^3$ presented in Figure 2.8 (d), which is almost as good as H but less expensive to implement. The high value given in the last row of Table 2.1 for this filter confirms the good agreement between (d_t°) and the daily component (d_t) obtained for this filter.

Let us summarise these results. Whilst most filters performed reasonably well if the storm did not last longer than 6 hours, the M-estimators including the composites showed the most consistent behaviour. As a result of their high breakdown points they recaptured the sea breeze cycle almost exactly even if the storms lasted for 12 hours. The median filter which achieved the highest degree of association, lost slightly if (g_t°) or (f_t°) were less regular, but would still provide a good initial estimate of the geostrophic component. Furthermore, it has the property of preserving sudden jumps in the mean level of the filtered time series. This could actually be an advantage, since the local region may come quite suddenly under the influence of a low depression, which would imply a sudden change in the characteristics of the underlying stochastic process.

2.4 Multivariate Filter Study

In this section the robustness concepts are adapted to time series in $\mathcal{R}^k$ ($k > 1$). Given a sample $\{\mathbf{x}_1, \ldots, \mathbf{x}_n\}$ of not necessarily independent observations in $\mathcal{R}^k$ the aim is as in the last section to estimate the *centre of the data* denoted by τ. By applying the ρ-function introduced in section 2.2 to the magnitude of the *influence vector*, which is defined as the difference between τ and the observation $\mathbf{x}_j$, the concept of M-estimation can be extended in a natural way to multivariate situations. Referring to a record (v_t, θ_t) of wind speed and direction of the series $(\mathbf{w}_t)$, this means that the ρ-function is applied only to the speed v_t, leaving the direction θ_t unchanged. Defining $\psi(\mathbf{x}) = d\rho(\mathbf{x})/d\mathbf{x}$, the M-estimate of τ is then obtained as the solution of

$$\sum_{j=1}^{n} \psi(\mathbf{x}_j - \hat{\tau}) = \mathbf{0} \quad , \quad \hat{\tau} \in \mathcal{R}^k .$$

Let us illustrate this procedure. The ordinary median of a sample $\{x_1, \ldots, x_n\} \subseteq \mathcal{R}$ is defined as the value $x \in \mathcal{R}$ which minimises the function

$$L(x) = \sum_{j=1}^{n} |x_j - x| . \tag{3}$$

Following Brown (1983) and Breckling & Chambers (1988) one could therefore define a *spatial median* as the vector $\mathbf{x} \in \mathcal{R}^k$ that minimises the function

$$L_k(\mathbf{x}) = \sum_{j=1}^{n} \|\mathbf{x}_j - \mathbf{x}\| , \tag{4}$$

and distinguish between

(i) the *restricted* median N_r where the minimum is taken over the sample $\{\mathbf{x}_1, \ldots, \mathbf{x}_n\} \subseteq \mathcal{R}^k$, and

(ii) the *unrestricted* median N where the minimum is taken over $\mathcal{R}^k$.

Note that the spatial median defined this way differs from the *spherical median* proposed by Fisher (1985) and Brown (1985). Suppose that $\mathbf{x}_j$ and $\mathbf{x}$ are unit vectors, and let α_j be the angle between $\mathbf{x}_j$ and $\mathbf{x}$. Then (4) can also be written in the form

$$L_k(\mathbf{x}) = \sum_{j=1}^{n} 2 \sin(\alpha_j/2) .$$

Whilst the spatial median minimises the function $L_k(\mathbf{x})$, the spherical median minimises the expression $\sum_{j=1}^{n} \alpha_j$. In the case where $\mathbf{x}$ is constrained to be a unit vector, it may be more natural to define *distance* in terms of arc lengths. However, in this chapter the spatial median is generally used for multivariate data which do not necessarily lie on the unit sphere. Furthermore, if the dispersion of the vectors $\mathbf{x}_j$ is sufficiently small, then the two medians are clearly in reasonable agreement. For the actual calculation of the spatial median the reader is referred to appendix 2.6.2.

Figure 2.9. Moving spatial median of wind records in $(\mathbf{w}_t)$ *from 3 December 1971, 1200 h to 5 December 1971, 1200 h.*

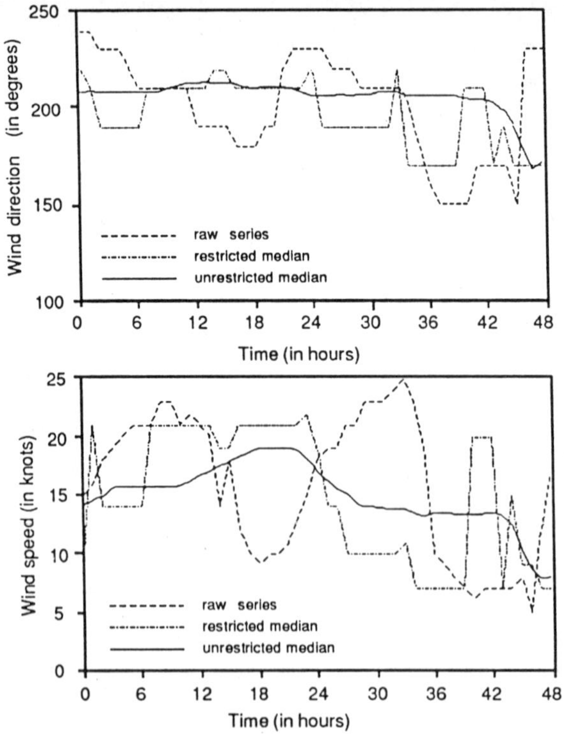

Figure 2.9 compares the restricted with the unrestricted median of the wind records in $(\mathbf{w}_t)$ using a window of width 24. Although the restricted median is not as smooth it is easy to determine and therefore useful as an initial estimate for the iterative procedures which are usually required for the calculation of the unrestricted median or any other M-estimator. Having defined a spatial median it is obvious how analogues of the trimmed mean $T(\alpha)$ or the skipped estimator $S(f)$ could be set up. In the case of $T(\alpha)$ the proportion α furthest away from the median is removed from the data, whilst for $S(f)$ all those observations are trimmed which are further away than a specified distance.

Another decision which has to be made in this section concerns the data representation; that is, whether to perform the algebra in a cartesian, polar or hyperbolic framework, as the initial decomposition and thus the subsequent analysis will be slightly different in each case. The three representations are briefly discussed and compared in appendix 2.6.1. Filters which are based on the polar and hyperbolic representations are denoted by the prefixes p and h, respectively. Further, to reduce the implementation costs we will, as in section 2.3, connect two or more estimators in series using formulae (2).

As in the univariate filter study the mean M turned out to be highly non-robust and very sensitive towards extreme outliers, whereas the M-estimators such as those suggested by Hampel, Andrews and Olshen gave the best results in terms of recapturing

Figure 2.10. Decomposition of a series of wind speeds and directions, $\delta = 3$.

(a) Simulated series ($\mathbf{w}_t$) and ($\mathbf{d}_t^\circ$). *(b) Mean filter M.*

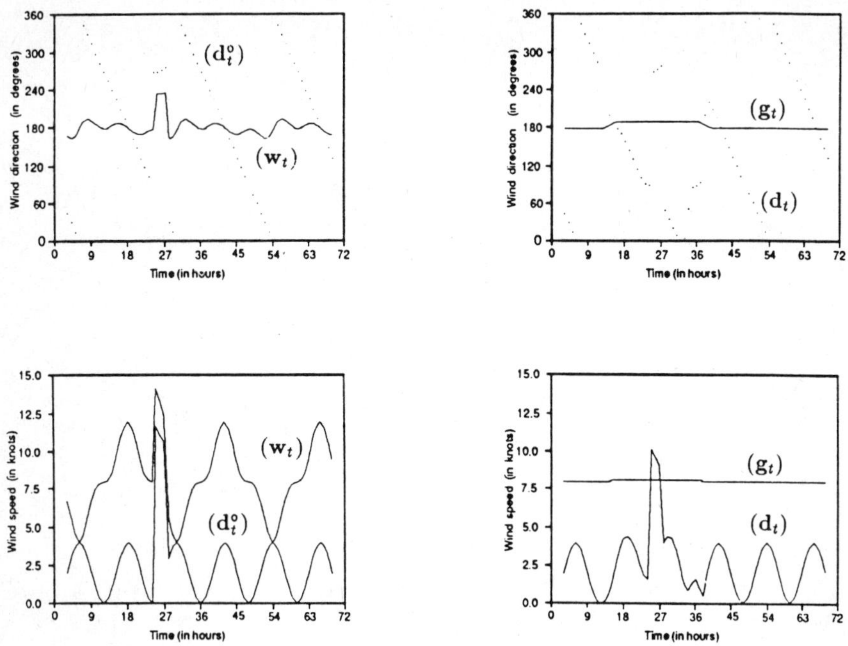

the components of the simulated series. The performance of various filters in a two-dimensional situation is illustrated in Figures 2.10 to 2.12. For the comparison the following components were selected ($t = 0, \ldots, 72$):

(i) the geostrophic component $\mathbf{g}^\circ(t)$ which is identical to $(8,\ 180°)$,

(ii) the land and sea breeze cycle

$$\mathbf{b}^\circ(t) = (4 \sin^2 \frac{2\pi}{24}\, t,\ 90° - 15°t)\ ,$$

(iii) the storm component

$$\mathbf{e}_\delta^\circ(t) = \begin{cases} (12,\ 270°) & \text{if } t \in (24,\ 24+\delta\,] \\ (0,\ 0°) & \text{otherwise} \end{cases}\ , \text{ and}$$

(iv) the short-term fluctuations

$$\mathbf{f}^\circ(t) = (0,\ 0°)\ .$$

The definition of $\mathbf{b}^\circ(t)$ means that wind direction moves anticlockwise in a full circle once every 24 hours, while wind speed is peaking twice a day corresponding to a land breeze at 0600 h from the north and a sea breeze at 1800 h from the south. As for a scale parameter we chose the median deviation (1) based on wind speeds only.

Figure 2.11. Decomposition of a series of wind speeds and directions, $\delta = 6$.

(a) Simulated series ($\mathbf{w}_t$) and ($\mathbf{d}_t^\circ$).

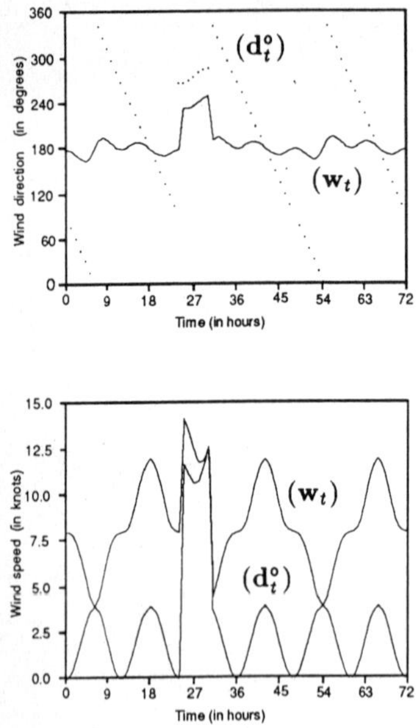

(b) Mean filter M.

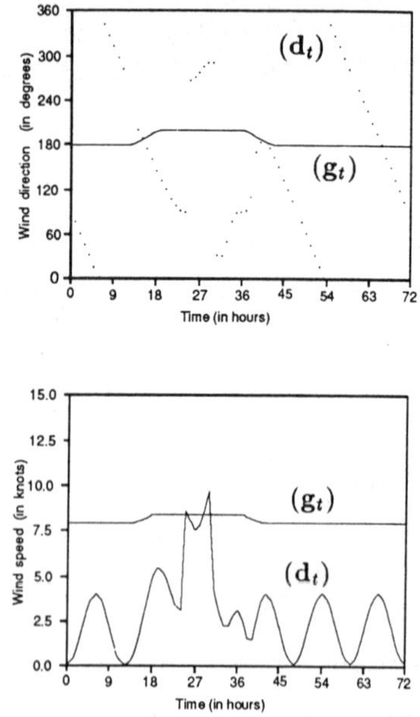

(c) Median filter N.

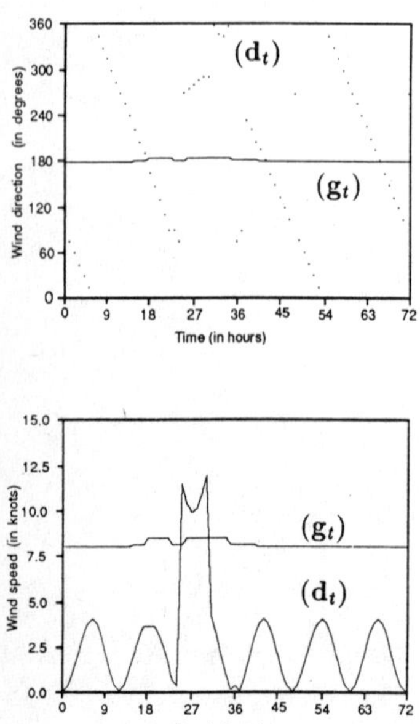

(d) Hyperbolic mean hM.

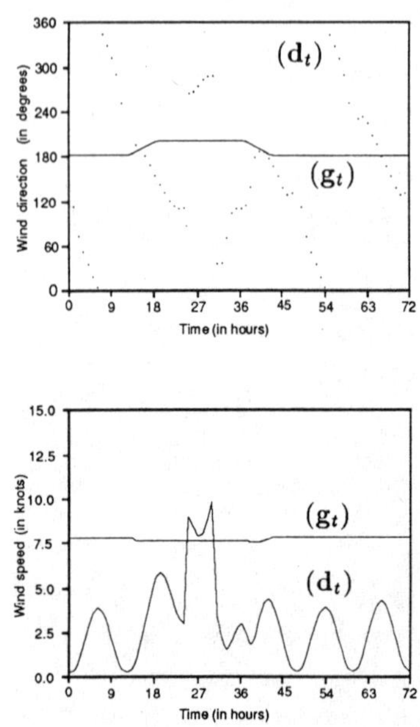

Figure 2.11. Decomposition of a series of wind speeds and directions, $\delta = 6$.

(e) Composite $(5H)^3$.

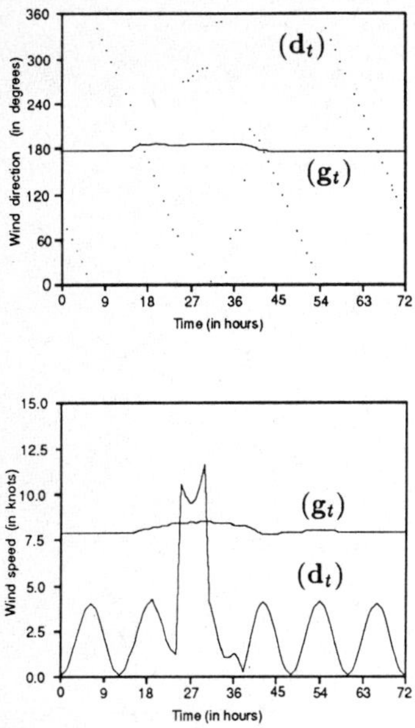

(f) Hyperbolic $(h5H)^3$.

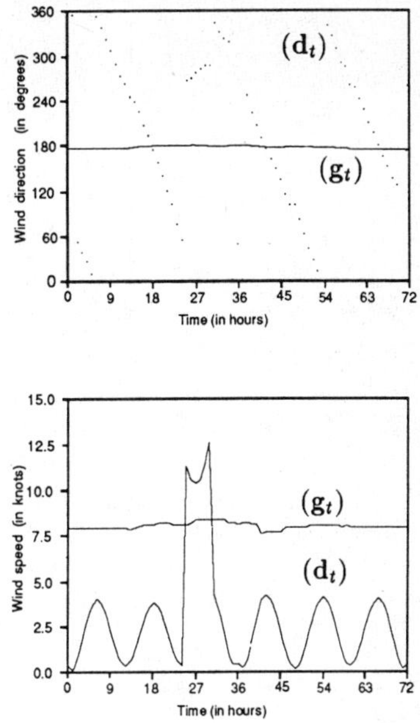

(g) Composite $(5H)^2$.

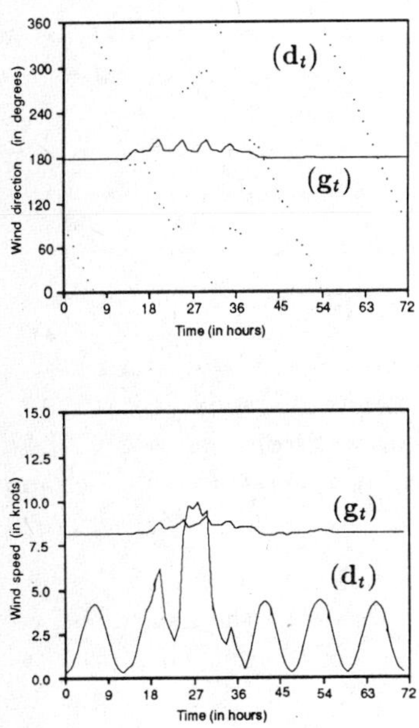

(h) Hampel's $H = H(1.7, 3.4, 8.5)$.

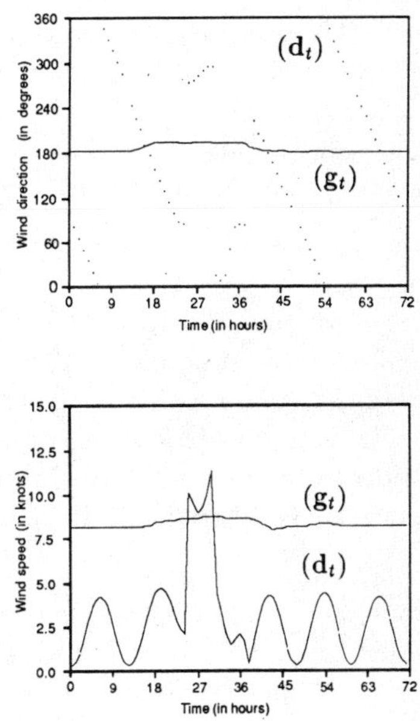

Figure 2.11. Decomposition of a series of wind speeds and directions, $\delta = 6$.

(i) Polar mean pM.

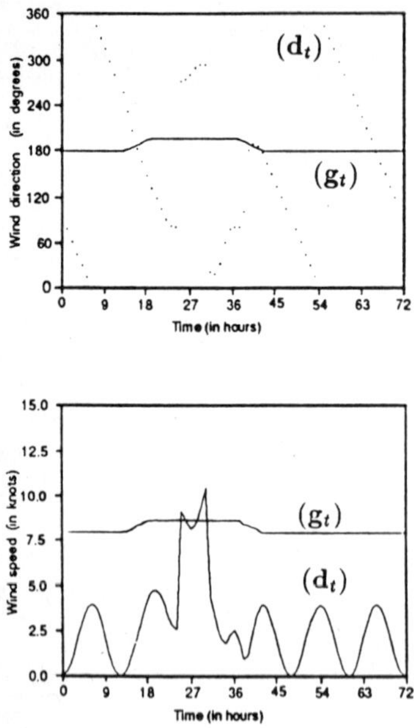

(j) Polar $(p5H)^3$.

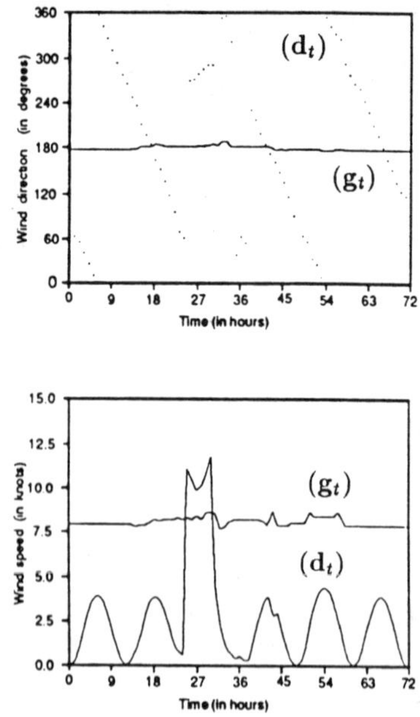

The simulated series $(\mathbf{w}_t)$ and $(\mathbf{d}_t)$ are depicted in diagram (a) of Figures 2.10 to 2.12. Note that each diagram contains one plot for wind direction and one for wind speed. The direction in $(\mathbf{d}_t)$ as well as the speed in $(\mathbf{w}_t)$ clearly exhibit three cycles which are interrupted only for the period of the storm. Referring to Figures 2.10 (a), 2.11 (a) and 2.12 (a), it is noted that the wind in $(\mathbf{d}_t)$ just after the storm continues to turn anticlockwise and to blow from a northerly to westerly direction.

However, when applying the mean filter the wind following the storm tends to come from a north-easterly direction and to turn clockwise until the sea breeze is peaking at about 4200 h, as shown in diagram (b) of Figures 2.10 to 2.12. Of course, the longer the storm lasts, the more pronounced is the distortion. In case of a 6-hour storm duration the land breeze at 3000 h is highly distorted but still noticeable, while in case of a 10-hour storm duration it is virtually eliminated. It is further seen from Figure 2.12 (b) that both, direction and speed in $(\mathbf{d}_t)$ exhibit only two instead of the three cycles imposed on the geostrophic component by $(\mathbf{b}_t^\circ)$.

Table 2.2 shows the degree of association between the generated and the filtered daily component. As in the last section, the comparison is based on the measure of vector correlation discussed in chapter 7, with the 24 observations of the daily component just before and the 24 observations just after the storm stacked into one 96-dimensional vector. The numbers presented in Table 2.2 clearly show that the performance of the mean deteriorates much faster than that of the other filters as the storm duration

Figure 2.12. Decomposition of a series of wind speeds and directions, $\delta = 10$.

(a) Simulated series $(\mathbf{w}_t)$ *and* $(\mathbf{d}_t^\circ)$.

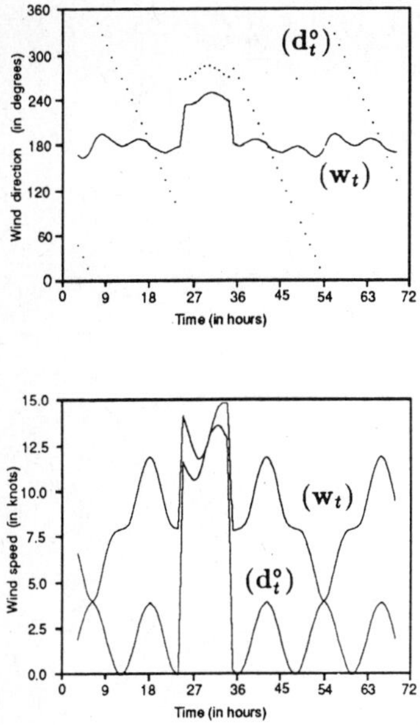

(b) Mean filter M.

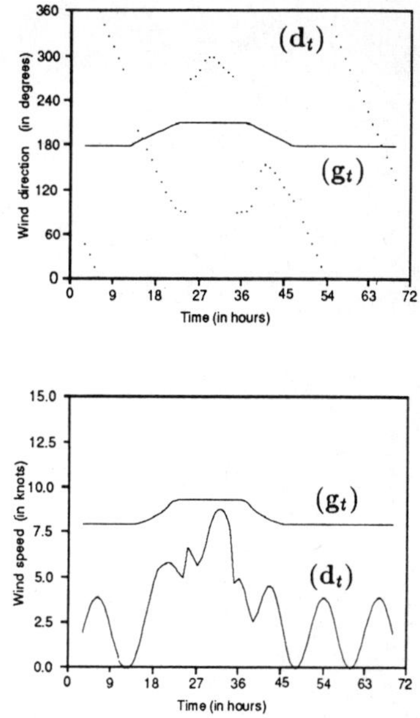

(c) Median filter N.

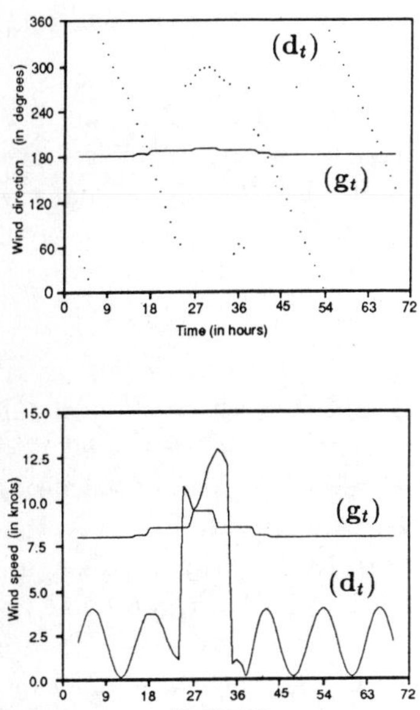

(d) Hyperbolic mean hM.

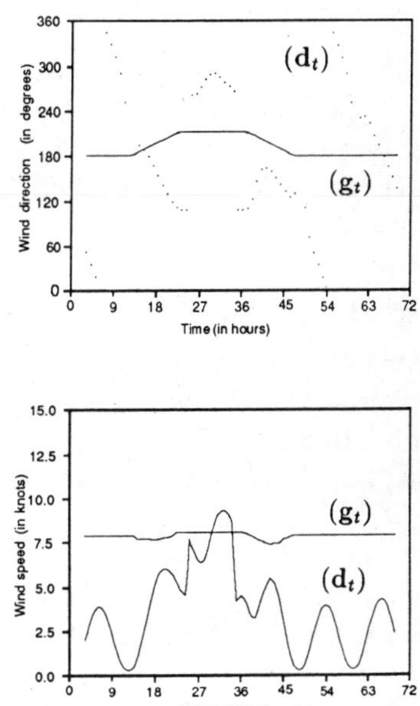

Figure 2.12. Decomposition of a series of wind speeds and directions, $\delta = 10$.

(e) Composite $(5H)^3$. *(f) Hyperbolic $(h5H)^3$.*

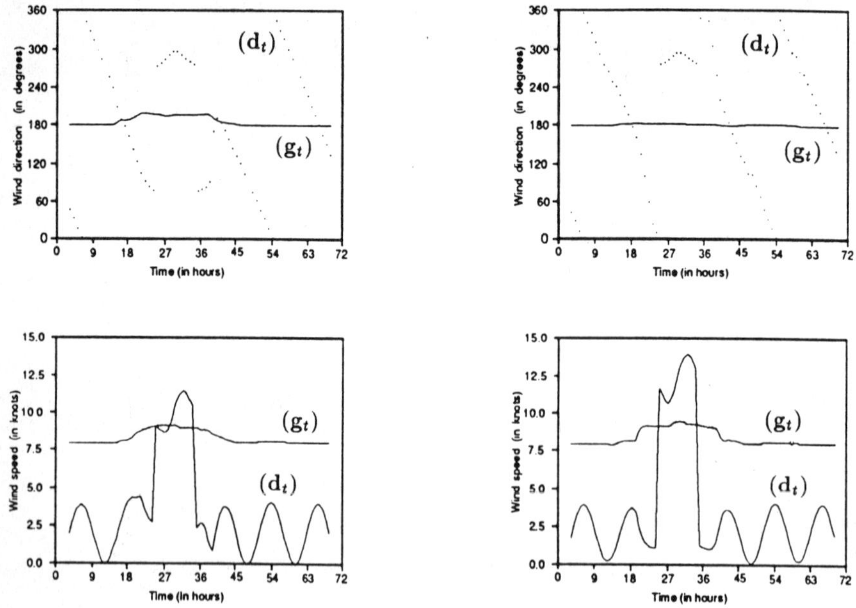

increases. Note, however, that the performance of the mean filter M relative to $(5H)^3$, for example, is not as bad as it was in section 2.3. This is because the component $e_\delta(t)^\circ$ is not as dominant as it was in the univariate simulation study.

Both the hyperbolic mean hM depicted in diagram (d) of Figures 2.11 and 2.12, and the polar mean pM depicted in Figure 2.11 (i) demonstrate a similar behaviour to the cartesian mean, even though the numbers presented in Table 2.2 indicate that they are slightly superior. Further note that the speed of $(\mathbf{g}_t)$ in Figure 2.11 (d) actually decreases during the storm. This is a consequence of the hyperbolic data representation as explained in appendix 2.6.1 and illustrated in Figure 2.17.

As in the univariate filter study, the clearly best performance is shown by the one-step M-estimators. From Figures 2.11 (c), 2.12 (c) and 2.11 (h) it is seen that the median and Hampel's 3-part descending estimator recapture the generated daily component $(\mathbf{d}_t^\circ)$ almost exactly. This is also reflected in the high degree of association, listed in Table 2.2, between the generated and filtered daily component. However, their performance is not greatly reduced when using the composite procedures $(5N)^2$ and $(5H)^2$ instead, judging by the numbers in Table 2.2 and the plot in Figure 2.11 (g). Indeed, a comparison of Figures 2.11 (e) and (g) shows, that if an additional $(5H)$-filter is used for smoothing the series produced by $(5H)^2$, the result is almost as good as for the corresponding one-step H-filter.

Table 2.2. Degree of association between the generated and reconstructed daily components.

Representation	δ	M	N_r	$(5N)^2$	$(5H)^2$	$(5H)^3$
Cartesian	3	0.93	1.00	0.99	0.97	0.97
	6	0.77	0.98	0.97	0.93	0.95
	10	0.72	0.98	0.97	0.87	0.88
Polar	6	0.82			0.93*	0.97*
Hyperboloid	6	0.80	0.97	0.94	0.98*	0.99*
	10	0.78	0.93	0.94	0.97*	0.97*

The asterisk indicates use of the cartesian median as an initial estimate.

The same is true for the composite filters based on a polar or hyperbolic representation. Their behaviour is illustrated in Figure 2.11 (j), and Figures 2.11 (f) and 2.12 (f), respectively. However, both representations are very sensitive to directional variations, besides the fact that the hyperbolic representation is not scale invariant. The use of these representations therefore had to be restricted to the addition of vectors pointing in similar directions since the results were not reliable otherwise. In fact, if used in conjunction with the cartesian median N, as an initial estimate and as centre for the calculation of the median deviation (1), both the hyperbolic and polar representation produced almost perfect results as indicated by the numbers in the last column of Table 2.2. However, if the geostrophic or residual components were less regular they both proved less successful than the cartesian representation.

In summarising, it can be said that the geostrophic component was recaptured in virtually all cases where bounded influence curves were used. The composite M-estimators were almost as good as the corresponding one-step filters. Even though the polar and hyperbolic representations produced satisfactory results they turned out to be sensitive to changes in the geostrophic component. The performance of the mean was clearly inferior to that of the other filters if the storm lasted longer than 6 hours. However, the distortion caused by the storm was confined mainly to the period between the peaks of the land and sea breeze circulation at 1800 h and at 4200 h.

2.5 Application to Wind Series

In order to test the different filters on real data, let us return to the wind record of 3 to 5 December 1971 plotted in Figure 2.14 (a). The filter used for the decomposition of the wind series $(\mathbf{w}_t)$ comprises a filter of the type described in the last section and a smoothing filter based on a moving average. This is done because the geostrophic and residual components are not as regular as in the previous Monte-Carlo studies. As for the smoothing operation we chose the filter S which converts an input series $(\mathbf{y}_t)$ into an output series $(\mathbf{z}_t)$ according to

$$\mathbf{z}_t = \tfrac{1}{8}\,\mathbf{y}_t + \sum_{j=1}^{6} \tfrac{7-j}{48}\,(\mathbf{y}_{t-j} + \mathbf{y}_{t+j})\,.$$

The weights applied to $(\mathbf{y}_t)$ are shown in Figure 2.13. The reader is reminded that all algebraic operations are performed within the data representation chosen for the particular filter, unless it is noted otherwise.

Figure 2.13. Weights of the smoothing filter S.

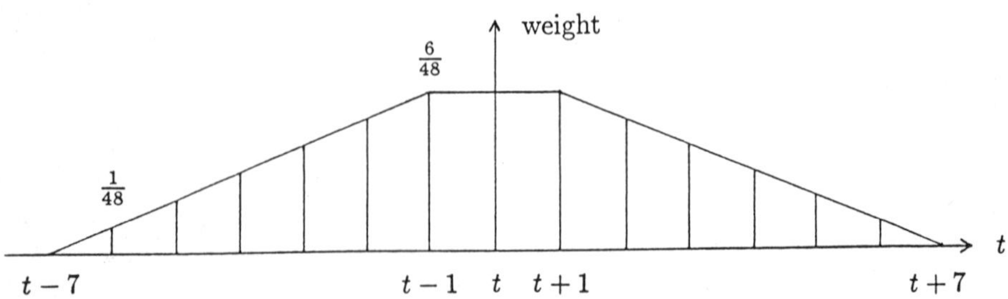

The performance of four different filters using the wind record $(\mathbf{w}_t)$ of 3 to 5 December 1971 is illustrated in Figure 2.14. First observe that all filters produce a similar anticlockwise cycle of the direction in $(\mathbf{d}_t)$, with an easterly land breeze in the morning and a south-westerly sea breeze in the late afternoon. However, the speed components of $(\mathbf{d}_t)$ obtained for the different filters exhibit a quite distinct behaviour. Instead of the three cycles depicted in Figure 2.14 (a) we would expect six peaks in the speed component of $(\mathbf{d}_t)$ corresponding to the three land and sea breeze cycles. Figures 2.14 (c) and (d) provide marginal evidence that the smoothed median NS and the composite M-estimator $(5H)^2S$ actually produce such a pattern, except for the sea breeze on 3 December 1971. However, as the three cycles reflecting the 24-hour periodicity are still noticeable in the speed component of $(\mathbf{d}_t)$ obtained by the median, the composite M-estimator seems to be slightly superior. Indeed, given the less regular geostrophic component the median turns out to be not as successful as in the previous Monte-Carlo studies. Also note the similarity of $(5H)^2S$ with the mean filter M depicted in Figure 2.14 (b).

Figure 2.14. Decomposition of the wind series ($\mathbf{w}_t$), *3 to 5 December 1971.*

(a) Wind series ($\mathbf{w}_t$).

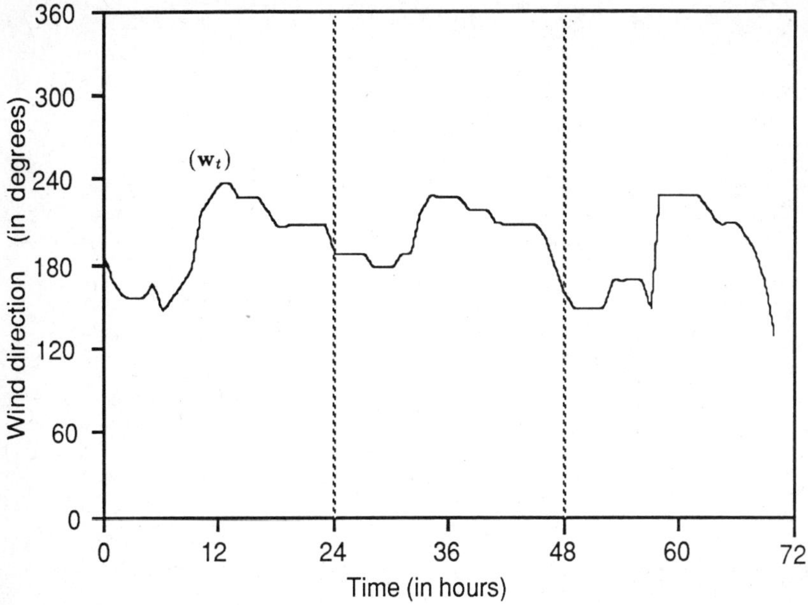

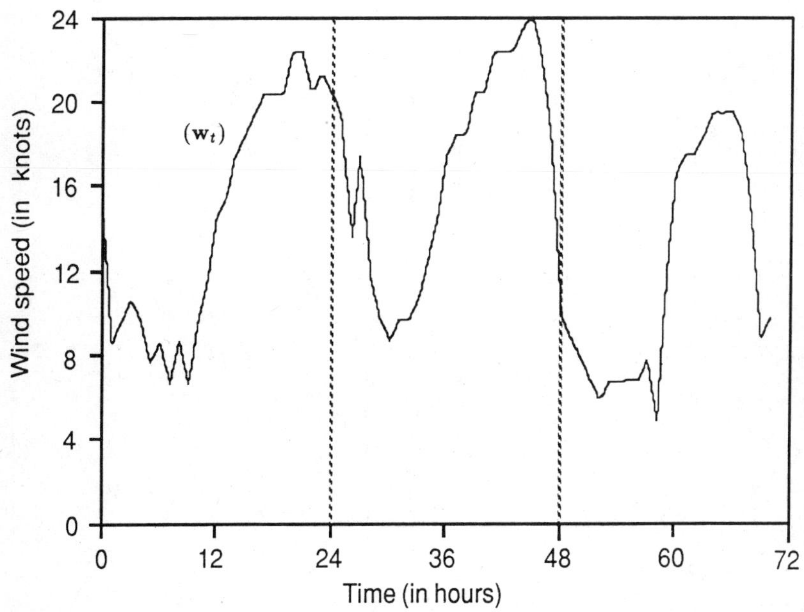

Figure 2.14. Decomposition of the wind series ($\mathbf{w}_t$)*, 3 to 5 December 1971.*

(b) Mean filter M.

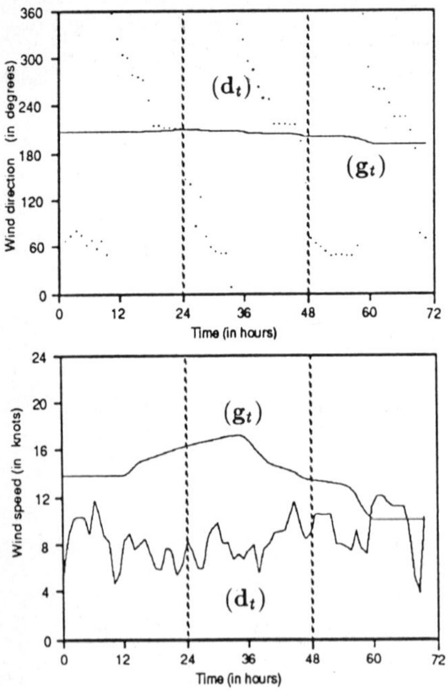

(c) Smoothed median NS.

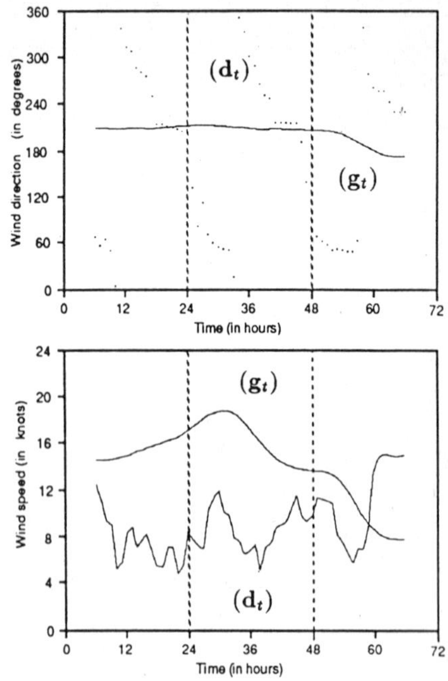

(d) Smoothed composite $(5H)^2S$*.*

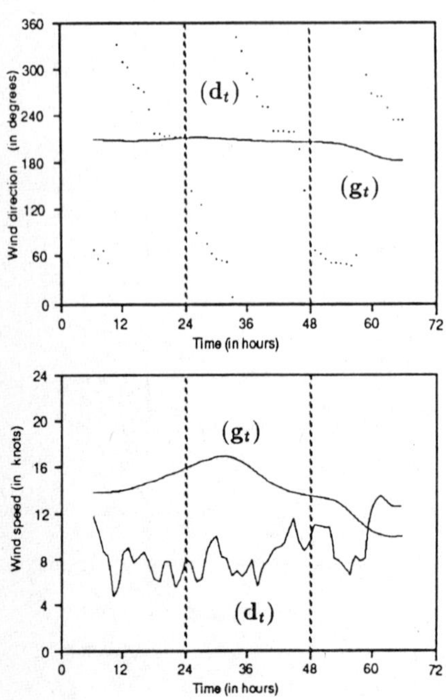

(e) Smoothed hyperbolic $(h5H)^2S$*.*

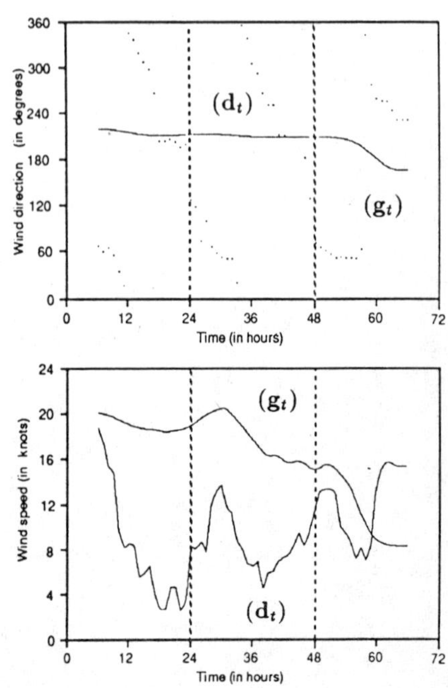

The effect the data representation can have on the decomposition is illustrated in Figure 2.14 (e). Whereas in case of a cartesian representation most filters eliminate the pronounced 24-hour periodicity, this cycle may be retained if a polar or hyperbolic representation is used, due to the estimated high mean level of the geostrophic component. Both these representations emphasize the direction more than the speed, thus reducing the influence of the dominant vector by more than a cartesian representation. This can be a vital point when adding vectors with different lengths and directions, as demonstrated in Figure 2.18.

Let us define the *mean sum of squares* of a time series $(\mathbf{x}_t)_{t=1\ldots T}$ as

$$s_{\mathbf{x}}^2 = \frac{1}{T} \sum_{t=1}^{T} \|\mathbf{x}_t\|^2 .$$

It is noted that the expression $s_{\mathbf{x}}^2$ can be physically interpreted as the *power* of the series $(\mathbf{x}_t)$. The ratios $q_{\mathbf{g}} = s_{\mathbf{g}}^2/s_{\mathbf{w}}^2$ and $q_{\mathbf{d}} = s_{\mathbf{d}}^2/s_{\mathbf{w}}^2$ provide some indication of how well the wind series $(\mathbf{w}_t)$ is explained by the geostrophic component $(\mathbf{g}_t)$, and of the relative importance of the daily fluctuation $(\mathbf{d}_t)$ about $(\mathbf{g}_t)$. Table 2.3 gives the results for the data presented in Figure 2.14 (a). Usually, the smaller the ratio $q_{\mathbf{d}}/q_{\mathbf{g}}$ and the smaller the sum $q_{\mathbf{d}} + q_{\mathbf{g}}$, the better the filter is.

Table 2.3. Mean sum of squares relative to the wind series $(\mathbf{w}_t)$ for the geostrophic component $(\mathbf{g}_t)$ and the daily component $(\mathbf{d}_t)$.

Representation	Ratio	M	N_r	$(5N)^2S$	$(5H)^2S$	$(5H)^2S^*$
Cartesian	$q_{\mathbf{g}}$	0.737	0.772	1.022	0.732	
	$q_{\mathbf{d}}$	0.268	0.317	0.305	0.289	
Hyperboloid	$q_{\mathbf{g}}$	0.517	1.016	1.041	1.011	0.726*
	$q_{\mathbf{d}}$	0.292	0.414	0.397	0.378	0.288*

The asterisk indicates use of the cartesian median as an initial estimate.

The results listed in Table 2.3 confirm the conclusions drawn from Figure 2.14. The high values for $q_{\mathbf{g}}$ in case of a hyperbolic representation reflect the high mean level of the geostrophic component. An exception is the 0.517 for the mean filter due to the *concave* behaviour of the hyperbolic representation, as explained in appendix 2.6.1. Apart from this particular filter the smallest values are obtained for the cartesian mean and almost identically for the composite $(5H)^2S$. The cartesian median, on the other hand, bears some similarity with the hyperbolic filters.

Although $(5H)^2S$ is a better estimator in general, for most of the time the mean filter M did not result in a noticeably different decomposition. The reason is that in the real data series the departures from the mean flow, as determined by the geostrophic, and land and sea breeze components, are far less pronounced than in the simulated series of sections 2.3 and 2.4. Since the use of $(5H)^2S$ brought only a small improvement over M for a large increase in computing time, it was felt that its application was not justified under the given circumstances. It was therefore decided to use a mean filter for the initial decomposition and to experiment with the filter M_α $(0 \leq \alpha \leq 1)$ defined for an input series $(\mathbf{w}_t)$ as follows:

$$\mathbf{g}_t = \frac{1}{24+48\alpha}\left(\frac{\alpha}{2}\left(\mathbf{w}_{t-36}+\mathbf{w}_{t+36}\right)+\alpha\sum_{s=13}^{35}\left(\mathbf{w}_{t-s}+\mathbf{w}_{t+s}\right)\right.$$
$$\left.+\frac{1+\alpha}{2}\left(\mathbf{w}_{t-12}+\mathbf{w}_{t+12}\right)+\sum_{s=-11}^{11}\mathbf{w}_{t+s}\right).$$

Figure 2.15. Weights of the mean filter M_α.

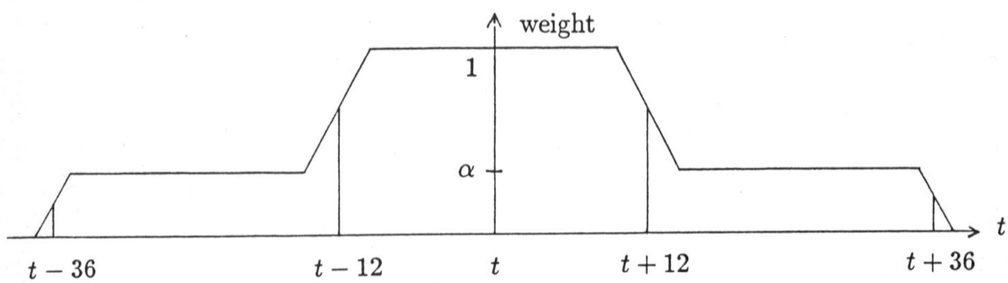

The weights applied to $(\mathbf{w}_t)$ according to this filter are shown in Figure 2.15. The difference between this class of filters and the mean filter M considered so far is in the additional smoothing covering three consecutive cycles. This implies that the distortions in $(\mathbf{g}_t)$ caused by the storms are distributed over a wider range of observations and therefore are less noticeable. The few missing values in the wind record $(\mathbf{w}_t)$ were estimated by linear interpolation, separately for speeds and directions. The daily component $(\mathbf{d}_t)$ which we refer to in subsequent chapters is defined as the residual series after subtraction of the geostrophic component $(\mathbf{g}_t)$, obtained by the filter $M_{.5}$, from $(\mathbf{w}_t)$.

Table 2.4 shows a few characteristics of the original wind series $(\mathbf{w}_t)$, and of the geostrophic component $(\mathbf{g}_t)$ and the daily component $(\mathbf{d}_t)$. These include the mean wind directions $\overline{\theta}_\mathbf{w}$ and $\overline{\theta}_\mathbf{g}$ as well as the mean wind speeds $\overline{v}_\mathbf{w}$ and $\overline{v}_\mathbf{g}$ of $(\mathbf{w}_t)$ and $(\mathbf{g}_t)$, respectively. Also presented are the mean directional resultant length $r_\mathbf{w}$, the sum of squares $s^2_\mathbf{w}$ of $(\mathbf{w}_t)$, and the ratios $q_\mathbf{g}$ and $q_\mathbf{d}$ for various filters M_α defined above $(\alpha = 0.1, 0.3, 0.5)$.

The absence of a predominant wind direction in winter explains the small value of $r_\mathbf{w}$, as opposed to summer when southerlies prevail. This is reflected also in the mean

Table 2.4. Speed, direction and sum of squares characteristics of the three series $(\mathbf{w}_t)$, $(\mathbf{g}_t)$ *and* $(\mathbf{d}_t)$.

Season	$\overline{\theta}_{\mathbf{w}}$	$\overline{v}_{\mathbf{w}}$	$r_{\mathbf{w}}$	$s^2_{\mathbf{w}}$	Filter	$\overline{\theta}_{\mathbf{g}}$	$\overline{v}_{\mathbf{g}}$	$q_{\mathbf{g}}$	$q_{\mathbf{d}}$
Winter 1971	91°	9.3	0.12	124	$M_{.1}$	131°	6.7	0.51	0.29
					$M_{.3}$	163°	6.0	0.39	0.39
					$M_{.5}$	200°	5.7	0.35	0.46
Summer 1971/72	180°	12.4	0.56	184	$M_{.1}$	181°	9.8	0.60	0.33
					$M_{.3}$	181°	9.5	0.56	0.35
					$M_{.5}$	181°	9.4	0.54	0.38

direction $\overline{\theta}_{\mathbf{g}}$ of the geostrophic component $(\mathbf{g}_t)$: whereas in winter it varies depending on the filter used for the decomposition, in summer it remains virtually unchanged. Note that the mean speed in summer 1971/72 was higher than in winter 1971, although usually speeds were higher in winter than in summer.

Further, as α increases the ratio $q_{\mathbf{g}}$ decreases while $q_{\mathbf{d}}$ increases, due to the more pronounced smoothing. Whereas in summer the values for $q_{\mathbf{g}}$ and $q_{\mathbf{d}}$ were roughly 0.55 and 0.35, respectively, in winter they varied more widely with an average of about 0.4 for both $q_{\mathbf{g}}$ and $q_{\mathbf{d}}$. This indicates that the geostrophic component in summer is a slightly better predictor of the actual wind than in winter. It is also noted that the mean sum of squares $s^2_{\mathbf{d}}$ was generally higher in summer, which can be attributed mainly to the stronger sea breeze circulation during this time of the year.

The fact that the wind speeds in $(\mathbf{g}_t)$ were higher in summer 1971/72 than in winter 1971 is also evident from the scattergrams in Figure 2.16 which show the simultaneous distribution of wind speed and direction in the geostrophic component obtained for $M_{.5}$. In summer the wind comes almost exclusively from an easterly to south-westerly direction, whereas in winter the direction is spread more uniformly over the full circle. The distribution of the wind records in $(\mathbf{g}_t)$ in winter exhibits an almost circular ridge of uniform height. Comparing the diagrams with those of Figure 1.5 it is observed that the filtering of $(\mathbf{w}_t)$ has enhanced this characteristic and at the same time confined the spread of wind direction in summer much more clearly to the south to south-easterly sector.

Figure 2.16. Frequency of wind speed and direction in $(\mathbf{g}_t)$.

(a) Winter 1971.

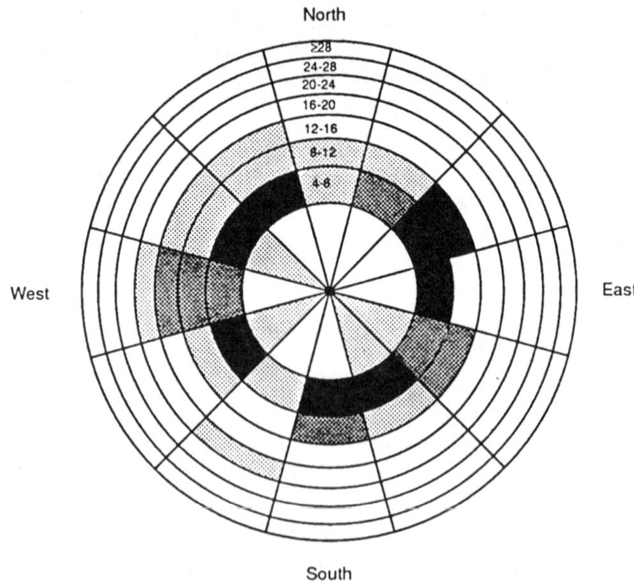

(b) Summer 1971/72.

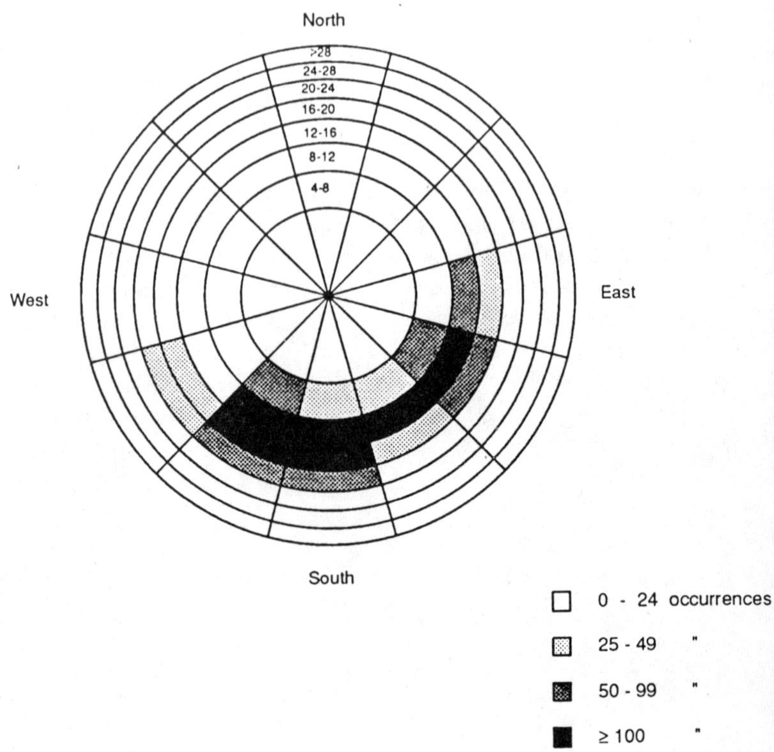

2.6 Appendix: Mathematical Details

2.6.1 Cartesian, polar and hyperbolic data representation

In the case of the given time series the data would commonly be represented in cartesian coordinates. However, this approach may be inconvenient for the following three reasons:

(i) Most wind patterns are usually considered in terms of speed and direction, so that a polar representation (which retains these aspects) seems more appropriate.
(ii) In many situations one may be interested only in the directional aspect. It is therefore desirable that the restriction of our analysis to the directional component follows in a natural way.
(iii) A bivariate normal distribution does not provide an appropriate model for the simultaneous record of wind speed and direction, since densities are required that can exhibit a circular ridge of not necessarily uniform height. Referring to Figure 2.16 (a), for example, we notice a higher frequency of winds with speeds between 4 and 8 knots irrespective of the direction.

In contrast with previous studies, the directional aspect is therefore strongly emphasized. However, different data representations have occasionally been associated with different ways of combining vectors of speed and direction (for example, cf. Jensen 1980). This will now be explained in detail.

The data consist of hourly records of wind speed and direction, and are viewed as elements of the set $\mathcal{S} = \{(v,\theta) \in \mathcal{R}^2 \mid -\pi < \theta \leq \pi,\, v \geq 0\}$. By means of a mapping $\kappa : \mathcal{S} \to \mathcal{R}^k$ $(k \in \mathcal{N})$ the data are represented as elements of a Euclidean space $\mathcal{R}^k$ furnished with two operations, an addition and a scalar multiplication. The mapping κ, the two operations and k are specified below for the three different data representations. The sample space defined as the image $\kappa(\mathcal{S})$ is usually a proper subset of $\mathcal{R}^k$ and may not be closed under the two operations. Given (v,θ), $(u,\eta) \in \mathcal{S}$ and α, $\beta \in \mathcal{R}$ the question therefore is how to interpret the resultant $\alpha\kappa(v,\theta) + \beta\kappa(u,\eta)$ as a record of wind speed and direction.

(a) *Cartesian representation.* Here $k = 2$, while the mapping κ is given by

$$\kappa : (v,\theta) \mapsto \begin{pmatrix} v\cos\theta \\ v\sin\theta \end{pmatrix}.$$

Vector addition and scalar multiplication are defined as usual. Suppose that

$$\begin{pmatrix} r\cos\zeta \\ r\sin\zeta \end{pmatrix} = \alpha \begin{pmatrix} v\cos\theta \\ v\sin\theta \end{pmatrix} + \beta \begin{pmatrix} u\cos\eta \\ u\sin\eta \end{pmatrix}$$

then r and ζ are interpreted as *resultant speed* and *resultant direction*, respectively. Further, the *mean speed* is defined as $w_c = r/[\,|\alpha| + |\beta|\,]$ while the *mean direction* ζ_c is identical with the resultant direction ζ.

(b) Polar representation. In this case we have $k = 3$ and

$$\kappa : (v, \theta) \mapsto \begin{pmatrix} \cos\theta \\ \sin\theta \\ v \end{pmatrix} .$$

Here ordinary vector addition is complemented by the scalar multiplication

$$\alpha \begin{pmatrix} x \\ y \\ z \end{pmatrix} = \begin{pmatrix} \alpha x \\ \alpha y \\ |\alpha| z \end{pmatrix} \qquad x, y \in \mathcal{R}\,, \quad z \in \mathcal{R}_0^+ \,.$$

If

$$\begin{pmatrix} r\cos\zeta \\ r\sin\zeta \\ w \end{pmatrix} = \alpha \begin{pmatrix} \cos\theta \\ \sin\theta \\ v \end{pmatrix} + \beta \begin{pmatrix} \cos\eta \\ \sin\eta \\ u \end{pmatrix}$$

then the ***resultant speed*** and ***direction*** are given by $wr/[\,|\alpha| + |\beta|\,]$ and ζ, respectively. The ***mean direction*** is $\zeta_p = \zeta$, while the ***mean speed*** is defined as $w_p = wr/[\,|\alpha| + |\beta|\,]^2$. Further, r and $r/[\,|\alpha| + |\beta|\,]$ are commonly called ***directional resultant length*** and ***mean directional resultant length***, respectively.

(c) Hyperbolic representation. Here we have $k = 3$ and, putting $v = \sinh z$,

$$\kappa : (v, \theta) \mapsto \begin{pmatrix} v\cos\theta \\ v\sin\theta \\ \sqrt{v^2+1} \end{pmatrix} = \begin{pmatrix} \sinh z \;\; \cos\theta \\ \sinh z \;\; \sin\theta \\ \cosh z \end{pmatrix} .$$

Vector addition and scalar multiplication are defined as for the polar representation. Suppose that

$$\begin{pmatrix} r\cos\zeta \\ r\sin\zeta \\ w \end{pmatrix} = \alpha \begin{pmatrix} v\cos\theta \\ v\sin\theta \\ \sqrt{v^2+1} \end{pmatrix} + \beta \begin{pmatrix} u\cos\eta \\ u\sin\eta \\ \sqrt{u^2+1} \end{pmatrix} .$$

Whilst ***resultant direction*** and ***mean direction*** are both given by $\zeta_h = \zeta$, ***resultant speed*** and ***mean speed*** are defined as r and $w_h = r/[w^2 - r^2]^{1/2}$, respectively. This representation is closely related to the ***hyperboloid distribution*** which provides a model for the simultaneous record (v, θ) of wind speed and direction. In many ways the hyperboloid distribution behaves like the von Mises distribution. In particular, the direction θ conditional on the speed v follows a von Mises distribution. However, if v and θ are supposed to be independent, then θ is forced to be uniformly distributed. For a detailed discussion of the hyperboloid distribution the reader is referred to Jensen (1980).

To illustrate the effect of the data representation let us consider the convex combination of the two vectors $\mathbf{v} = (0.5, 120°)$ and $\mathbf{u} = (1, 0°)$ plotted in Figure 2.17, with α varying between 0 and 1, and $\beta = 1 - \alpha$. Mean speed and direction are calculated below for each of the three data representations.

(a) Cartesian representation

$$w_c = \tfrac{1}{2}\sqrt{4 - 10\alpha + 7\alpha^2}$$
$$\zeta_c = \arctan(\alpha\sqrt{3}\,,\, 4 - 5\alpha)$$

(b) Polar representation

$$w_p = (1 - \tfrac{1}{2}\alpha)\sqrt{1 - 3\alpha + 3\alpha^2}$$
$$\zeta_p = \arctan(\alpha\sqrt{3}\,,\, 2 - 3\alpha)$$

(c) Hyperbolic representation

$$w_h = \sqrt{\frac{4 - 10\alpha + 7\alpha^2}{4 + 2\alpha(1-\alpha)(2\sqrt{10} - 3)}}$$
$$\zeta_h = \arctan(\alpha\sqrt{3}\,,\, 4 - 5\alpha) = \zeta_c$$

Figure 2.17. Convex combination in different data representations.

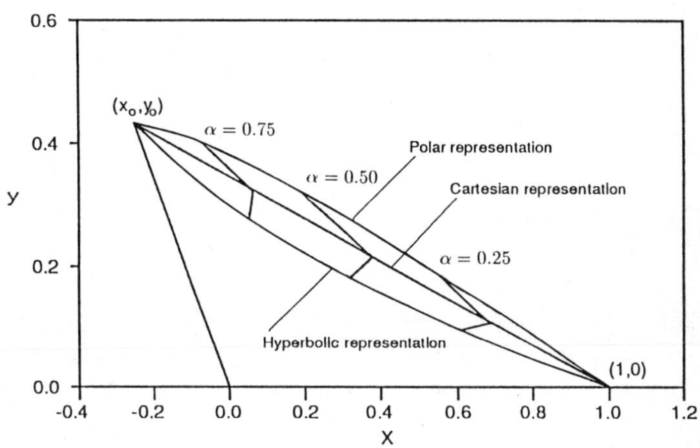

First observe that the curve is convex for the polar representation, linear for the cartesian representation and concave for the hyperbolic representation. Further, it is easy to show that for each value of α between 0 and 1 both w_p and w_h are smaller than w_c. In case of a cartesian representation the algebra is dominated by the *largest* vector, whereas in case of a polar representation the mean vector is more strongly determined by the *directions* of the two vectors. This can be seen in Figure 2.17 by comparing the results for the three data representations for $\alpha = 0.25$, $\alpha = 0.5$ and $\alpha = 0.75$. It follows that algebraic operations using a cartesian representation are more stable, especially when dealing with extremely different speeds and directions. This is even more obvious in Figure 2.18 where we form the mean of $\mathbf{v} = (1, \theta)$ and $\mathbf{u} = (10, 0°)$, and vary θ over the full circle.

In case of a polar representation the mean vector becomes $\mathbf{0}$ if the vectors $\mathbf{v}$ and $\mathbf{u}$ point into opposite directions. This illustrates that it is the direction and not the speed which is the dominant aspect of this representation. For the cartesian representation the situation is reversed, with speed being the dominant aspect. Finally, note that the mean directions for the cartesian and hyperbolic representation always coincide.

Figure 2.18. Mean of two vectors in different data representations.

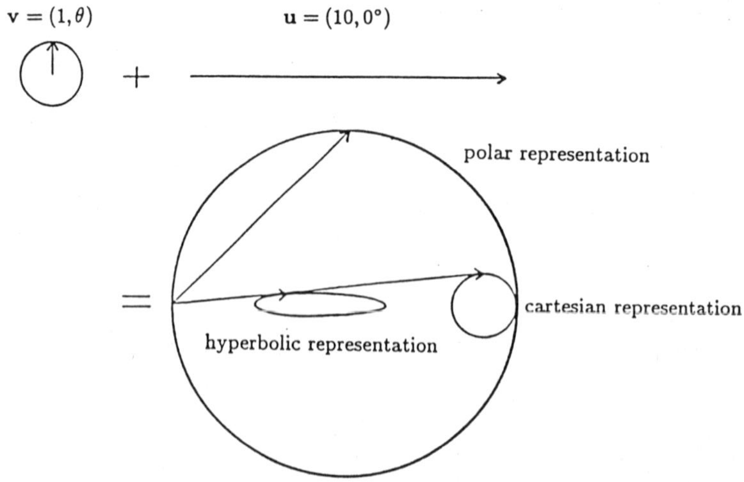

2.6.2 The spatial median

Given a sample $\{\mathbf{x}_1, \ldots, \mathbf{x}_n\}$ of observations in $\mathcal{R}^k$ $(k > 1)$, we will now present a procedure for calculating the spatial median as defined by (4). Let $\mathbf{x} = (x_1, \ldots, x_k)'$ be the spatial median and $\mathbf{x}_0$ be an approximation, for example the restricted median. We then have

(5) $$\left(\frac{\partial L_k}{\partial x_i}\right)_{i=1\ldots k} [\mathbf{x}] = \mathbf{0} \quad ,$$

assuming that the derivative exists in a neighbourhood of $\mathbf{x}$ including $\mathbf{x}_0$. Expanding the left-hand side of (5) into a Taylor series about $\mathbf{x}_0$ gives

$$\left(\frac{\partial L_k}{\partial x_i}\right)_{i=1\ldots k} [\mathbf{x}] =$$
$$\left(\frac{\partial L_k}{\partial x_i}\right)_{i=1\ldots k} [\mathbf{x}_0] + \left(\frac{\partial^2 L_k}{\partial x_i \partial x_j}\right)_{i,j=1\ldots k} [\mathbf{x}_0]\ (\mathbf{x} - \mathbf{x}_0) + o(\|\mathbf{x} - \mathbf{x}_0\|^2)\ .$$

This leads to the following correction term

$$(6) \qquad (\mathbf{x} - \mathbf{x}_0) = -\left(\frac{\partial^2 L_k}{\partial x_i \partial x_j}\right)^{-1}_{i,j=1\ldots k} [\mathbf{x}_0] \cdot \left(\frac{\partial L_k}{\partial x_i}\right)_{i=1\ldots k} [\mathbf{x}_0]$$

provided the inverse of the matrix exists. Repeated application of (6) then yields the spatial median.

If $\|\cdot\|$ denotes the L_2-norm in $\mathcal{R}^k$ then (4) reduces to

$$(7) \qquad L_k(\mathbf{x}) = \sum_{m=1}^{n} \left(\sum_{i=1}^{k} (x_i - x_{mi})^2 \right)^{1/2} \quad , \mathbf{x} \in \mathcal{R}^k ,$$

where x_{mi} denotes the ith coordinate of $\mathbf{x}_m$. Note that (7) corresponds to bounding the length of the influence vector. Let us now concentrate on the case $k = 2$. Writing

$$r_m(x,y) = \sqrt{(x - x_m)^2 + (y - y_m)^2} \qquad (x,y) \in \mathcal{R}^2$$

yields

$$L(x,y) = L_2(x,y) = \sum_{m=1}^{n} r_m(x,y)$$

and hence

$$\frac{\partial L}{\partial x} = \sum_{m=1}^{n} \frac{x - x_m}{r_m(x,y) + \epsilon}$$

$$\frac{\partial L}{\partial y} = \sum_{m=1}^{n} \frac{y - y_m}{r_m(x,y) + \epsilon}$$

$$\frac{\partial^2 L}{\partial x^2} = \sum_{m=1}^{n} \frac{(y - y_m)^2}{r_m^3(x,y) + \epsilon}$$

$$\frac{\partial^2 L}{\partial x \partial y} = -\sum_{m=1}^{n} \frac{(x - x_m)\,(y - y_m)}{r_m^3(x,y) + \epsilon}$$

$$\frac{\partial^2 L}{\partial y^2} = \sum_{m=1}^{n} \frac{(x - x_m)^2}{r_m^3(x,y) + \epsilon}$$

with $\epsilon = 0$. To ensure numerical stability, however, ϵ was set to 0.1. For the same reason the term $\epsilon \cdot \mathbf{I}_2$ was added to the matrix in (6), where $\mathbf{I}_2$ denotes the identity matrix of order 2, before inverting it. That is, the inverse in (6) was replaced by

$$\left[\begin{pmatrix} \frac{\partial^2 L}{\partial x^2} & \frac{\partial^2 L}{\partial x \partial y} \\ \frac{\partial^2 L}{\partial x \partial y} & \frac{\partial^2 L}{\partial y^2} \end{pmatrix} + \epsilon\, \mathbf{I}_2 \right]^{-1} [\mathbf{x}_0] \qquad \text{with} \quad \epsilon = 0.1 \,.$$

CHAPTER 3

THE GEOSTROPHIC COMPONENT

In this chapter the geostrophic component ($\mathbf{g}_t$) *is analysed and related to the synoptic pressure field. A simple model is proposed to predict the geostrophic wind from the configuration of low and high pressure systems, and to establish the extent of association between this wind, which describes a meteorological phenomenon, and the geostrophic component* ($\mathbf{g}_t$) *obtained in the last chapter by filtering the raw data series* ($\mathbf{w}_t$)*. It is shown that the model is fairly robust towards its parameterization and that it achieves a good agreement between* ($\mathbf{g}_t$) *and the predicted geostrophic wind despite its simplicity. It can therefore be assumed that* ($\mathbf{g}_t$) *captures the main characteristics of the geostrophic wind and that the initial decomposition is physically meaningful. We then go on to specify particular synoptic configurations in terms of wind speed and direction, which will be referred to as synoptic states. The objective is to find regularities in the passage of high and low pressure systems, and to establish basic weather patterns. It is demonstrated that a complete partition of* ($\mathbf{g}_t$) *into the synoptic states roughly reflects the seasonal weather patterns as outlined in section 1.2.*

Following an explanation of the meteorological background in section 3.1, the model for prediction of the geostrophic wind is introduced in section 3.2. The agreement between the estimated geostrophic wind and ($\mathbf{g}_t$) *is demonstrated in section 3.3, while the partition of* ($\mathbf{g}_t$) *into sequences of synoptic patterns is discussed in section 3.4. Appendix 3.5 presents the mathematical detail on the derivation of the geostrophic wind equation.*

3.1 The Geostrophic Wind

The horizontal wind field that is associated with synoptic scale systems is approximated by the geostrophic wind to within 10 to 15 per cent in midlatitudes (Holton 1979, p.38). *Geostrophic wind* is an explicit meteorological term which describes the basic component of the global wind as resulting from the configuration of high and low pressure systems. In the present study the geostrophic wind will be estimated on an 6-hourly basis using the central pressure and location of all high and low pressure systems relevant to the weather in the Fremantle region. This series $(\mathbf{h}_t)$ of estimated geostrophic winds captures a meteorological phenomenon and therefore has an immediate physical interpretation. We then compare $(\mathbf{h}_t)$ with the geostrophic component $(\mathbf{g}_t)$ constructed from the wind series $(\mathbf{w}_t)$ in chapter 2, in order to confirm that $(\mathbf{g}_t)$ can in fact be interpreted as a series of synoptic scale winds.

Despite the apparent complexity of atmospheric systems as depicted on synoptic weather charts, the pressure and velocity distributions are approximated by rather simple force balances. In particular, the geostrophic wind is obtained as

(1)
$$\mathbf{u}_g = \mathbf{k} \times \frac{1}{\varrho f} \operatorname{grad} p \, ,$$

where $\times$ denotes the ordinary vector product, $\mathbf{k}$ the vertical unit vector, ϱ the atmospheric density, f the Coriolis parameter and p the atmospheric pressure. According to formula (1) the geostrophic wind has no vertical velocity component and is directly related to the *pressure gradient force* which is defined as $-\varrho^{-1} \operatorname{grad} p$. However, this is usually sufficient to develop a basic understanding of atmospheric disturbances. Note that the Coriolis parameter is positive in the Northern and negative in the Southern Hemisphere and hence, that the geostrophic wind is directed along the isobars with low pressure to the left in the Northern Hemisphere and to the right in the Southern Hemisphere.

3.2 Estimation of the Geostrophic Wind

3.2.1 The pressure gradient model

According to formula (1) it is sufficient to estimate the horizontal pressure gradient, grad p, in order to determine the geostrophic wind $\mathbf{u}_g$. Let d_j denote the distance of the jth pressure system S_j from Fremantle, p_j its central pressure, and c_j a system specific parameter ($j \in \mathcal{N}$). Assuming that the relative influence of a pressure system decreases as the square of the distance increases, the following empirical model of the atmospheric pressure is proposed:

(2) $$p = \left[\sum_j \frac{c_j}{d_j^2}\, p_j\right] \Big/ \left[\sum_j \frac{c_j}{d_j^2}\right].$$

That is, the atmospheric pressure at Fremantle is obtained as a simple weighted average. In order to make the model more realistic in case of a single influential pressure system, and at the same time to ensure general stability of the model, a *prior* pressure system S_0 is included with a central pressure that is equal to standard sea level pressure $p_0 =$ 1013.25 mb and $c_0/d_0^2 = 1$. Hence,

(3) $$p = p_0 + \left[\sum_{j\geq 1} \frac{c_j}{d_j^2}\,(p_j - p_0)\right] \Big/ \left[1 + \sum_{j\geq 1} \frac{c_j}{d_j^2}\right].$$

Figure 3.1. Atmospheric pressure between a high and low pressure system according to equation (3).

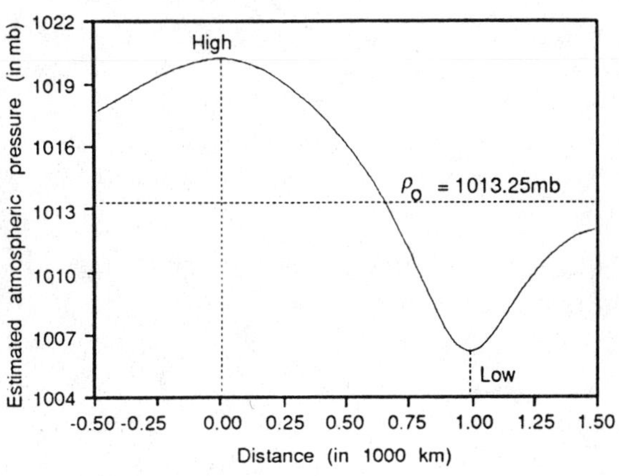

Figure 3.1 shows the estimated atmospheric pressure p given a high with $p_1 =$ 1020.25 mb at $x = 0$, and a low with $p_2 = 1006.25$ mb at $x = 1$. For an observer positioned at x we thus have $d_1 = |x|$ and $d_2 = |1 - x|$. The parameters c_1 and c_2 were set to 0.48 and 0.12, respectively. Notice on the one hand, the much further reaching influence of the high, and on the other hand, that the strongest gradients occur near the centre of the low.

It has implicitly been assumed that the influence of a pressure system depends only on the distance d_j from the centre and not on the direction η_j under which the centre appears. In case of a single influential pressure system, the isobars would therefore form concentric circles around the centre, with (3) reducing to

$$p = p_0 + \frac{p_1 - p_0}{1 + d_1^2/c_1} .$$

Figure 3.2. Estimated atmospheric pressure p as a function of the distance d_1 from the centre of the pressure system.

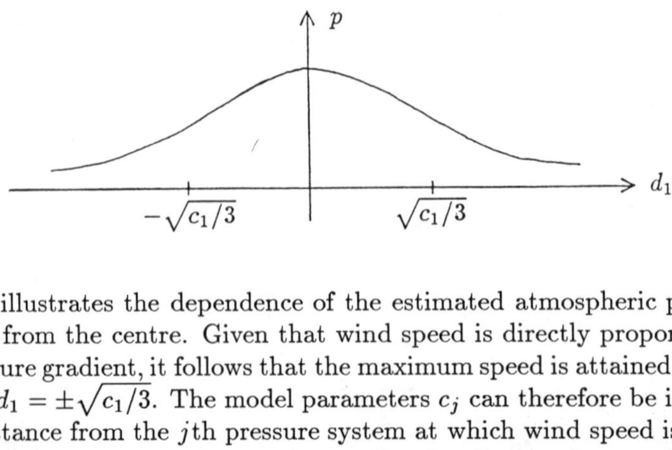

Figure 3.2 illustrates the dependence of the estimated atmospheric pressure p on the distance d_1 from the centre. Given that wind speed is directly proportional to the horizontal pressure gradient, it follows that the maximum speed is attained at the points of inflection at $d_1 = \pm\sqrt{c_1/3}$. The model parameters c_j can therefore be interpreted in terms of the distance from the jth pressure system at which wind speed is expected to be a maximum. Hence, the smaller c_j, the smaller the distance between the centre of the system and the point at which the maximum speed is attained. It can therefore be expected that the system parameters are relatively small for lows and relatively large for highs.

Equation (3) suggests that the pressure gradient may be most conveniently expressed in terms of polar coordinates (d, η). Since

$$\text{(4)} \qquad \text{grad } p = \begin{pmatrix} \cos\eta & -\sin\eta \\ \sin\eta & \cos\eta \end{pmatrix} \begin{pmatrix} \frac{\partial p}{\partial d} \\ \frac{1}{d}\frac{\partial p}{\partial \eta} \end{pmatrix} ,$$

the formula for predicting the geostrophic wind is obtained as follows:

$$\text{(5)} \qquad \mathbf{u}_g = \mathbf{k} \times \frac{1}{\varrho f} \cdot \frac{2}{1 + \sum_{j=1}^{n} d_j^{-2} c_j} \begin{pmatrix} \sum_{j=1}^{n} d_j^{-3} c_j (p_j - p) \cos\eta_j \\ \sum_{j=1}^{n} d_j^{-3} c_j (p_j - p) \sin\eta_j \end{pmatrix} .$$

Hence, the component contributed by the jth pressure system S_j to the geostrophic wind $\mathbf{u}_g$ is characterized by a speed that is proportional to $(p_j - p)/d_j$ and a direction perpendicular to η_j. Furthermore, the relative importance of this component is given by the weight $\left[d_j^{-2} c_j\right] / \left[1 + \sum_{j=1}^{n} d_j^{-2} c_j\right]$. To estimate the geostrophic wind it is therefore sufficient that for each pressure system S_j one calculates the distance d_j from Fremantle and the direction η_j under which it appears. Note that p_j, d_j and η_j are measurable quantities, while the system parameters c_j have to be estimated. For a derivation of formula (5) the reader is referred to appendix 3.5.

3.2.2 Determination of η_j and d_j

The information available on the atmospheric pressure field comprises all pressure systems relevant to the wind at Fremantle. Associated with each pressure system is a triple (p_j, ψ_j, ϕ_j) of central pressure, and of longitude and latitude of the centre. The time series containing this information will subsequently be referred to as $(\mathbf{p}_t)$.

The data in $(\mathbf{p}_t)$ are presented in Table 3.1 for the period of 1 to 8 December 1971, which corresponds to the synoptic charts depicted in Figure 1.3. At 0000 h on 1 December we have a single influential high with central pressure 1016 mb south-east of Fremantle at $119°E$ of longitude and $35°S$ of latitude. While moving in an easterly direction, this system remains relevant to the local wind until 1200 h on 3 December. At the same time the heat low over the north-west of the continent becomes important. It is first recorded at 0600 h on 1 December with a central pressure of 1004 mb, and is almost stationary until the 6 December when it starts to move in a south to south-easterly direction. The data listed in the last three columns of Table 3.1 correspond to a persistent high over the eastern Indian Ocean. During the period considered two other high pressure systems emerged, the first lasting from 5 to 7 December and the second beginning on 8 December. For convenience both their records are presented in the first three columns of Table 3.1.

Taking the centre of the earth as origin, the points on the surface are most conveniently represented in spherical coordinates. That is

$$\mathbf{r} = R_0 \begin{pmatrix} \cos\phi \ \cos\psi \\ \cos\phi \ \sin\psi \\ \sin\phi \end{pmatrix}$$

where $R_0 = 6378$ km is the earth's radius, ψ is the longitude and ϕ is the latitude of the point on the surface. Let $\mathbf{r}_0$ and $\mathbf{r}_j$ denote the position vectors of Fremantle and the centre of the jth pressure system, respectively ($\psi_0 = 116°$ and $\phi_0 = -32°$). The *great circle* distance between these two points is then given by

(6) $$d_j = R_0 \ \delta_j$$

Table 3.1. Location and central pressure of all pressure systems between 90° E and 140° E of longitude, and 22° S and 46° S of latitude for the period of 1 to 8 December 1971.

	ψ_1	ϕ_1	p_1	ψ_2	ϕ_2	p_2	ψ_3	ϕ_3	p_3
1.12.71 0000 h	119°	−35°	1016						
0600 h	119°	−35°	1018	120°	−24°	1004			
1200 h	121°	−35°	1020	120°	−24°	1002			
1800 h	123°	−35°	1020	120°	−24°	1002			
2.12.71 0000 h	124°	−35°	1019	121°	−24°	1004	101°	−30°	1019
0600 h	125°	−36°	1018	122°	−23°	1004	102°	−30°	1019
1200 h	125°	−36°	1016	121°	−23°	1004	103°	−31°	1020
1800 h	126°	−35°	1015	121°	−23°	1003	103°	−31°	1020
3.12.71 0000 h	127°	−34°	1014	121°	−23°	1004	102°	−31°	1020
0600 h	128°	−34°	1014	121°	−23°	1005	102°	−31°	1020
1200 h	130°	−35°	1014	120°	−23°	1004	102°	−32°	1020
1800 h				120°	−24°	1003	102°	−33°	1020
4.12.71 0000 h				119°	−23°	1004	102°	−32°	1019
0600 h				119°	−24°	1004	102°	−32°	1019
1200 h				120°	−25°	1003	102°	−32°	1020
1800 h				121°	−26°	1002	103°	−31°	1020
5.12.71 0000 h				122°	−24°	1004	104°	−31°	1020
0600 h	118°	−36°	1017	123°	−24°	1004	105°	−32°	1019
1200 h	119°	−37°	1018	121°	−24°	1004	105°	−32°	1020
1800 h	120°	−37°	1019	120°	−24°	1003	105°	−32°	1020
6.12.71 0000 h	121°	−37°	1020	120°	−24°	1004	105°	−33°	1020
0600 h	123°	−37°	1020	119°	−24°	1004	104°	−32°	1020
1200 h	125°	−38°	1020	120°	−25°	1003	103°	−32°	1021
1800 h	127°	−38°	1020	119°	−27°	1002	103°	−32°	1020
7.12.71 0000 h	129°	−38°	1020	118°	−29°	1001	103°	−32°	1019
0600 h	130°	−37°	1020	117°	−30°	1000	103°	−32°	1018
1200 h				118°	−31°	998			
1800 h				119°	−31°	997			
8.12.71 0000 h				120°	−32°	999			
0600 h	100°	−30°	1020	122°	−33°	1000			
1200 h	100°	−30°	1022	125°	−33°	1000			
1800 h	100°	−30°	1024	128°	−33°	1002			

where δ_j is the angle between $\mathbf{r}_0$ and $\mathbf{r}_j$. That is,

$$\cos\delta_j = \cos\phi_0\,\cos\phi_j\,\cos(\psi_0 - \psi_j) + \sin\phi_0\,\sin\phi_j\ . \tag{7}$$

Figure 3.3. Relationship between the different angles.

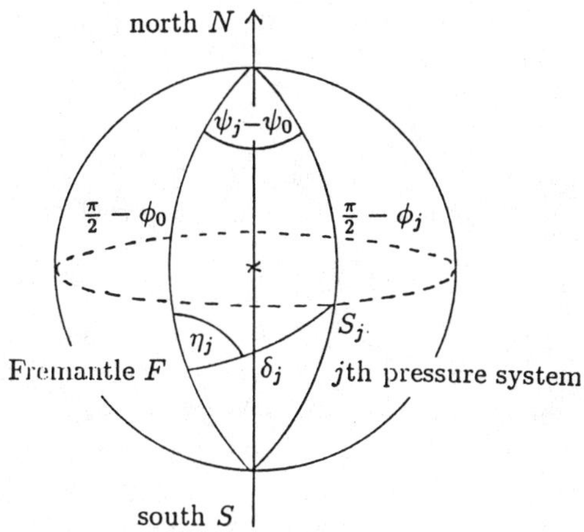

Only those pressure systems which lie between 90°E and 140°E of longitude, and 22°S and 46°S of latitude were considered because the pressure configuration beyond this region can be regarded as irrelevant for the wind at Fremantle. This implies $\cos\delta_j > 0$ and hence, $0° \leq \delta_j < 90°$ with $\cos\delta_j = 1$ if and only if $\psi_0 = \psi_j$ and $\phi_0 = \phi_j$. Equation (7) is also known as the *fundamental formula of spherical trigonometry* (Smart 1962, p.7). Associated with (7) is the triangle FNS_j sketched in Figure 3.3, which shows the relationship between δ_j and the different longitudes and latitudes relevant to our discussion. Now, instead of this triangle let us consider NFS_j. By complete analogy with (7) we then obtain:

$$\sin\phi_j = \sin\delta_j\cos\phi_0\cos\eta_j + \cos\delta_j\sin\phi_0$$

and thus

$$\cos\eta_j = \frac{\sin\phi_j - \cos\delta_j\sin\phi_0}{\sin\delta_j\cos\phi_0}\ . \tag{8}$$

If the pressure system S_j is located on the same latitude as Fremantle then $\cos\eta_j > 0$ unless $\delta_j = 0$. Formula (8) thus suggests that S_j lies in an either east-north-easterly or in a west-north-westerly direction, even though it should appear exactly east or west of Fremantle. The problem is that the distance d_j is always measured along great circles. However, none of the circles of equal latitude, except for the equator, corresponds to such a great circle. We therefore perform a transformation where ϕ_0 is subtracted from all latitudes, whilst the longitudes are left unchanged. This is, as if the complete synoptic configuration is shifted in a northerly direction such that Fremantle comes to lie on the equator. Formula (8) will therefore be replaced by

$$\cos\eta_j = \frac{\sin(\phi_j - \phi_0)}{\sin\delta_j} . \tag{9}$$

Whereas $\sin\delta_j$ is always positive, $\sin(\phi_j - \phi_0)$ can be positive or negative depending on whether the jth pressure system appears north or south of Fremantle. Given the relationship

$$\eta_j = \begin{cases} \arccos(\cos\eta_j) & \text{if } \psi_j \geq \psi_0 \\ 2\pi - \arccos(\cos\eta_j) & \text{otherwise} \end{cases} \tag{10}$$

the distance d_j can be calculated from (7) and (6), while η_j can be calculated using (7), (9) and (10).

Finally, the system parameters c_j in (5) have to be determined. In case of a single influential pressure system S_1 we had $c_1 = 3d_1^2$ where d_1 is the distance from the centre at which the strongest gradients are experienced. Although in general d_1 decreases with p_1, in this study the parameterization will be simplified by introducing two global parameters d_L and d_H for lows and highs, respectively, and defining $c_L = 3d_L^2$ and $c_H = 3d_H^2$.

3.2.3 Sensitivity of the pressure gradient model

In this section we investigate the sensitivity of the pressure gradient model introduced in section 3.2.1 to changes in the following parameters:

(i) the high pressure system parameter c_H,
(ii) the low pressure system parameter c_L,
(iii) the central pressure p_j of the jth pressure system S_j,
(iv) its direction η_j from Fremantle, and
(v) its distance d_j from Fremantle.

In particular, we will study the influence of these parameters on the atmospheric pressure p evaluated according to (3), and on the wind speed u and the direction η of the geostrophic wind as predicted from (5). Since c_H and c_L are parameters that are specific to the model, it is important to know how sensitive the results are to changes in these parameters and thus, with what accuracy they should be estimated. Furthermore, the error associated with measuring p_j, η_j and d_j was not always negligible. Their inaccurate representation on weather charts results from the fact that the pressure gradient, especially near the centre of a high, can be extremely flat. It was therefore decided to include these parameters in the sensitivity analysis.

Table 3.2. Central pressure, distance and direction of the pressure systems S_1 to S_5 used in the sensitivity analysis.

System	Type	Central pressure p_j	Distance d_j	Direction η_j
S_1	High	1018 mb	300 km	180°
S_2	Low	1006 mb	600 km	330°
S_3	High	1018 mb	900 km	30°
S_4	Low	1006 mb	1200 km	90°
S_5	High	1018 mb	1500 km	270°

The study described in this section is based on an atmospheric pressure configuration which encapsulates up to five hypothetical pressure systems. Having examined the pressure gradient model in a situation where the winds are determined by a single influential pressure system S_1, the complexity of the configuration will gradually be increased by successively introducing the systems S_2 to S_5 listed in Table 3.2 above. Eventually this leads to the situation sketched in Figure 3.4. The different pressure configurations constructed in this way can be regarded as typical of the data in the series $(\mathbf{p}_t)$ presented in Table 3.1.

In each of the situations where the wind is determined by the systems S_1 to S_n $(n = 1, \ldots, 5)$ we will study the variation of the output variables p, u and η as

(i) d_H changes from 200 km to 600 km,
(ii) d_L changes from 50 km to 400 km,
(iii) p_n changes from 1000 mb to 1024 mb,
(iv) η_n varies over the full circle, and
(v) d_n changes from 100 km to 1500 km.

Figure 3.4. Pressure configuration in sensitivity analysis.

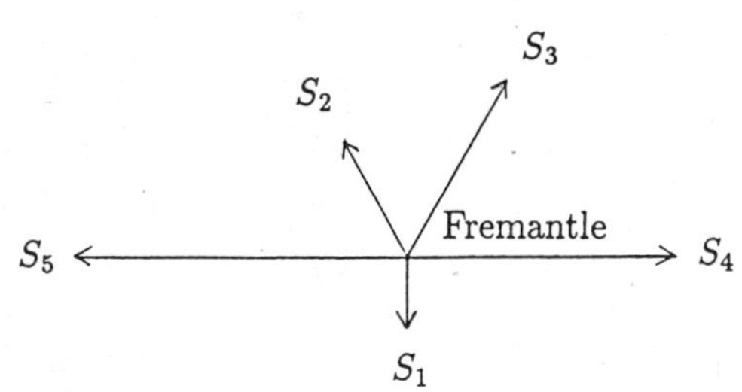

As for the *standard values* of the global system parameters we chose $d_H = 400$ km and $d_L = 200$ km. Whereas a change of d_H or d_L affects the influence of all pressure systems, p_n, η_n and d_n are related to the nth pressure system only. Since in this case the information about S_1 to S_{n-1} remains unaltered the effect of changing these variables becomes less noticeable as n increases.

Figure 3.5 illustrates the impact of a change in p_n and d_n on the predicted pressure p and the wind speed u of the geostrophic wind for $n = 1$. Observe that the change in p is slightly less than the change in p_1. Further, as d_1 increases from 100 km to 1000 km, p only falls from 1018 mb to 1015 mb, consistent with the observation that the pressure gradient is usually flat near the centre of a high. Even wind speed varies only between 8 and 18 knots given the same range of d_1-values. However, if p_1 is altered the change in wind speed becomes more marked. In particular, if p_1 falls to 1001 mb we obtain a speed of 80 knots. This suggests that the pressure gradient model overestimates wind speed, a fact that will be further discussed in section 3.3 below.

Figure 3.5. Atmospheric pressure p and wind speed u as a function of central pressure p_1 and distance d_1 ($n = 1$).

(a) p versus p_1.

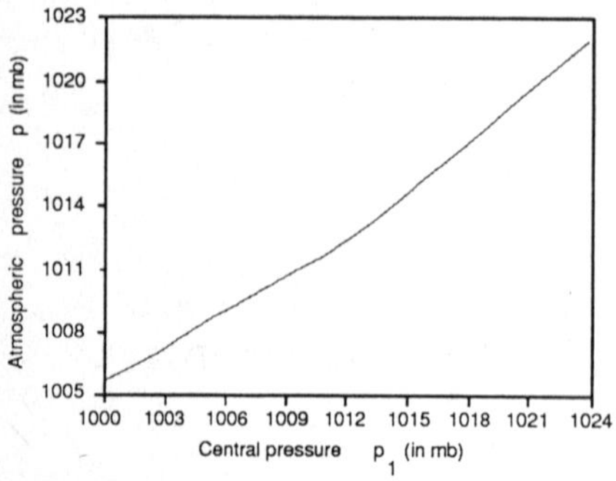

(b) u versus p_1.

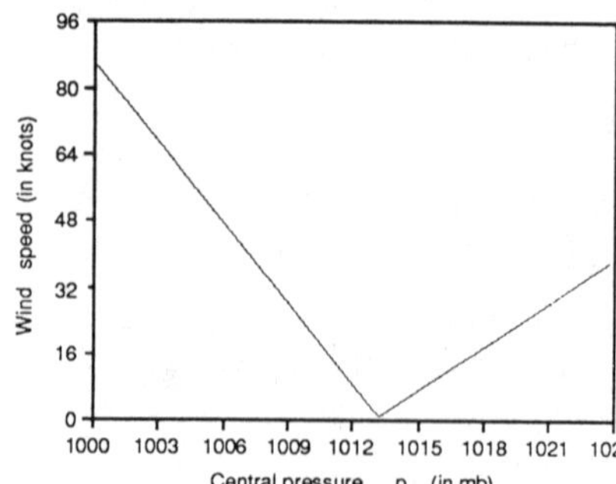

(c) p versus d_1.

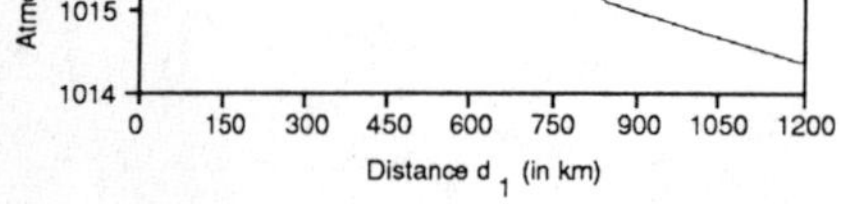

(d) u versus d_1.

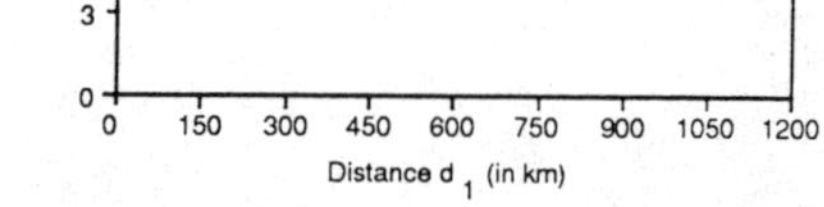

Studying the effect of a variation of the two system parameters d_H and d_L on the speed u of the geostrophic wind, it turns out that the percentage change in u is of about the same magnitude as the percentage change in d_H and d_L, as is illustrated in Figure 3.6.

Figure 3.6. The percentage change in wind speed u as a function of the percentage change in the two system parameters d_H and d_L ($n = 4$).

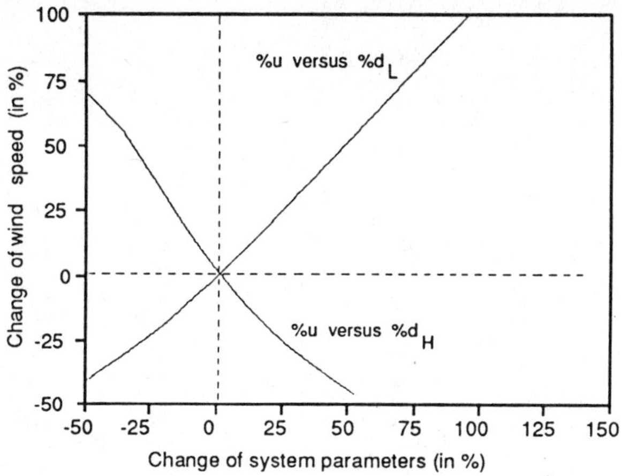

The wind direction η, however, is only little affected, even if the input variables, including d_H and d_L, vary over the full range specified above. Changing d_H from 200 km to 600 km or d_L from 50 km to 400 km does not alter η by more than 3°, whatever the value of n. As far as wind direction η is concerned, the model thus proves to be fairly robust towards its parameterization. However, the sensitivity of wind speed u with respect to changes in d_L or d_H suggests that the system parameters have to be carefully chosen, even though the values $d_H = 400$ km and $d_L = 200$ km appear to be in the right order of magnitude from a meteorological point of view (Haltiner 1971, chapter 3).

3.2.4 Application

Now, using the pressure gradient model (3), the geostrophic wind can be predicted for each pressure configuration in $(\mathbf{p}_t)$, where it is implicitly assumed that the vertical component of the pressure gradient can be ignored. The series of predicted geostrophic winds will subsequently be referred to as $(\mathbf{h}_t)$. Table 3.3 presents the values of this series for the period of 1 to 8 December 1971. Note that at 0000 h on 1 December when a single influential high was recorded south-east of Fremantle, we obtain a wind of 5.3 knots from the north-east. This seems to conform well with the isobars shown in the first diagram of Figure 1.3. On 7 and 8 December the predicted wind is strengthening and turning from an easterly to a southerly direction. From Figure 1.3 it can be seen that a trough had moved through the region at the time, and that strong to moderate south-westerlies, which were associated with a depression south of the continent, prevailed. The pressure gradient model thus serves to quantify the synoptic pressure field such that it can be related to the geostrophic component $(\mathbf{g}_t)$, and to validate the decomposition of $(\mathbf{w}_t)$ described in chapter 2.

Table 3.3. Estimated geostrophic wind for the period of 1 to 8 December 1971.

Day	Time of day	No. of systems n	Pressure p	Speed u	Direction η
1.12.71	0000 h	1	1015.2	5.3	50°
	0600 h	2	1016.2	12.5	58°
	1200 h	2	1016.5	15.1	47°
	1800 h	2	1015.5	11.9	45°
2.12.71	0000 h	3	1015.4	5.8	54°
	0600 h	3	1015.0	3.2	77°
	1200 h	3	1014.6	3.2	135°
	1800 h	3	1014.3	3.8	151°
3.12.71	0000 h	3	1013.9	3.9	157°
	0600 h	3	1014.0	3.9	159°
	1200 h	3	1013.9	4.6	151°
	1800 h	2	1013.7	6.8	144°
4.12.71	0000 h	2	1013.7	5.1	145°
	0600 h	2	1013.5	6.0	140°
	1200 h	2	1013.4	8.0	143°
	1800 h	2	1013.3	9.9	150°
5.12.71	0000 h	2	1014.3	6.9	164°
	0600 h	3	1015.9	6.0	91°
	1200 h	3	1016.0	7.0	98°
	1800 h	3	1016.2	8.2	88°
6.12.71	0000 h	3	1016.5	8.5	82°
	0600 h	3	1015.9	7.2	82°
	1200 h	3	1015.2	7.2	107°
	1800 h	3	1013.7	14.0	115°
7.12.71	0000 h	3	1010.6	36.5	120°
	0600 h	3	1006.6	64.9	114°
	1200 h	1	1002.4	58.3	150°
	1800 h	1	1004.1	53.7	159°
8.12.71	0000 h	1	1006.7	38.5	180°
	0600 h	2	1011.0	21.4	191°
	1200 h	2	1012.9	11.1	188°
	1800 h	2	1014.1	7.3	187°

3.3 Comparison with the Geostrophic Component

3.3.1 Initial comparison

In this section the series $(\mathbf{h}_t)$ of predicted geostrophic winds will be compared with the raw data series $(\mathbf{w}_t)$ and with the geostrophic component $(\mathbf{g}_t^\alpha)$ $(\alpha = 0.1, 0.3, 0.5)$. The superscript α is to denote the output series obtained by using the mean filter M_α introduced in section 2.5 for the decomposition of $(\mathbf{w}_t)$. Let (u_t, η_t) be a pair of wind speed and direction in the geostrophic wind series $(\mathbf{h}_t)$ and (v_t, θ_t) be a record in either $(\mathbf{w}_t)$ or $(\mathbf{g}_t^\alpha)$. After forming the directional differences

$$\xi_t = \theta_t - \eta_t \text{ (mod } 360°)$$

and the speed differences

$$x_t = v_t - u_t$$

for each pair of corresponding winds, we obtain the deviation diagrams in Figure 3.7 which compare $(\mathbf{h}_t)$ with $(\mathbf{g}_t^{.5})$. The histograms clearly indicate a bias towards positive direction differences and towards negative speed differences. Also note that the spread of the direction differences is much wider in winter than in summer. The geostrophic component $(\mathbf{g}_t^{.5})$ in summer is therefore, after a constant and season specific rotation, a better predictor of the synoptic scale wind, as captured by $(\mathbf{h}_t)$, than it is in winter. The speed differences, on the other hand, exhibit a pronounced lower tail which is represented in Figure 3.7 by the spike at -18 knots. This indicates that the pressure gradient model will have the tendency to overestimate strong and gusty winds in particular.

The mean directional deviations $\overline{\xi}$ and the mean speed differences $\overline{x}$ are listed in Table 3.4 for the four comparisons $(\mathbf{h}_t)$ with $(\mathbf{w}_t)$ and $(\mathbf{h}_t)$ with $(\mathbf{g}_t^\alpha)$ for $\alpha = 0.1, 0.3, 0.5$. Most noticeable are the huge directional deviations $\overline{\xi}$ in summer. This phenomenon had been observed before in section 1.2, where Figure 1.4 showed a predominantly southerly wind for the period of 1 to 8 December 1971, while the corresponding weather charts in Figure 1.3 indicated a synoptic scale flow from an easterly direction. This turning of the local wind can be attributed to temperature gradients that emerge near the coast and are much stronger in summer than in winter. For a detailed discussion of this phenomenon the reader is referred to Holton (1979, section 3.4).

The fact that the pressure gradient model tends to overestimate wind speed is also reflected by the difference between $s^2_{\mathbf{g}}$, and the mean sum of squares $s^2_{\mathbf{h}} = 156$ in winter and $s^2_{\mathbf{h}} = 355$ in summer. As expected, the more the original series $(\mathbf{w}_t)$ is smoothed, the more the seasonal mean speed $\overline{v}$ and the mean sum of squares $s^2_{\mathbf{g}}$ are reduced. The mean speed difference $\overline{x}$ is therefore smallest if $(\mathbf{h}_t)$ is compared with the raw data series $(\mathbf{w}_t)$. In winter the smoothing also results in a shift of the mean wind direction $\overline{\theta}$, which is due to the fact that wind direction has an almost uniform distribution during this time of the year. The mean direction is therefore barely defined, though the mean directional deviation is almost unaffected.

Figure 3.7. Histogram of the direction differences ξ_t and of the speed differences x_t (non-adjusted).

(a) Winter 1971. *(b) Summer 1971/72.*

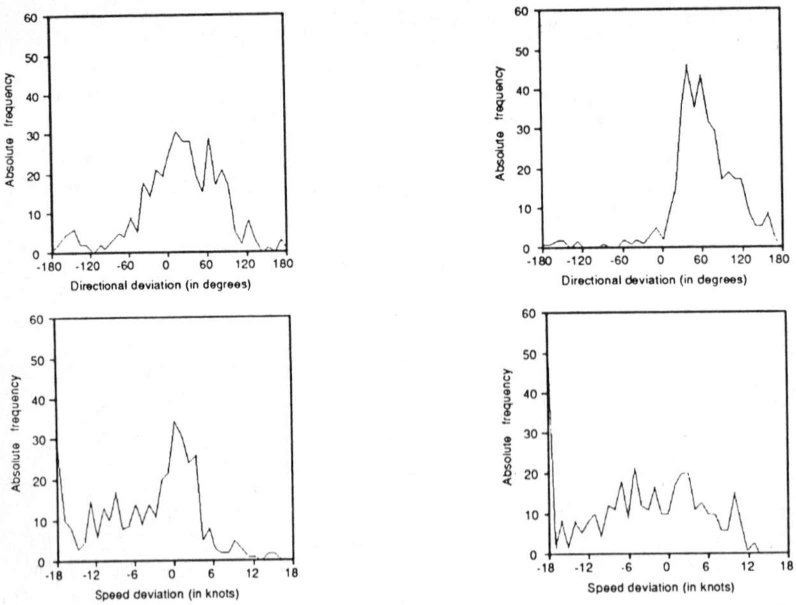

Figure 3.8. Histogram of the direction differences ξ_t and of the speed differences x_t (adjusted).

(a) Winter 1971. *(b) Summer 1971/72.*

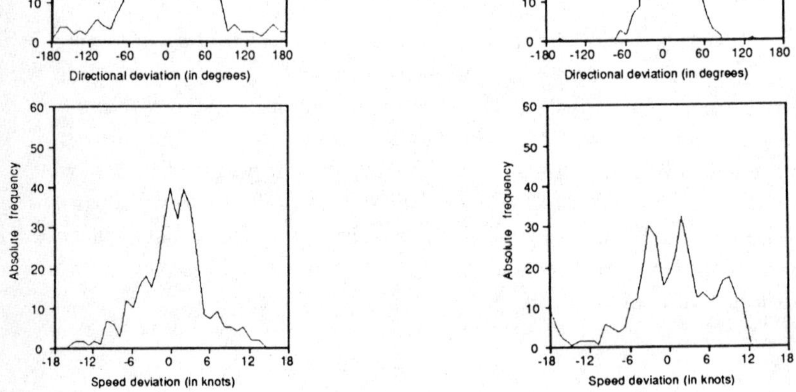

Table 3.4. Summary statistics for the comparison of the (non-filtered) predicted geostrophic wind ($\mathbf{h}_t$) *with the raw and filtered wind series* ($\mathbf{w}_t$) *and* ($\mathbf{g}_t^\alpha$), *resp.*

Season	Series	Mean wind direction $\overline{\theta}$	Mean direction difference $\overline{\xi}$	Mean wind speed $\overline{v}$	Mean speed difference $\overline{x}$	Mean sum of squares $s_{\mathbf{g}}^2$
Winter 1971	($\mathbf{g}_t^{.5}$)	200°	22.6°	5.7	-4.4	44
	($\mathbf{g}_t^{.3}$)	163°	23.4°	6.0	-4.1	50
	($\mathbf{g}_t^{.1}$)	131°	24.3°	6.7	-3.4	64
	($\mathbf{w}_t$)	91°	24.3°	9.3	-0.8	124
Summer 1971/72	($\mathbf{g}_t^{.5}$)	181°	64.6°	9.4	-5.5	99
	($\mathbf{g}_t^{.3}$)	181°	64.4°	9.5	-5.4	103
	($\mathbf{g}_t^{.1}$)	181°	63.8°	9.8	-5.1	110
	($\mathbf{w}_t$)	180°	56.3°	12.4	-2.5	184

3.3.2 Adjustment of ($\mathbf{h}_t$)

The discussion in the last section suggests that the same filter that is used for the decomposition of ($\mathbf{w}_t$) is also applied to ($\mathbf{h}_t$), and that the resultant series $\mathbf{h}_t^\alpha = M_\alpha(\mathbf{h}_t)$ is compared with the corresponding series $\mathbf{g}_t^\alpha = M_\alpha(\mathbf{w}_t)$ for $\alpha = 0.1,\ 0.3,\ 0.5$. Summary statistics describing this comparison and similar to those presented in Table 3.4 are given below in Table 3.5. On average, the mean direction differences are about 18° in winter and 65° in summer, while the mean wind speeds differ by a factor of around 1.2 in winter and 1.3 in summer. In contrast with the initial comparison conducted in section 3.3.1, the ratio of the mean speeds is about the same irrespective of whether we compare the raw series ($\mathbf{h}_t$) and ($\mathbf{w}_t$) or the filtered series ($\mathbf{h}_t^\alpha$) and ($\mathbf{g}_t^\alpha$).

The predicted geostrophic winds are now adjusted such that their mean speed and direction coincide with the corresponding values for the geostrophic component ($\mathbf{g}_t^\alpha$). This adjustment reflects the effect of temperature gradients and other local phenomena not captured by the pressure gradient model (3). Let (u_t, η_t) be a record of either ($\mathbf{h}_t$) or ($\mathbf{h}_t^\alpha$) ($\alpha = 0.1,\ 0.3,\ 0.5$). For all t, the speed u_t is multiplied by the constant factor $\overline{v}/\overline{u}$, while the direction η_t is rotated by $\overline{\xi}$. The deviation diagrams illustrating the agreement between the adjusted series ($\mathbf{h}_t^{.5}$) and ($\mathbf{g}_t^{.5}$) are presented in Figure 3.8. In particular

Table 3.5. Summary statistics for the comparison of the filtered series $(\mathbf{h}_t^\alpha)$ *and* $(\mathbf{g}_t^\alpha)$, *and of* $(\mathbf{h}_t)$ *and* $(\mathbf{w}_t)$.

Season	Series	Mean direction difference $\bar{\xi}$	Mean wind speed $\bar{u}$	Ratio of mean speeds $\bar{u}/\bar{v}$	Mean sum of squares $s^2_{\mathbf{h}}$
Winter 1971	$(\mathbf{h}_t^{.5})$	17.4°	6.8	1.20	66
	$(\mathbf{h}_t^{.3})$	17.1°	7.1	1.18	73
	$(\mathbf{h}_t^{.1})$	19.3°	7.8	1.14	89
	$(\mathbf{h}_t)$	24.3°	10.1	1.08	156
Summer 1971/72	$(\mathbf{h}_t^{.5})$	65.3°	11.9	1.27	194
	$(\mathbf{h}_t^{.3})$	64.8°	12.2	1.28	207
	$(\mathbf{h}_t^{.1})$	63.8°	12.8	1.30	236
	$(\mathbf{h}_t)$	56.3°	14.9	1.20	355

the plots of directional deviation provide ample evidence of a good agreement between the two series. Further, comparing Figure 3.8 with Figure 3.7, it is observed that the adjustment of $(\mathbf{h}_t)$ has resulted in centering the modes of the speed and direction differences at 0, and in reducing their spread at the same time.

Figure 3.9. Schematic representation of prevailing synoptic situation ($\mathbf{v}$ *in* $(\mathbf{g}_t^\alpha)$ *and* $\mathbf{u}$ *in* $(\mathbf{h}_t^\alpha)$).

(a) Winter pattern.

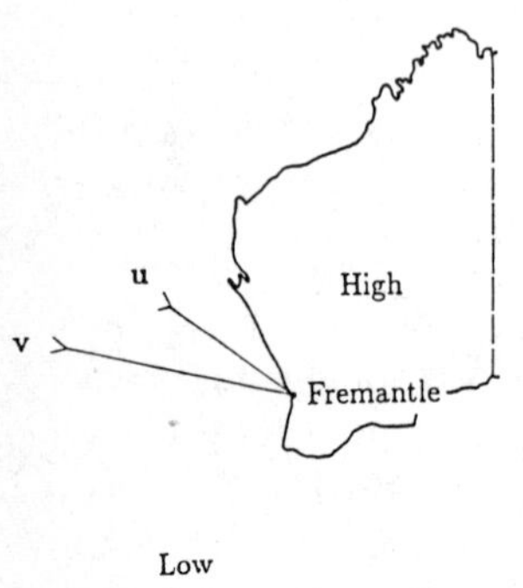

(b) Summer pattern.

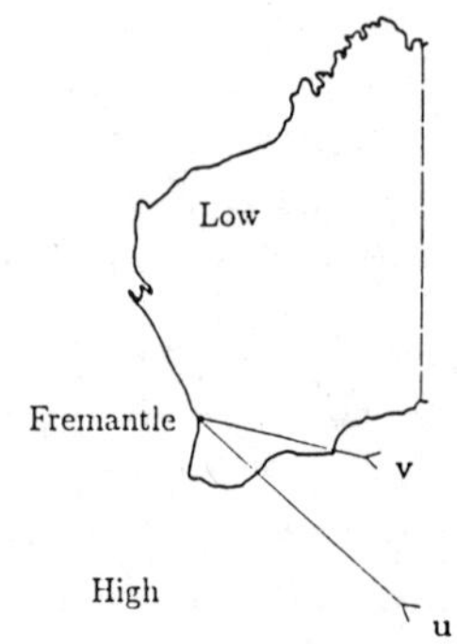

When comparing the speed and direction differences simultaneously, it is revealed that positive directional deviations in winter are generally associated with negative speed differences and *vice versa.* In summer, however, the situation is reversed as positive (negative) directional deviations seem to correspond to positive (negative) speed differences. This result can be explained as follows. The pressure gradient model tends to overestimate the speed of synoptic scale easterlies, which occur more frequently in summer, and to underestimate the speed of westerlies, which predominate in winter. On the other hand, it also tends to underestimate the frequency of both easterlies and westerlies, as the high pressure systems in particular extend much further in an easterly than in a northerly direction and usually are far from being circular. The diagrams in Figure 3.9 above are to illustrate this situation.

3.3.3 Conclusion

Despite the simplistic model (3) for predicting the geostrophic wind, the results are quite remarkable. Using a ratio of sums of squares as a measure for the variation in $(\mathbf{g}_t^{.5})$ that is explained by $(\mathbf{h}_t^{.5})$, it is found that $(\mathbf{h}_t^{.5})$ accounts for 71 per cent of the variation in $(\mathbf{g}_t^{.5})$ in winter and 84 per cent in summer. It thus seems justified to interpret the series $(\mathbf{g}_t) = (\mathbf{g}_t^{.5})$ as the *synoptic scale component* of the wind series $(\mathbf{w}_t)$. This result will now be established.

The residual series obtained after removal of the predicted geostrophic wind $(\mathbf{h}_t^{\alpha})$ from the geostrophic component $(\mathbf{g}_t^{\alpha})$ will be denoted by $(\mathbf{q}_t^{\alpha})$ $(\alpha = 0.1, 0.3, 0.5)$. Let $\mathbf{v}_t$ be a record in $(\mathbf{g}_t^{\alpha})$ and $\mathbf{u}_t$ be the corresponding record in the adjusted series $(\mathbf{h}_t^{\alpha})$. We then define $\mathbf{q}_t^{\alpha} = \mathbf{v}_t - \beta \mathbf{u}_t$ where $\beta = \max\{0, v_t/u_t \cos \xi_t\}$. The factor β thus guarantees that the residual vector $\mathbf{q}_t^{\alpha}$ has minimum length and at the same time does not point into the opposite direction to $\mathbf{v}_t$. See Figure 3.10 for a graphical illustration.

Figure 3.10. Definition of the residual component $(\mathbf{q}_t^{\alpha})$.

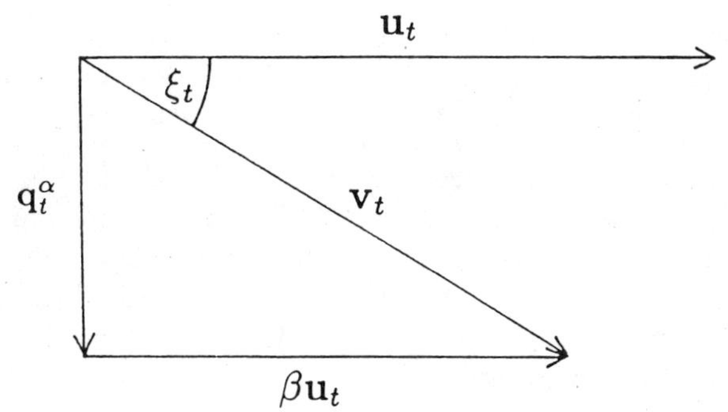

Table 3.6. Agreement between the geostrophic component $(\mathbf{g}_t^\alpha)$ *and the predicted geostrophic wind* $(\mathbf{h}_t^\alpha)$.

Season	Comparison	Mean sum of squares $s_{\mathbf{q}}^2$	Relative amount of explanation ρ
Winter 1971	$(\mathbf{g}_t^{;5})$ & $(\mathbf{h}_t^{;5})$	12.9	0.71
	$(\mathbf{g}_t^{;3})$ & $(\mathbf{h}_t^{;3})$	13.8	0.72
	$(\mathbf{g}_t^{;1})$ & $(\mathbf{h}_t^{;1})$	17.1	0.73
	$(\mathbf{w}_t)$ & $(\mathbf{h}_t)$	41.5	0.67
Summer 1971/72	$(\mathbf{g}_t^{;5})$ & $(\mathbf{h}_t^{;5})$	16.3	0.84
	$(\mathbf{g}_t^{;3})$ & $(\mathbf{h}_t^{;3})$	18.6	0.82
	$(\mathbf{g}_t^{;1})$ & $(\mathbf{h}_t^{;1})$	24.1	0.78
	$(\mathbf{w}_t)$ & $(\mathbf{h}_t)$	68.1	0.63

With $s_{\mathbf{q}}^2$ denoting the mean sum of squares in $(\mathbf{q}_t^\alpha)$, define the parameter $\rho = (s_{\mathbf{g}}^2 - s_{\mathbf{q}}^2)/s_{\mathbf{g}}^2$ which provides an indication of the variation in the geostrophic component $(\mathbf{g}_t^\alpha)$ that is explained by the predicted geostrophic wind $(\mathbf{h}_t^\alpha)$. It is noted that the parameter ρ would be 0.25 if the two series were completely unrelated and equal to 1 if they were identical. Values of $s_{\mathbf{q}}^2$ and ρ are given in Table 3.6 for winter 1971 and summer 1971/72. The difference between the ρ-values for the *raw* series on the one hand, and the *filtered* series on the other hand, is much larger in summer than it is in winter. This result reflects the different strength of the sea breeze circulation which is present in the $(\mathbf{w}_t)$ but not in $(\mathbf{g}_t^\alpha)$. Finally, it is noted that throughout the year the best agreement between $(\mathbf{h}_t^\alpha)$ and $(\mathbf{g}_t^\alpha)$ is achieved for moderate and strong geostrophic winds. However, whereas in summer the best results are obtained for winds from the northerly and easterly sectors, in winter this is if the prevailing wind comes from a south-easterly to westerly direction.

When estimating the system parameters d_H and d_L, a grid search was performed where both d_H and d_L were altered in steps of 50 km so as to maximize ρ. The highest amount of explanation was achieved for $d_H = 400$ km and $d_L = 200$ km. However, as could be expected on the grounds of the sensitivity analysis conducted in section 3.2.3, the ρ-values hardly changed when d_H and d_L were varied. A refinement of the search grid was therefore felt to be unnecessary.

Figure 3.11. Synoptic charts for 8 to 15 June 1971 (0700 h). Winter pattern. Source: Bureau of Meterology.

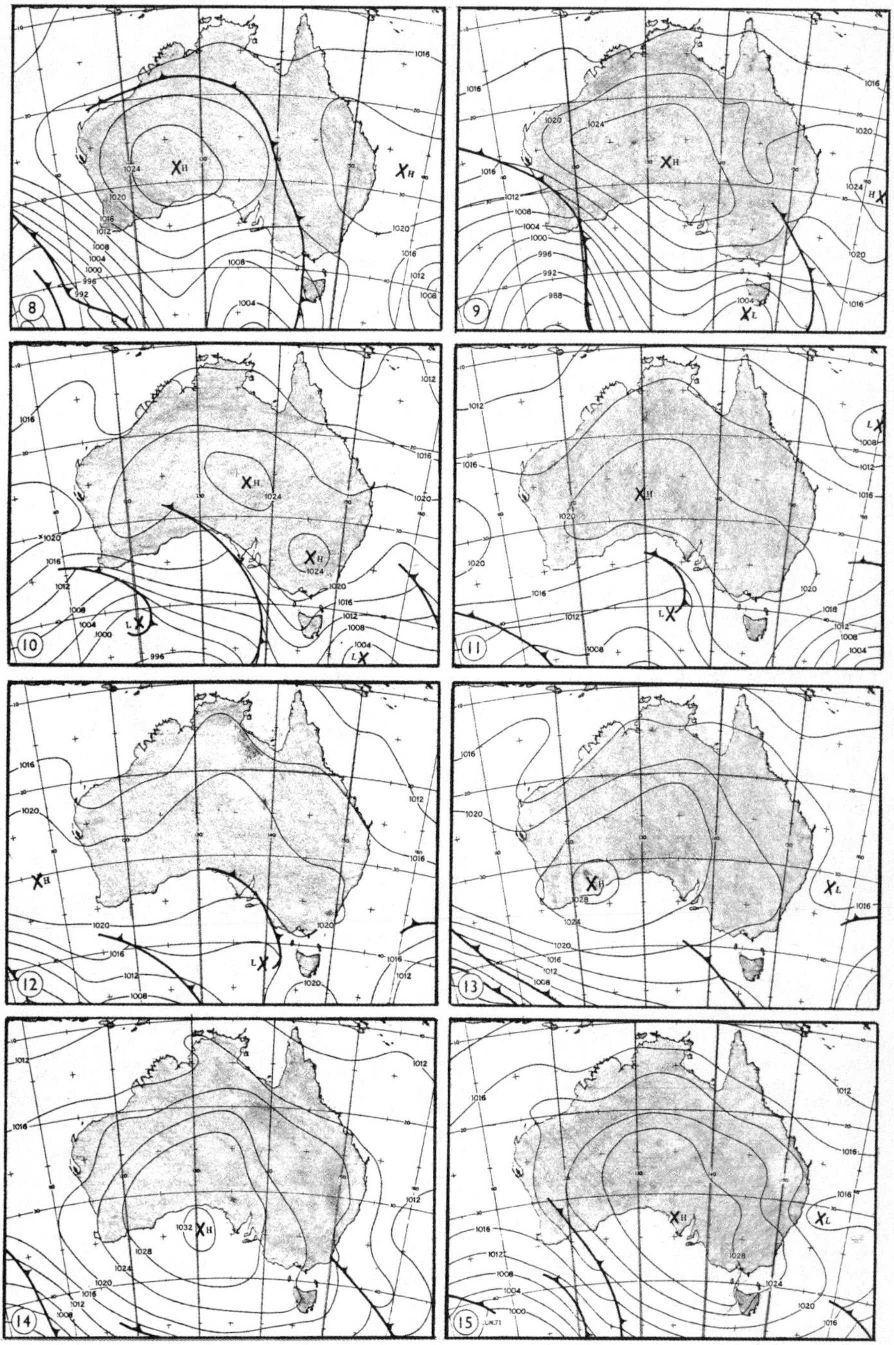

3.4 Synoptic States

In the last section it was confirmed that the geostrophic component ($\mathbf{g}_t$) can be interpreted as a series of synoptic scale winds. Using this sequence of geostrophic winds, the objective in this section is to establish and quantify the seasonal patterns described in section 1.2. Certain synoptic configurations are defined in terms of wind speed and direction, which can be related to particular atmospheric pressure distributions and be conceived as *synoptic states.* In order to detect regularities in the passage of high and low pressure systems, and to determine the major seasonal wind patterns, the geostrophic component ($\mathbf{g}_t$) will then be partitioned into these states and the transition probabilities be examined.

3.4.1 Predominant winter pattern

The winter months in the Fremantle region are characterized by a high pressure belt lying across the continent and a series of low pressure systems approaching from the south-west. A sequence of synoptic charts, covering the period of 8 to 15 June 1971 and typical for this time of the year, is presented in Figure 3.11. It is noted that the atmospheric pressure field on 15 June is similar to that on 8 June, which suggests a kind of *periodic* pattern.

Figure 3.12. Sequence of synoptic configurations in winter.

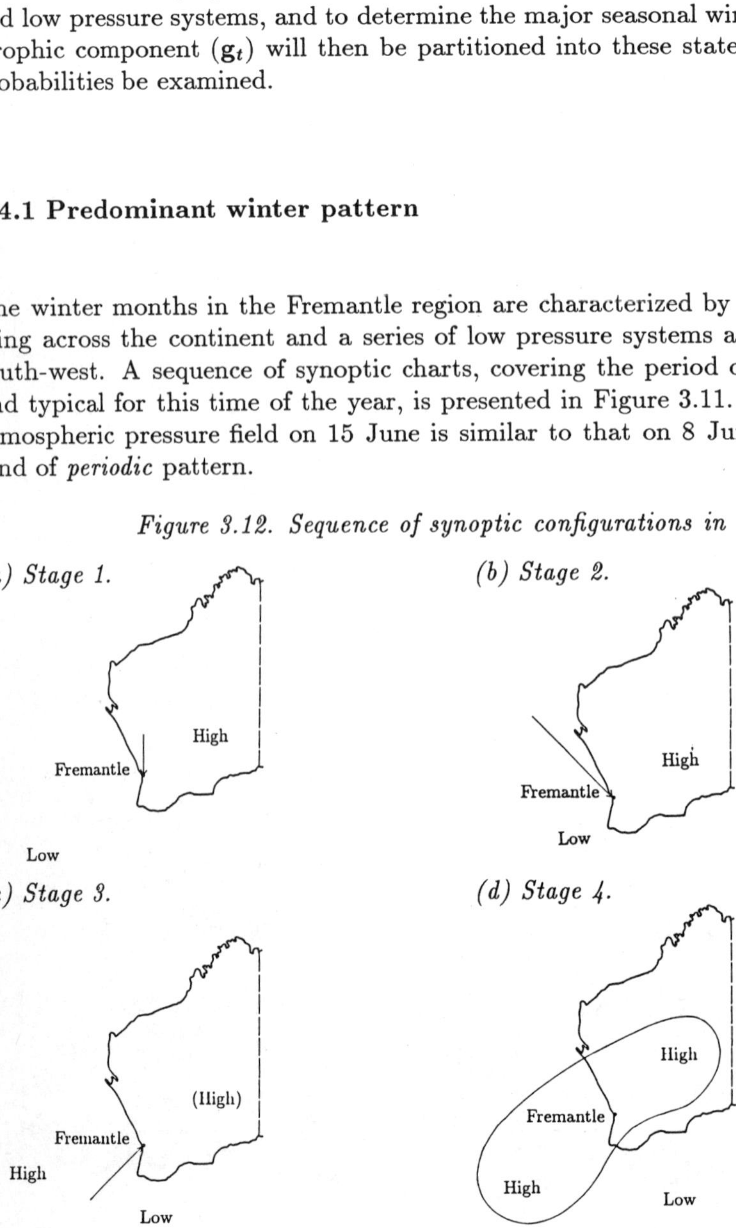

On 14 June light northerlies dominate the pattern, which is sketched in Figure 3.12 (a) and is referred to as stage 1. As the low moves in a north-easterly direction, the wind begins to strengthen and to turn north-westerly. This corresponds to a pressure configuration as shown on the synoptic chart of 8 June and which is denoted as stage 2 of the winter pattern. When on 10 June the low moves in an eastward direction south of the continent, the wind becomes south-westerly. Determined by the high and low pressure configuration the wind also tends to moderate. Following this third stage of the winter pattern, a configuration as depicted in diagram (d) of Figure 3.12 emerges. That is, as the low moves further to the east, Fremantle comes under the influence of the anticyclonic belt which extends from the eastern Indian Ocean right across Western Australia. Small gradients and therefore light winds with no distinct direction are associated with this situation which is found on 12 June.

Figure 3.13. Schematic winter pattern.

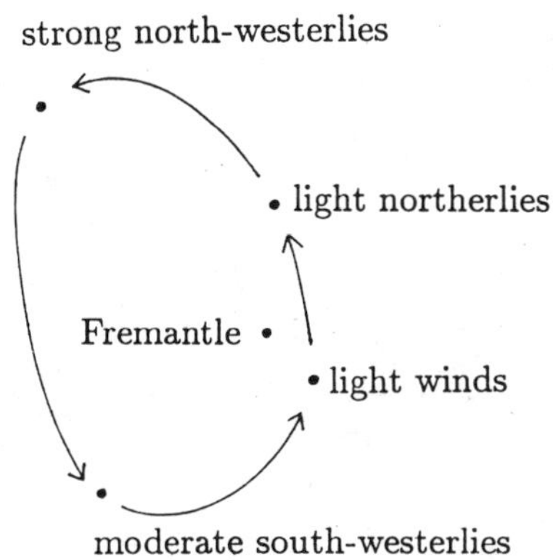

The pattern described in the last paragraph can be summarised as follows. For most of the time the wind moves through the full circle in an anticlockwise direction, with the strongest winds from the north-west and the lightest winds from the south-east. As it swings through the east to the north, the speed gradually increases and reaches a maximum when the wind turns north-westerly. A schematic representation of this pattern is given in Figure 3.13.

3.4.2 Predominant summer pattern

Synoptic charts for the period of 1 to 7 February 1972 are presented in Figure 3.14. They reveal a pattern that is typical for the Fremantle region in summer. Note that the atmospheric pressure field emerging on 7 February is similar to that on 1 February and hence, that there is a similar 7- to 8-day *period* as in winter. The dominant feature

Figure 3.14. Synoptic charts for 1 to 7 February 1972 (0700 h). Summer pattern. Source: Bureau of Meterology.

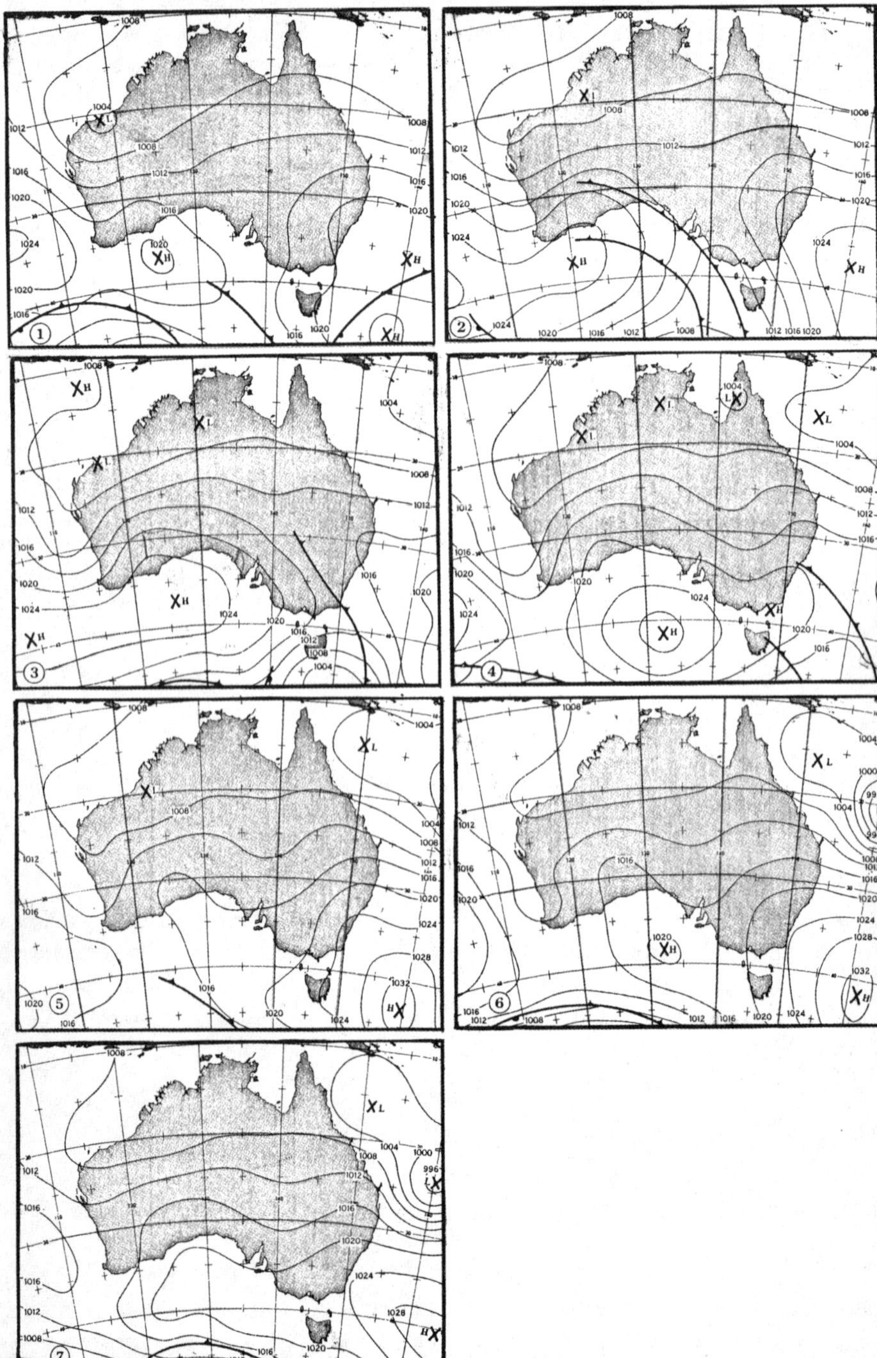

Figure 3.15. Sequence of synoptic states in summer.

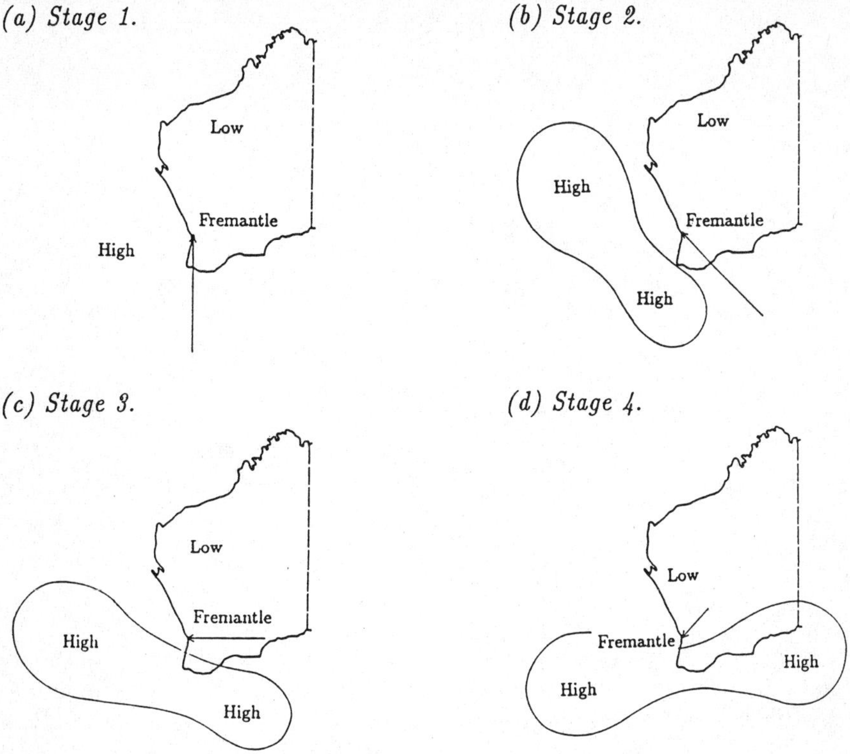

during the summer months is the high pressure system over the eastern Indian Ocean and the heat low over the north-west of the continent.

Such a pattern that is associated with freshening winds from a southerly direction is recorded on 7 February and corresponds to the first diagram in Figure 3.15. As the high extends further to the east it produces strong winds from the south-east along its northern edge. This pattern is found on 1 February and is represented in Figure 3.15 by the second diagram. On 2 and 3 February when the high pressure ridge reaches right across into the Bight the wind tends to moderate and to blow from a predominantly easterly direction. Under the influence of the low pressure system in the north a trough begins to develop in a southerly direction, splitting an eastern cell away from the high pressure ridge, as happening on 4 February. These last two stages of the summer pattern are stylized in diagrams (c) and (d) of Figure 3.15. At this stage the wind field across the Fremantle area is dominated by light north-easterlies but as the trough moves across the coast, the wind swings back to a southerly direction, producing again the configuration captured in Figure 3.15 (a).

As a summary of this cycle we obtain the schematic pattern shown in Figure 3.16. Strong southerlies are followed by moderate easterlies and light north-easterlies. As the trough passes through the region the wind strengthens and swings back to a southerly direction.

Figure 3.16. Schematic summer pattern.

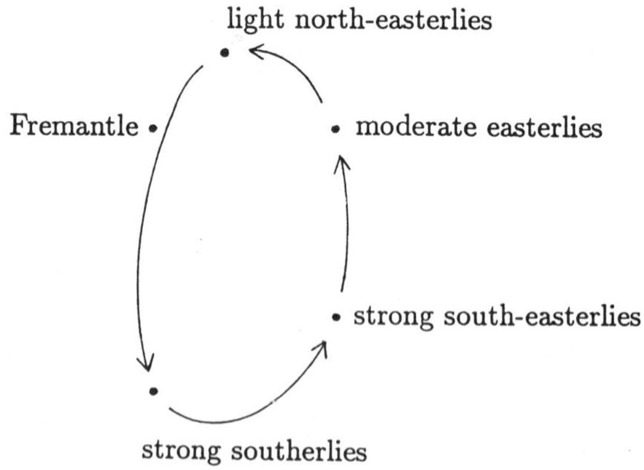

Both synoptic patterns plotted in Figures 3.13 and 3.16 are reflected in the histograms of geostrophic wind speed and direction in Figure 2.16. Whereas in summer strong to moderate winds from the southerly and easterly sectors almost exclusively dominate the pattern, in winter strong westerlies and light easterlies tend to prevail.

3.4.3 Definition of synoptic states

In order to identify the synoptic patterns described in sections 3.4.1 and 3.4.2 in the geostrophic component $(\mathbf{g}_t)$, it will be necessary to group and classify the wind records such that they can be related to the atmospheric pressure field. Eight classes are required to distinguish between the four basic wind directions on the one hand, and low and high speeds on the other hand. An additional class will be included for calms, since they cannot with certainty be associated with any particular wind direction. In order to study the development of large scale pressure configurations and to determine the sequence of corresponding wind patterns, the following *synoptic states* are proposed, with (v_t, θ_t) denoting a record in $(\mathbf{g}_t)$:

LN: light winds from a northerly direction
$(-45° < \theta_t \leq 45°$ and 1.5 knots $\leq v_t <$ 6 knots),

SN: strong winds from a northerly direction
$(-45° < \theta_t \leq 45°$ and $v_t \geq$ 6 knots),

LE: light winds from an easterly direction
($45° < \theta_t \leq 135°$ and 1.5 knots $\leq v_t < 6$ knots),
SE: strong winds from an easterly direction
($45° < \theta_t \leq 135°$ and $v_t \geq 6$ knots),
LS: light winds from a southerly direction
($135° < \theta_t \leq 225°$ and 1.5 knots $\leq v_t < 6$ knots),
SS: strong winds from a southerly direction
($135° < \theta_t \leq 225°$ and $v_t \geq 6$ knots),
LW: light winds from a westerly direction
($225° < \theta_t \leq 315°$ and 1.5 knots $\leq v_t < 6$ knots),
SW: strong winds from a westerly direction
($225° < \theta_t \leq 315°$ and $v_t \geq 6$ knots), and
0: calms with no particular direction and $v_t < 1.5$ knots.

It is pointed out that these states are exhaustive and that no two states can occur at the same time. A complete classification of the wind records in $(\mathbf{g}_t)$ will therefore result in a *partition* of the geostrophic component. In order to reduce the number of repeated transitions between alternating states, a tolerance was conceded for both, directional and speed changes.

Given the discussion in sections 3.4.1 and 3.4.2 the following sequence of synoptic states can be expected in winter:

$$(11) \qquad LE \to LN \to SN \to SW \to \left\{ \begin{array}{c} LW \\ SS \end{array} \right\} \to LS \to LE \, .$$

That is, light easterlies (*LE*) are followed by light northerlies (*LN*), when the wind is expected to strengthen (*SN*). However, it continues to turn anticlockwise, thus leading to strong westerlies (*SW*). Then, as the wind weakens, it swings to a southly direction (*LS*) and finally back to *LE*. Since the geostrophic component $(\mathbf{g}_t)$ is extremely smooth, the change from *SW* to *LS* will be either via light westerlies (*LW*) or via strong southerlies (*SS*). The dominant states in this sequence are expected to be *SW* and to a smaller extent *LE*, corresponding to the extratropical cyclones and the high pressure belt, respectively.

Quite different characteristics are expected in summer. The pattern sketched in Figure 3.16 translates into the following cycle of synoptic states:

$$(12) \qquad SS \to SE \to LE \to \left\{ \begin{array}{c} LS \\ 0 \to LS \end{array} \right\} \to SS \, .$$

As the wind moves anticlockwise from a southerly direction (*SS*) to an easterly direction (*SE*), it gradually moderates (*LE*). Then it quickly swings back, possibly through *0* and *LS*, and again produces strong southerlies (*SS*). Presumably, the most important state in summer is *SS*, while *LN*, *SN*, *LW* and *SW*, which are associated with winds from the northerly and westerly sectors, are not expected to play a significant role.

If the sequence of synoptic states was modelled as a simple Markov chain, 81 transition probabilities would have to be estimated. Of course, we are mainly interested in those transitions where the states actually change. Hence, there could be a severe estimation problem, since together only 136 transitions were recorded in the summer seasons of 1971/72, 1973/74, 1975/76 and 1977/78, with the corresponding figure for the winter seasons being 330. However, given the smoothness of the geostrophic component ($\mathbf{g}_t$) certain transitions, for example from a strong westerly into a strong easterly or a light southerly, are highly unlikely. This reduces the number of transition probabilities to be estimated to 32. Since northerlies, and to some extent westerlies, are rarely recorded in summer, this figure could be brought further down to 20 or even 12 for this time of the year. Nevertheless, given the ratio of actual transitions over the number of parameters to be estimated, the results presented in this section are regarded as indicative only. It is therefore suggested that the procedure set out in this study be applied to the full data set spanning over a period of 16 years from 1971 to 1986.

3.4.4 Empirical winter pattern

The number of transitions between synoptic states is given in Table 3.7 with the labels on the left and on the top denoting the states before and after the transition, respectively. It is noted that the numbers are based on the winter seasons of 1971, 1973, 1975 and 1977, and that the transitions represent hourly changes. This and the smoothness of the geostrophic component ($\mathbf{g}_t$) also explain the dominance of the diagonal in the Table.

Most of the figures in Tables 3.8 to 3.11 can be derived from those in Table 3.7. The first of these Tables presents the number of transitions into the different synoptic states, while Table 3.9 gives the frequency of each state, which simply is the number presented in the last column of Table 3.7 divided by the total of these numbers. The average amount of time the wind remained in a particular state is given in Table 3.10. Apart from the effect associated with the first pattern in a season, this is the same as the ratio of the total time spent in a state and the number of transitions into that state. Table 3.11, finally, gives the one-step transition probabilities conditional on being in a particular state and leaving it.

From Table 3.9 it follows that the four wind directions are almost equally represented in the winter pattern. Nevertheless, the clearly most important state is strong westerlies (*SW*) which account for almost 20 per cent of the time and exhibit a distinctly longer duration (cf. Table 3.10). Judging by the frequencies listed in Table 3.9, other common patterns include light easterlies (*LE*) due to frequent transitions into this state, and strong easterlies (*SE*) due to its longer duration. Together, westerlies and easterlies account for 56 per cent of the time, compared with only 37 per cent accounted for by northerlies and southerlies. This again reflects the tendency of synoptic scale pressure systems to extend much further in a west-easterly direction than in a south-northerly direction, thus giving rise for frequent westerlies and easterlies relative to southerlies and northerlies.

Table 3.7. Number of transitions between synoptic states in winter (based on observations of the years 1971, 1973, 1975 and 1977).

	LN	*SN*	*LE*	*SE*	*LS*	*SS*	*LW*	*SW*	*0*	Total
LN	963	12	6	.	.	.	19	.	10	1010
SN	9	614	.	.	.	.	.	13	.	636
LE	13	.	1238	20	9	.	.	.	14	1294
SE	.	9	19	1063	.	1	.	.	.	1092
LS	.	.	21	.	1201	8	2	.	8	1240
SS	.	.	.	9	9	422	.	.	.	440
LW	15	.	.	.	7	.	918	16	10	966
SW	.	1	.	.	.	9	19	1715	.	1744
0	10	.	12	.	14	.	6	.	556	598

Table 3.8. Number of transitions into synoptic states from other states.

LN	*SN*	*LE*	*SE*	*LS*	*SS*	*LW*	*SW*	*0*	Total
47	22	58	29	39	18	46	29	42	330

Table 3.9. Relative frequency of synoptic states in winter (in %).

LN	*SN*	*LE*	*SE*	*LS*	*SS*	*LW*	*SW*	*0*	Total
11	7	14	12	14	5	11	19	7	100

Table 3.10. Average duration of synoptic states in winter (in hours).

LN	*SN*	*LE*	*SE*	*LS*	*SS*	*LW*	*SW*	*0*	Across states
21.5 (15.4)	28.9 (17.1)	22.3 (17.8)	37.7 (31.4)	31.8 (24.8)	24.4 (22.9)	21.0 (13.7)	60.1 (67.8)	14.2 (12.1)	27.0 (28.4)

The numbers in brackets are standard deviations.

Table 3.11. Transition probabilities conditional on a change of synoptic states in winter (based on 4 years' observations).

	LN	*SN*	*LE*	*SE*	*LS*	*SS*	*LW*	*SW*	*0*	Total
LN	.	0.26 (0.06)	0.13 (0.05)	.	.	.	0.40 (0.07)	.	0.21 (0.06)	1.00
SN	0.41 (0.10)	.	.	.	.	.	.	0.59 (0.10)	.	1.00
LE	0.23 (0.06)	.	.	0.36 (0.06)	0.16 (0.05)	.	.	.	0.25 (0.06)	1.00
SE	.	0.31 (0.09)	0.66 (0.09)	.	.	0.03 (0.03)	.	.	.	1.00
LS	.	.	0.53 (0.08)	.	.	0.21 (0.06)	0.05 (0.04)	.	0.21 (0.06)	1.00
SS	.	.	.	0.50 (0.12)	0.50 (0.12)	.	.	.	.	1.00
LW	0.31 (0.07)	.	.	.	0.15 (0.05)	.	.	0.33 (0.07)	0.21 (0.06)	1.00
SW	.	0.03 (0.03)	.	.	.	0.31 (0.09)	0.66 (0.09)	.	.	1.00
0	0.24 (0.07)	.	0.29 (0.07)	.	0.33 (0.07)	.	0.14 (0.05)	.	.	1.00

The numbers in brackets are estimated standard deviations.

Apart from strong westerlies (*SW*), the longest durations were recorded for strong easterlies (*SE*), light southerlies (*LS*) and strong northerlies (*SN*). This suggests that these states more than others are associated with particular synoptic configurations. In fact, a short state duration can be an indication of a mere transition between two states that are unlikely to be transformed into each other directly, given the smoothing procedure performed in the last chapter. Hence, the shorter the duration of a particular state, the less likely it reflects a *typical* synoptic configuration. Strong westerlies (*SW*) are surely associated with extratropical cyclones approaching the continent from the south-west, while light easterlies (*LE*) tend to occur more often under the influence of the high pressure belt. Further, if the wind comes from a westerly direction then it is more likely strong than light, while for all other directions the converse is true.

It is also noted that there are about twice as many transitions into the state representing light winds as there are transitions into the state that is associated with strong winds from the same direction. On the other hand, the average duration for the latter states is noticeably longer. This merely reflects the fact that light winds are more susceptible to fluctuations and thus state changes. The lower the speed, the more quickly the wind usually swings around. For example, strong westerlies (*SW*), having the highest mean speed, exhibit a markedly longer duration than any other pattern, even though the difference may not be statistically significant.

To establish the general synoptic circulation, we now turn to the transition probabilities given in Table 3.11. It follows that strong northerlies (*SN*) are most likely followed by strong westerlies (*SW*), which in turn are most likely followed by light westerlies (*LW*). Similarly, strong southerlies (*SS*) lead to light southerlies (*LS*) or strong easterlies (*SE*), both of which precede light easterlies (*LE*). The most likely transitions (with a probability ≥ 0.4) are:

strong northerly (*SN*)	$\rightarrow$	strong westerly (*SW*)	or light northerly (*LN*)
light northerly (*LN*)	$\rightarrow$	light westerly (*LW*)	
strong westerly (*SW*)	$\rightarrow$	light westerly (*LW*)	
strong southerly (*SS*)	$\rightarrow$	strong easterly (*SE*)	or light southerly (*LS*)
light southerly (*LS*)	$\rightarrow$	light easterly (*LE*)	
strong easterly (*SE*)	$\rightarrow$	light easterly (*LE*)	

That is, there is a complete analogy between northerlies and westerlies on the one hand, and southerlies and easterlies on the other hand. When the wind comes from the north-westerly sector it tends to move to a light westerly (*LW*), while light easterlies (*LE*) seem to form an absorbing state for winds from the south-easterly sector. However, there is also a good chance that westerlies are followed by southerlies, and easterlies are followed by northerlies. This suggests that the wind moves in a predominantly anticlockwise direction and further, that it strengthens as it turns from the easterly to the westerly sector, and conversely, that it moderates as it swings back from the westerly to the easterly sector.

Figure 3.17. Flow of synoptic states in winter.

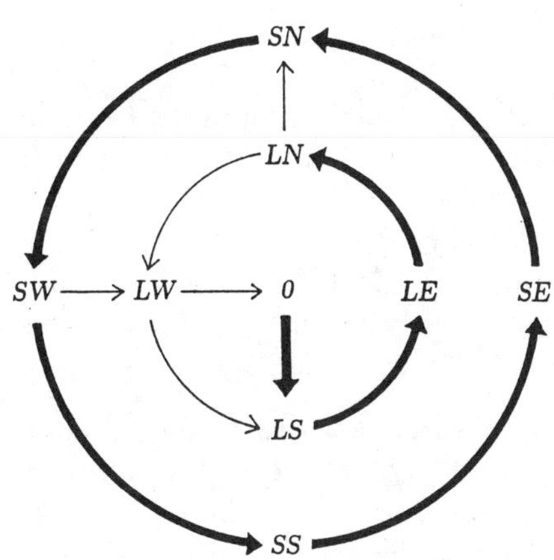

These findings are even more evident in Figure 3.17 which shows the *net transitions* defined as the number of transitions from state 1 to state 2 minus the number of transitions in the opposite direction. A light arrow corresponds to 3-5 net transitions, whilst a thick arrow represents more than 5 net transitions. The diagram in Figure 3.17 gives a very clear picture of the path that is typical of the wind in winter. While moving around the full circle in an anticlockwise direction, it moderates as it swings through the south-west and strengthens as it turns north-westerly. Given that there are hardly any records of a strong wind turning clockwise, this anticlockwise rotation is particularly pronounced for winds with a speed of more than 6 knots.

In fact, anticlockwise rotating winds occurred for about 69 per cent of the time. This in turn reflects a predominantly southerly influence, as a low passing south of Fremantle can generally be associated with strong winds from the north-west turning south-westerly, whereas a high passing south would cause light southerlies swinging anticlockwise through the east to a northerly direction.

It is readily seen that Figure 3.17 exhibits similar characteristics to the diagram in Figure 3.13 and roughly reflects the predominant synoptic pattern as described in section 3.4.1. The only unexpected result is the weakening of westerlies before turning southerly. Although the evidence is not particularly strong, this could mean that the wind starts to moderate just after the cold front, which is associated with strong westerlies, has passed through the region. It may also be interesting to note that the average durations of the states which constitute the basic cycle (11) amount to a 7- to 8-day period, even though this is far from being conclusive, given the departures from this cycle on the one hand, and the large standard deviations on the other hand.

3.4.5 Empirical summer pattern

The results of classifying the geostrophic component ($\mathbf{g}_t$) in summer are listed below in Tables 3.12 to 3.16. Just like Tables 3.7 to 3.11 they give the number of transitions between synoptic states, the number of transitions into the different states, the percentage time for which these states occurred, the average number of hours the wind remained in a certain state and the one-step transition probabilities, respectively. As in the last section the transitions reflect hourly changes, and are based on the summer seasons of the years 1971/72, 1973/74, 1975/76 and 1977/78.

Strong southerlies (*SS*) clearly constitute the dominant state in summer, which occurs for more than two thirds of the time and has a much longer duration than any other pattern. Given an average duration of almost 7 days, it follows that the synoptic configuration must be extremely stable. Indeed, the atmospheric pressure field during this time of the year is characterized by a persistent high over the eastern Indian Ocean, a ridge extending from this high into the Bight and an almost stationary low over the north-west of the continent.

Table 3.12. Number of transitions between synoptic states in summer (based on observations of the years 1971/72, 1973/74, 1975/76 and 1977/78).

	LN	*SN*	*LE*	*SE*	*LS*	*SS*	*LW*	*SW*	*0*	Total
LN	34	.	.	.	.	.	3	.	.	37
SN	.	42	.	.	.	.	.	1	.	43
LE	1	.	628	2	13	.	.	.	4	648
SE	.	1	12	1067	.	12	.	.	.	1092
LS	.	.	6	.	592	19	1	.	2	620
SS	.	.	.	23	13	5931	.	1	.	5968
LW	.	.	.	.	4	.	160	5	.	169
SW	.	.	.	.	.	6	1	216	.	223
0	2	.	.	.	.	.	4	.	70	76

Table 3.13. Number of transitions into synoptic states from other states.

LN	*SN*	*LE*	*SE*	*LS*	*SS*	*LW*	*SW*	*0*	Total
3	1	18	25	30	37	9	7	6	136

Table 3.14. Relative frequency of synoptic states in summer (in %).

LN	*SN*	*LE*	*SE*	*LS*	*SS*	*LW*	*SW*	*0*	Total
0	1	7	12	7	67	2	3	1	100

Table 3.15. Average duration of synoptic states in summer (in hours).

LN	*SN*	*LE*	*SE*	*LS*	*SS*	*LW*	*SW*	*0*	Across states
12.3	43.0	35.9	43.7	20.7	161.3	18.8	31.9	12.7	63.4
-	-	(27.6)	(51.9)	(15.1)	(129.1)	(6.6)	(24.2)	(8.5)	(90.9)

The numbers in brackets are standard deviations.

Table 3.16. Transition probabilities conditional on a change of synoptic states in summer (based on 4 years' observations).

	LN	*SN*	*LE*	*SE*	*LS*	*SS*	*LW*	*SW*	*0*	Total
LN	.	.	.	.	.	.	1.00 -	.	.	1.00
SN	.	.	.	.	.	.	.	1.00 -	.	1.00
LE	0.05 (0.05)	.	.	0.10 (0.07)	0.65 (0.11)	.	.	.	0.20 (0.09)	1.00
SE	.	0.04 (0.04)	0.48 (0.10)	.	.	0.48 (0.10)	.	.	.	1.00
LS	.	.	0.21 (0.08)	.	.	0.68 (0.09)	0.04 (0.04)	.	0.07 (0.05)	1.00
SS	.	.	.	0.62 (0.08)	0.35 (0.08)	.	.	0.03 (0.03)	.	1.00
LW	.	.	.	.	0.44 (0.17)	.	.	0.56 (0.17)	.	1.00
SW	.	.	.	.	.	0.86 (0.13)	0.14 (0.13)	.	.	1.00
0	0.33 (0.19)	.	.	.	.	.	0.67 (0.19)	.	.	1.00

The numbers in brackets are estimated standard deviations.

However, strong easterlies (*SE*), light easterlies (*LE*) and light southerlies (*LS*) are also important, as indicated by their relative frequencies in Table 3.14. Although *SE* and *LE* have a duration not nearly as long as *SS*, they are still relatively persistent, especially if compared with the average durations in winter (cf. Table 3.10). This suggests that both *SE* and *LE* are a significant part of the general summer pattern, even though they reflect a similar synoptic configuration as *SS*. However, whereas the high over the Indian Ocean has a dominant influence on the wind pattern when strong southerlies (*SS*) prevail, *SE* and *LE* are evidence that the high pressure ridge is extending further into the Bight and that the influence of the low over the north-west increases.

An examination of the transition probabilities in Table 3.16 shows that strong southerlies (*SS*) are usually followed by strong easterlies (*SE*), which then either moderate (*LE*) or swing back to a southerly direction (*SS*). If we have light easterlies (*LE*), then the wind is most likely to return to *SS* via *LS*. The most likely transitions in summer (with a probability ≥ 0.4) are:

strong southerly (*SS*) $\rightarrow$ strong easterly (*SE*)
strong easterly (*SE*) $\rightarrow$ light easterly (*LE*) or strong southerly (*SS*)
light easterly (*LE*) $\rightarrow$ light southerly (*LS*)

light southerly (*LS*) → strong southerly (*SS*)
light westerly (*LW*) → strong westerly (*SW*) or light southerly (*LS*)
strong westerly (*SW*) → strong southerly (*SS*)

Hence, the following sequence of synoptic states seems to prevail in summer:

$$SS \rightarrow SE\ (\rightarrow LE \rightarrow LS\ )\ \rightarrow SS$$

Note that this cycle is almost identical with (12). If we plot the *net* transitions between synoptic states, just as in Figure 3.17, this result becomes even clearer.

Figure 3.18. Flow of synoptic states in summer.

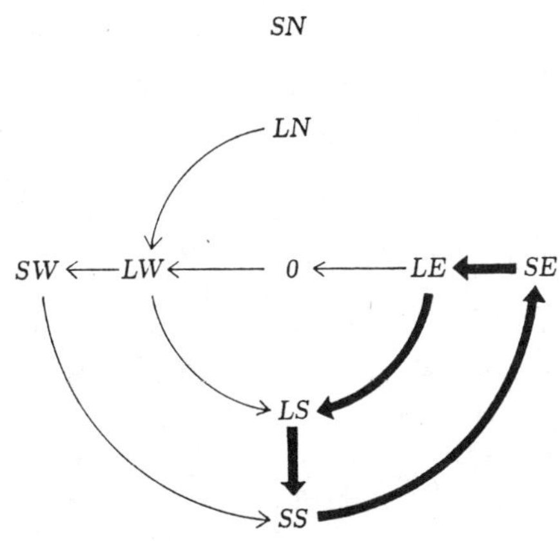

From Figure 3.18 it is obvious that strong southerlies usually are followed by strong easterlies. The wind then tends to moderate and, as it swings back to a southerly direction, strengthens again. Given the relatively large number of transitions into light southerlies (*LS*) and at the same time their shorter duration, *LS* seems to be more a transition state than a reflection of a particular synoptic configuration. This suggests that *SS*, *SE* and *LE* constitute the basic sequence of synoptic states in summer, just as could be expected from the discussion in section 3.4.2. The reader is reminded that the wind direction in the geostrophic component ($\mathbf{g}_t$) and the series ($\mathbf{h}_t$), which is a reflection of the atmospheric pressure field, differ by more than 60° in summer. That is, if the synoptic chart indicates winds from the north-easterly sector, they actually appear as easterlies in ($\mathbf{g}_t$).

As in winter, it is observed that the states representing strong winds usually have a longer duration than the corresponding states associated with light winds. Further, winds from the northerly and westerly sectors rarely appear in the record of the geostrophic component ($\mathbf{g}_t$) as is evident from Table 3.14. However, once in a while the wind turns from a light easterly through *0* and *LW* into a strong westerly. This sequence can be interpreted as follows. After the trough, which is associated with the low over the north-west, has passed through the region, the wind usually swings back from an easterly to a southerly direction, driven by the high over the Indian Ocean. Occasionally, however, the region may come under the influence of a depression passing south of the continent, with a pressure configuration similar to that in winter developing. Finally, it is noted that for 63 per cent of the time in summer we recorded an anticlockwise rotating wind, which again signifies a predominantly southerly influence.

In summarising, it can be said that *synoptic states* as discussed in this section 3.4 provide a useful device for examining and quantifying the sequence of synoptic scale winds as determined by the passage of high and low pressure systems.

3.5 Appendix: Derivation of the Geostrophic Wind Equation

In this section we will derive equation (5) which formed the basis for the prediction of the geostrophic wind. Consider a coordinate system with origin at Fremantle, and with horizontal and vertical axes in an easterly and northerly direction, respectively. The situation is sketched in Figure 3.19. Suppose that the jth pressure system S_j is centered at (x_j, y_j), which in polar coordinates is equivalent to (d_j, η_j), and let P be an arbitrary point at $(x, y) \equiv (d, \eta)$ in the neighbourhood of the origin. This ensures that the curvature of the earth can be ignored and that the formulae for plane triangles (Abramowitz & Stegun 1966, section 4.3.148) hold approximately. The distance between S_j and P is denoted by $d_j(x, y)$.

Figure 3.19.

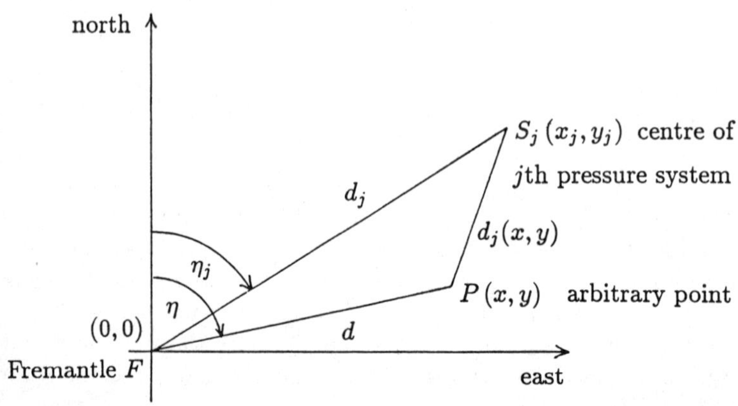

Putting $e_j(d,\eta) = d_j^2(x,y)$ and applying the cosine theorem to the triangle PFS_j yields

$$e_j(d,\eta) = d^2 + d_j^2 - 2d_j d\,\cos(\eta - \eta_j)\ .$$

In order to formulate equation (2) for an arbitrary point P, one needs to replace the square of the distance d_j by $e_j(d,\eta)$. Hence,

$$p(d,\eta) = \Big[\sum_j \frac{c_j}{e_j(d,\eta)}\, p_j\Big] \Big/ \Big[\sum_j \frac{c_j}{e_j(d,\eta)}\Big]\ . \tag{13}$$

Differentiation of (13) with respect to d and η then leads to

$$\frac{\partial p}{\partial d} = \left[-\sum_j \frac{e_{jd}(d,\eta)}{e_j^2(d,\eta)}\, c_j\, p_j \sum_k \frac{c_k}{e_k(d,\eta)} + \sum_j \frac{e_{jd}(d,\eta)}{e_j^2(d,\eta)}\, c_j \sum_k \frac{c_k\, p_k}{e_k(d,\eta)} \right] \Big/ \left[\sum_j \frac{c_j}{e_j(d,\eta)} \right]^2$$

and

$$\frac{1}{d}\frac{\partial p}{\partial \eta} = \left[-\sum_j \frac{e_{j\eta}(d,\eta)}{d\, e_j^2(d,\eta)}\, c_j\, p_j \sum_k \frac{c_k}{e_k(d,\eta)} + \sum_j \frac{e_{j\eta}(d,\eta)}{d\, e_j^2(d,\eta)}\, c_j \sum_k \frac{c_k\, p_k}{e_k(d,\eta)} \right] \Big/ \left[\sum_j \frac{c_j}{e_j(d,\eta)} \right]^2 ,$$

where

$$e_{jd}(d,\eta) = \frac{\partial e_j(d,\eta)}{\partial d} = 2d - 2d_j \cos(\eta - \eta_j)$$

and

$$d^{-1}\, e_{j\eta}(d,\eta) = \frac{\partial e_j(d,\eta)}{\partial \eta} = 2d_j \sin(\eta - \eta_j)\ .$$

In order to calculate the geostrophic wind at Fremantle, grad $p(d,\ \eta)$ needs to be evaluated at the origin; that is

(14) $$\operatorname{grad} p(d,\eta)\Big|_{(0,0^\circ)} = \begin{pmatrix} \frac{\partial p}{\partial d} \\ \frac{1}{d}\frac{\partial p}{\partial \eta} \end{pmatrix}\Bigg|_{(0,0^\circ)} .$$

Since

$$e_j(0,0^\circ) = d_j^2\ ,$$
$$e_{jd}(0,0^\circ) = -2d_j \cos\eta_j\ ,$$

and $$d^{-1} e_{j\eta}(0,0^\circ) = -2d_j \sin\eta_j\ ,$$

one has

$$\frac{\partial p}{\partial d}(0,0^\circ) = \frac{2\sum_j d_j^{-3} c_j p_j \cos\eta_j - 2\sum_j d_j^{-3} c_j p \cos\eta_j}{\sum_j d_j^{-2} c_j}$$

(15a) $$= \frac{2\sum_j d_j^{-3} c_j (p_j - p) \cos\eta_j}{\sum_j d_j^{-2} c_j}$$

and

(15b) $$\frac{1}{d}\frac{\partial p}{\partial \eta}(0,0^\circ) = \frac{2\sum_j d_j^{-3} c_j (p_j - p) \sin\eta_j}{\sum_j d_j^{-2} c_j}\ .$$

If (15) is substituted into (14) and then into (1), the geostrophic wind is obtained as

$$\mathbf{u}_g = \mathbf{k} \times \frac{1}{\varrho f} \cdot \frac{2}{1+\sum_{j=1}^{n} d_j^{-2} c_j} \begin{pmatrix} \sum_{j=1}^{n} d_j^{-3} c_j (p_j - p) \cos \eta_j \\ \sum_{j=1}^{n} d_j^{-3} c_j (p_j - p) \sin \eta_j \end{pmatrix} .$$

Note that according to (4), equation (14) does not depend on η, since

$$\operatorname{grad} p(d,\eta)\Big|_{(0,\eta)} = \begin{pmatrix} \cos\eta & -\sin\eta \\ \sin\eta & \cos\eta \end{pmatrix} \begin{pmatrix} \frac{\partial p}{\partial d}(0,\eta) \\ \frac{1}{d}\frac{\partial p}{\partial \eta}(0,\eta) \end{pmatrix} = \operatorname{grad} p(d,\eta)\Big|_{(0,0^\circ)} .$$

CHAPTER 4

THE LAND AND SEA BREEZE CYCLE

In this chapter we begin to investigate the daily component $(\mathbf{d}_t)$*, concentrating on all effects that have a 24-hour pattern and in particular, the land and sea breeze cycle. This phenomenon is commonly studied in the field of dynamic meteorology with the aim of predicting the actual wind on a short-term basis. However, the objective in this chapter is to determine the types of circulation that occur depending on the time of the year and the geostrophic wind. In contrast with the rather intricate physical approach, we therefore use statistical methods to study this circulation. After the removal of the land and sea breeze cycle from* $(\mathbf{d}_t)$ *the analysis will continue in chapter 5 with the residual series* $(\mathbf{r}_t)$ *and again be resumed in chapter 10.*

The physical basis of the land and sea breeze circulation which dominates the local weather for most of the year is discussed in section 4.1. In section 4.2 our approach is described using the generated data of chapter 2. By focussing on those days when the wind swings around anticlockwise in a full circle, it is possible to adjust for the use of a mean filter for the initial decomposition, and to define an ideal land and sea breeze cycle. These techniques are then applied to $(\mathbf{d}_t)$ *in section 4.3 where various empirical circulation patterns are established. Their seasonal characteristics and dependence on the geostrophic wind are examined.*

By focussing on $(\mathbf{d}_t)$ *rather than the raw data series* $(\mathbf{w}_t)$ *it is possible to detect a land and sea breeze cycle also in winter when it is generally believed not to exist. However, when studying the influence of the geostrophic component* $(\mathbf{g}_t)$ *on this circulation it is found that the sea breeze is unlikely to occur when westerlies prevail. This means, given the increased frequency of westerlies in winter, that the sea breeze registers on fewer days than in summer.*

4.1 The Nature of the Circulation

During the summer months the wind pattern in the Fremantle area is dominated by the interaction of synoptic scale easterlies with the land and sea breeze circulation. In fact, most of the meteorological monitoring in the region is aimed at the resolution of the relevant characteristics of this cycle. In the present chapter an empirical sea breeze pattern is derived, and its main features and seasonal characteristics are determined.

The coastline in the Fremantle area forms an almost straight line from north to south. Since further the adjacent plain is relatively featureless, except for some hills to about 80 m elevation north of Perth, the conditions are ideal for an investigation of the land and sea breeze cycle. Even the Darling Range which rises at about 25 to 30 km inland reaches 300 m elevation in only a few places. The blocking which occurs at the western edge of other continents is therefore almost absent. Since the wind direction can be very sensitive to mesoscale terrain properties (Hardy & Walton 1978), the conclusions drawn from this study may not apply in general. It would therefore be interesting to perform a similar analysis for different geographical regions and to compare the results with those presented here. For a discussion of the effect of topographic barriers on the sea breeze circulation the reader is also referred to Mahrer & Pielke (1977).

Figure 4.1. Northerly wind component versus easterly component, 4 and 5 December 1971.

(a) Wind series ($\mathbf{w}_t$).

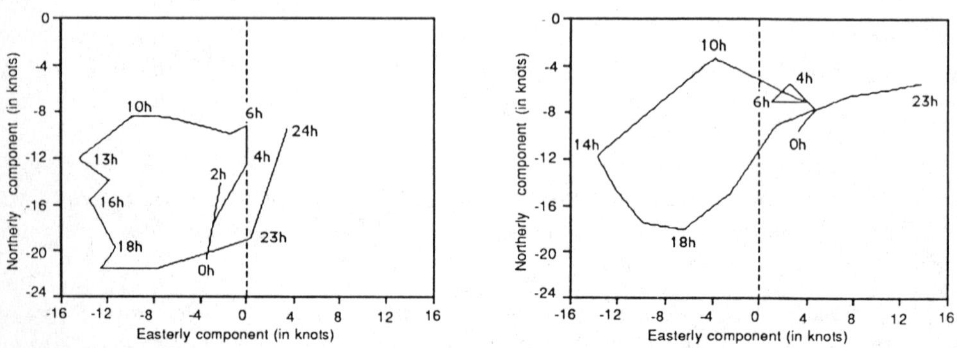

(b) Daily component ($\mathbf{d}_t$).

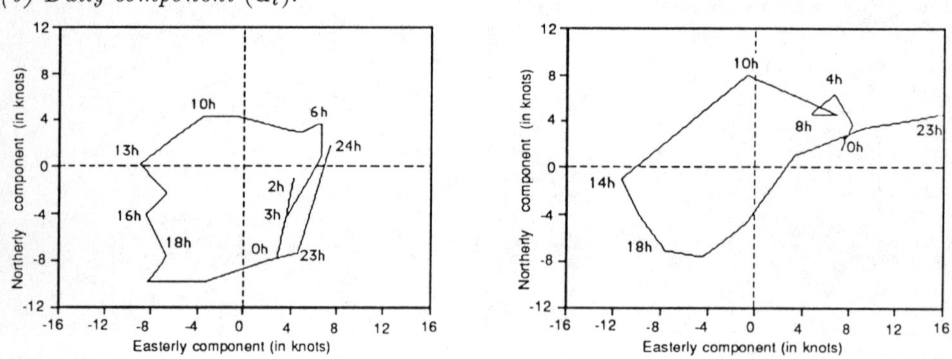

So far little reference was made to temperature gradients and their impact on the wind circulation. However, the surface wind is as much determined by temperature differences as by synoptic scale pressure configurations. During the day solar heating of the land increases the soil temperature, which causes an increase in the air temperature over the land and hence a reduction of the atmospheric density. Since the temperature over the adjacent ocean does not change as much, a temporary wind develops, blowing onshore from the ocean to the land. This wind is commonly referred to as *sea breeze*. At night this process is reversed. Heat radiates from the soil, cooling the land and thus decreasing the air temperature over the land. However, since the temperature over the sea remains almost constant, another temporary wind is generated, this time blowing offshore from the land to the sea and therefore called *land breeze*.

A typical path of the wind during two consecutive days in summer is shown in Figure 4.1 (a) where the time-of-day has been noted as a parameter along the trajectory. The vertical axis represents the northerly wind component with positive and negative values corresponding to winds from the northerly and southerly sector, respectively. In the same way, positive and negative values on the horizontal axis denote the easterly and westerly components. For example, at 1800 h on 4 December the two components result in a wind of 23 knots from 210°. Also note that the wind at 2400 h on 4 December is identical with that at 0000 h on 5 December.

Even though $(\mathbf{w}_t)$ is confined to the south-westerly sector on 4 and 5 December, the wind tends to describe a full circle on both days. The removal of the geostrophic component $(\mathbf{g}_t)$ results in centering these circles at the origin as shown in Figure 4.1 (b). Their shape, however, remains unchanged so that they can be thought of as being superimposed on the geostrophic wind. These circles also constitute the basic feature of the land and sea breeze pattern, as roughly depicted in Figure 4.2. For the definition of the mean wind direction and the mean directional resultant length the reader is referred to appendix 2.6.1.

Both, the mean direction and the mean directional resultant length exhibit a periodic circulation with a distinct land breeze at about 0700 h in the morning and a sea breeze at about 1700 h in the afternoon. The mean directional resultant length indicates that the wind direction changes less when either the land or the sea breeze is peaking. Note that wind speed has not been displayed in these diagrams, and that the circulation is identified solely by the wind direction moving anticlockwise in a full circle each day and by the speed with which the wind direction changes. The plots of the mean direction show that the land and sea breeze cycle is present throughout the year, something that was not obvious from the records in $(\mathbf{w}_t)$. Furthermore, the mean directional resultant seems to suggest that this cycle is more pronounced in autumn than in summer. Whereas in autumn there are two periods each day when the wind direction changes rapidly and two periods when the direction is almost constant, in summer the change of direction is slightly more uniform. In contrast with mean wind direction and mean directional resultant length, the plot of wind speed did not exhibit any contour which was similarly significant. This suggests that wind direction be more strongly emphasized in the analysis than wind speed.

Figure 4.2. Mean wind direction and mean directional resultant length for each hour of the day, calculated for the daily component ($\mathbf{d}_t$).

(a) Autumn 1971.

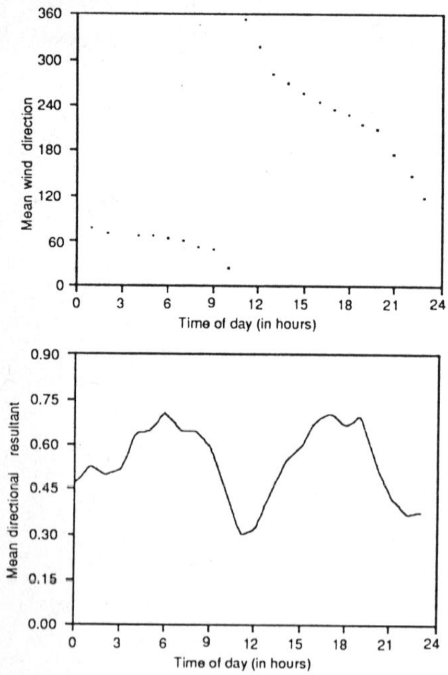

(b) Winter 1971.

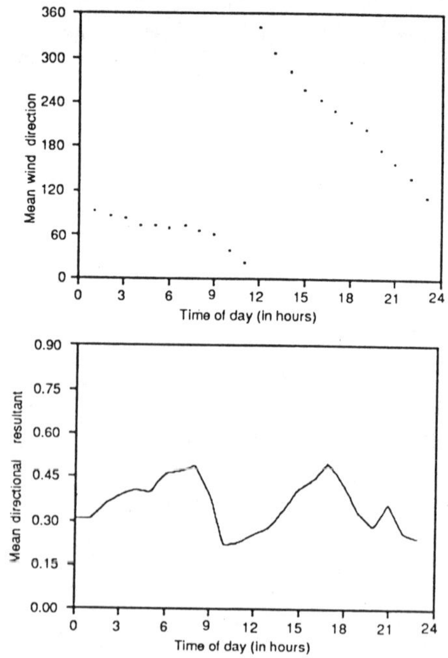

(c) Spring 1971.

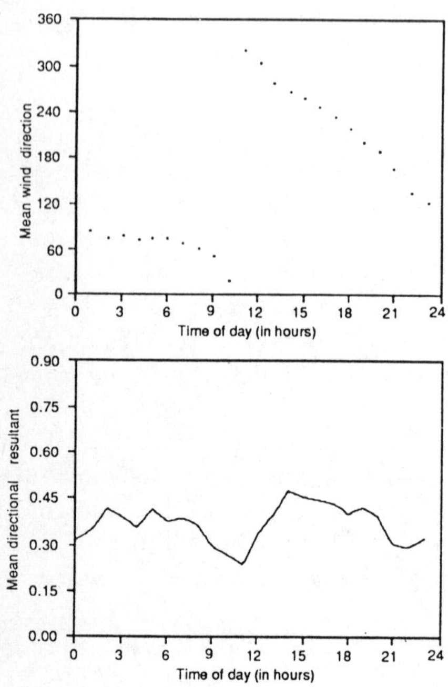

(d) Summer 1971/72.

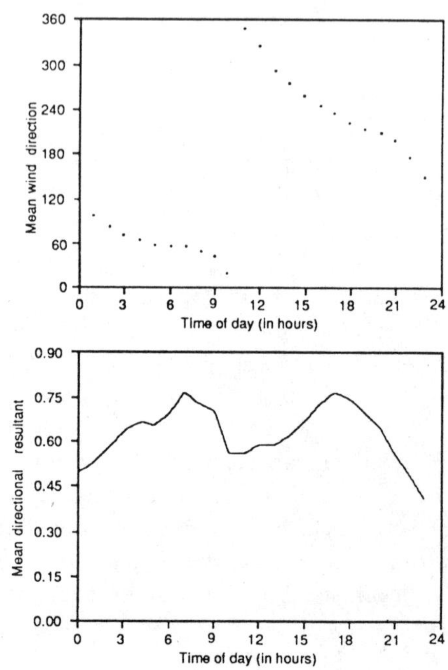

4.2 Statistical Approach

In the area of dynamic meteorology a number of models has been built to study the effect of synoptic scale forces and temperature fields on the development of the sea breeze circulation. See, for example, Fisher (1961), Estoque (1962), Neumann & Mahrer (1971), Pielke (1974), Mak & Walsh (1976), Physick (1976) and Rye (1980). All these models aim at a physical resolution of the *actual sea breeze*, which is particularly important for *short-term* prediction.

In this study, however, we are interested in categorizing the land and sea breeze cycle. We therefore follow a statistical approach in order to establish *types of circulation*, which is more relevant for *long-term* prediction. Central to our investigation is a statistical definition of the land and sea breeze cycle. By presenting this pattern in a form that is amenable to physical interpretation, further insight into the meteorological phenomenon is provided. For the use of statistical methods when exploring complex relationships the reader is also referred to Barndorff-Nielsen *et al.* (1985).

Figure 4.3 shows the daily component $(\mathbf{d}_t)$ for the period from 1 to 8 December 1971. Each day the wind direction typically describes a full anticlockwise circle, with a land breeze from the north-east in the morning and a sea breeze from the south-west in the afternoon. This pattern is most pronounced between 1 and 5 December when Fremantle was under the influence of a high pressure system as shown in Figure 1.3. The sea breeze on 6 and 7 December is hardly marked, and on 8 December not even the anticlockwise cycle of the wind direction is observed. Wind speed is peaking twice a day, although the speed pattern is not as distinct as the directional pattern. Also note the comparatively high wind speed at 2400 h on 5 December, at a time of day when the wind usually moderates. In order to quantify the circulation pattern the effect that is associated with any particular time of the day will now be estimated.

The discussion in the last paragraph suggests that the sea breeze not only has seasonal characteristics but also depends on factors such as the intensity of solar heating and the geostrophic wind. It therefore seems appropriate to derive a number of sea breeze patterns by means of a classification of all the days in a season. In this study we will distinguish between an *ideal* pattern based on the days which exhibit a distinct land and sea breeze cycle on the one hand, and a *complementary pattern* based on the remaining days on the other hand.

Let us rewrite the index t of the record $\mathbf{d}_t$ in the daily component $(\mathbf{d}_t)$ in terms of a day-index k and a time-of-day index m, that is $t = 24k + m$ $(m = 0, \ldots, 23)$. The two sea breeze patterns can then be associated with a set Δ of day-indices and its complement Δ^c, respectively. The sea breeze vector $\mathbf{b}_m(\Delta)$ $(m = 0, \ldots, 23)$ is now defined by averaging over all wind records which correspond to a particular time of the day, that is

$$\mathbf{b}_m(\Delta) = \frac{1}{|\Delta|} \sum_{k \in \Delta} \mathbf{d}_{24k+m} \qquad , \; m = 0, \ldots, 23 .$$

Figure 4.3 (a). Wind direction in $(\mathbf{d}_t)$ *from 1 to 8 December 1971.*

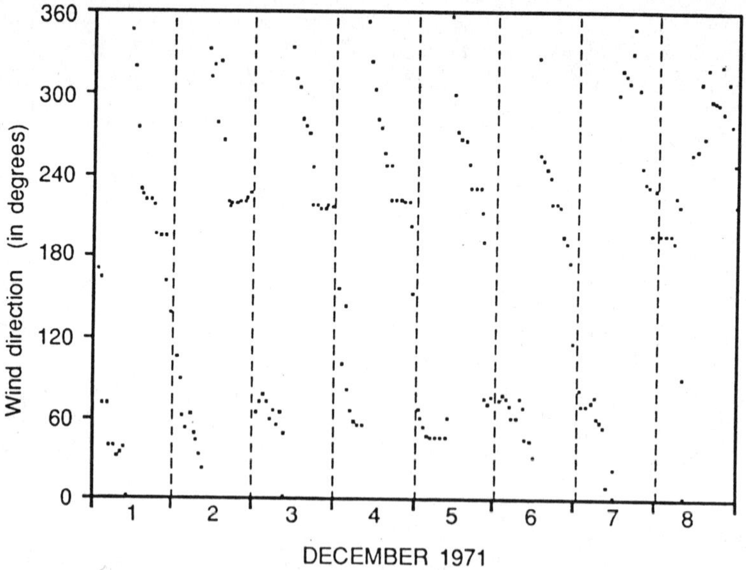

Figure 4.3 (b). Wind speed in $(\mathbf{d}_t)$ *from 1 to 8 December 1971.*

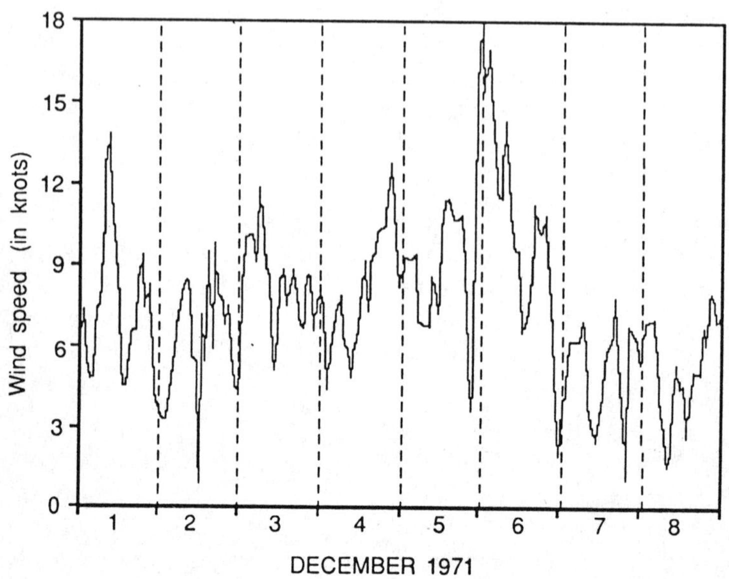

The remaining task therefore is to select the days with a distinct land and sea breeze cycle, and to determine the set Δ. Since for these days one would expect a short resultant vector

$$\mathbf{s}_k = \frac{1}{24} \sum_{m=0}^{23} \mathbf{d}_{24k+m} \quad ,$$

the ideal pattern will be based on all those days for which $\mathbf{s}_k$ is no further away from the mean vector $\overline{\mathbf{d}}$ of $(\mathbf{d}_t)$ than a given distance ϵ. That is,

$$(1) \qquad \Delta = \{ k \mid \| \mathbf{s}_k - \overline{\mathbf{d}} \| < \epsilon \} \; .$$

Figure 4.4. Admissible $\mathbf{s}_k$*-region for* $k \in \Delta$.

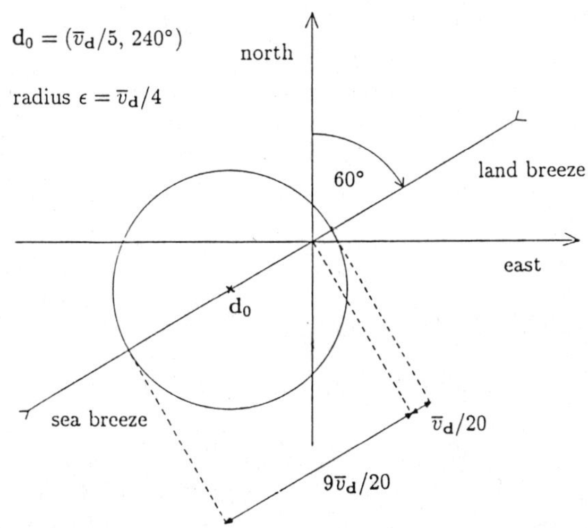

For all seasons, $\overline{\mathbf{d}}$ was found to be close to $\mathbf{d}_0 = (\overline{v}_\mathbf{d}/5, 240°)$ where $\overline{v}_\mathbf{d}$ is the seasonal mean speed. This roughly means that the land and sea breezes blow from directions of about 60° and 240°, respectively, and further that the ratio of their strengths is in the order of 2:3, even though this ratio will be different for the ideal and complementary patterns. As for the *critical distance* ϵ we chose the value $\overline{v}_\mathbf{d}/4$. The situation is broadly sketched in Figure 4.4. Whenever the day-vector $\mathbf{s}_k$ lies inside the circle the observations recorded on day k form part of the ideal pattern, whereas otherwise they contribute to the complementary pattern.

To illustrate this procedure let us return to the simulated series plotted in Figure 2.11 (a), where a 6-hour storm had been imposed on an ideal land and sea breeze cycle. Each cell in Table 4.1 contains mean wind speed, mean sum of squares and mean resultant length for each of the three days in the simulated series. Since the circulation pattern depends on the initial decomposition, the results are presented for the cases

Table 4.1. Summary values for the daily component (compare with Figures 2.11 (b), (c) and (e), $\delta = 6$).

		Mean M	Median N	Composite $(5H)^3$
Day 1	Mean speed	2.5	2.0	2.3
	Mean sum of squares	9.5	6.0	7.6
	Mean resultant length	1.02	0.26	0.72
Day 2	Mean speed	4.0	4.1	4.0
	Mean sum of squares	23.8	33.4	27.0
	Mean resultant length	0.96	2.59	1.79
Day 3	Mean speed	2.0	2.0	2.0
	Mean sum of squares	6.0	6.0	6.1
	Mean resultant length	0.00	0.00	0.04
Mean speed $\overline{v}_{\mathbf{d}}$		2.8	2.7	2.8
Critical distance $\epsilon = \overline{v}_{\mathbf{d}}/4$		0.71	0.68	0.69

where either a mean, a median or a composite 3-part descending M-filter was used for the decomposition. Values of the mean speed over the three days and the critical distance $\epsilon = \overline{v}_{\mathbf{d}}/4$ are given in the last two rows of Table 4.1. As for the centre of the data in this example we have $\mathbf{d}_0 = (0, 0°)$. Hence, in order to decide which days are to be taken as a basis for the derivation of the ideal circulation pattern, one simply has to compare the mean resultant lengths with ϵ according to formula (1), with $\overline{\mathbf{d}}$ replaced by $\mathbf{d}_0$.

In case the median filter N is used for the initial decomposition, the sea breeze pattern would be based on observations of the days 1 and 3, whilst the mean filter M and composite 3-part descending M-filter $(5H)^3$ preserve the cycle only on day 3. Note, however, that the mean resultant length obtained for $(5H)^3$ on day 1 is very close to the critical distance. Despite this difference the original sea breeze component $(\mathbf{b}_t^\circ)$ is almost exactly reconstructed in all three cases, as is illustrated in Figure 4.5 for the mean M and the median N. Nevertheless, the estimation of $(\mathbf{b}_t^\circ)$ would be more accurate if a median filter was used for the initial decomposition as it is based on two days' as opposed to one day's observations.

The residual series obtained after the removal of the land and sea breeze circulation is plotted in Figure 4.6. As expected the median filter N is best suited to recapture the storm component $(\mathbf{e}_t^\circ)$ imposed on $(\mathbf{b}_t^\circ)$. The mean filter M, on the other hand, results

Figure 4.5. Land and sea breeze pattern for simulated series, $\delta = 6$.

(a) Mean filter M.

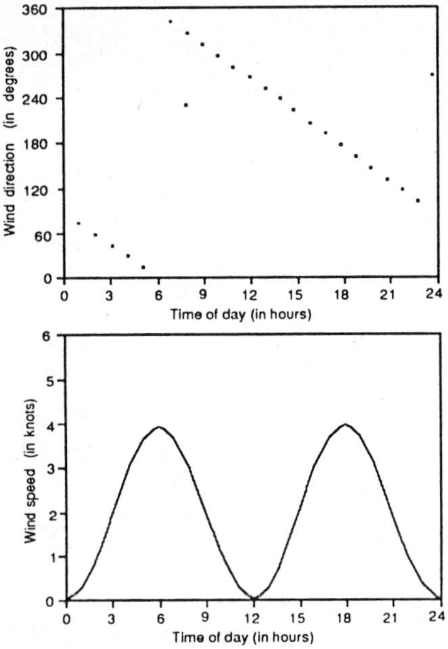

(b) Median filter N.

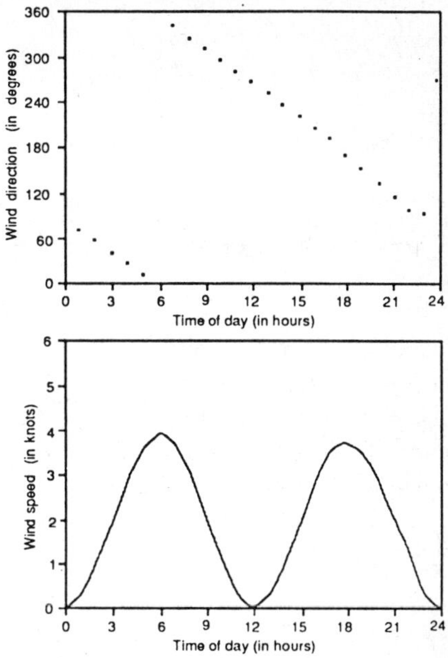

Figure 4.6. Residual series after removal of the land and sea breeze circulation.

(a) Mean filter M.

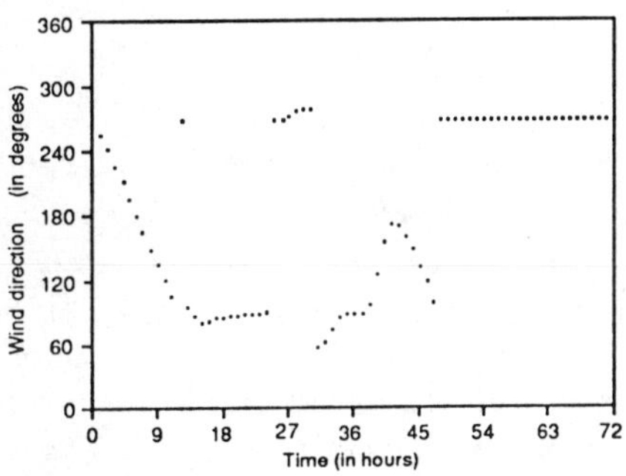

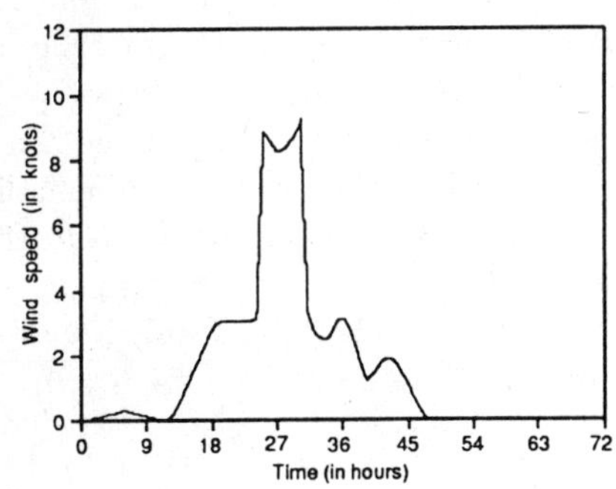

(b) Median filter N.

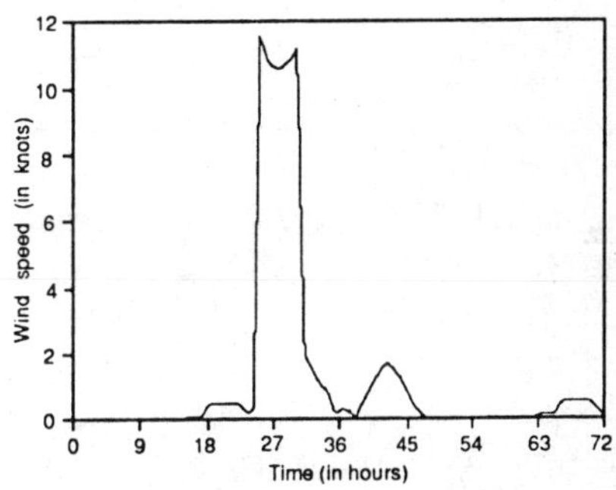

(c) Composite $(5H)^3$.

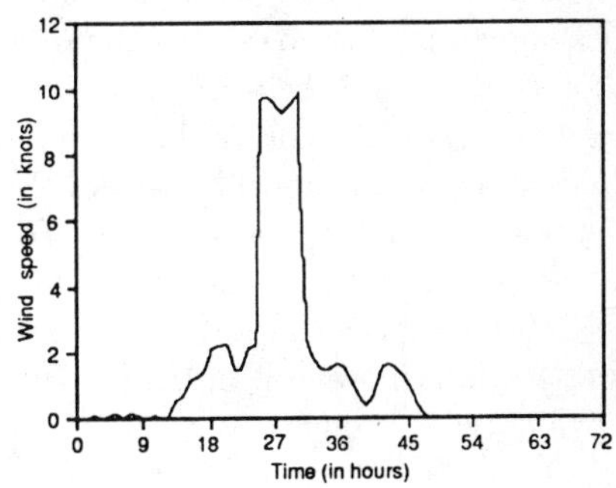

in a wind of about 3 knots from an easterly direction for a period of almost 10 hours just before and after the storm when it is supposed to be 0. How the sea breeze component is actually removed from the data will be explained in section 4.3.2 below.

4.3 Land and Sea Breeze Pattern

4.3.1 The structure of the circulation

The procedure outlined in the last section is now applied to the daily component ($\mathbf{d}_t$) of the wind series ($\mathbf{w}_t$). The land and sea breeze patterns obtained in summer 1971/72 are presented in Figure 4.7. Diagram (a) presents the *ideal* circulation based on those days which exhibit a distinct land and sea cycle, whilst diagram (b) shows the *complementary* pattern based on the remaining days. The *average* pattern based on all the days in summer 1971/72 is depicted in diagram (c).

Figure 4.7 (a) shows that there is a distinct land breeze from the north-east with speeds of around 7 knots between 0700 h and 0900 h in the morning. As the breeze weakens the wind starts to swing anticlockwise at a faster rate. Between 1400 h and 1900 h in the afternoon the sea breeze dominates the pattern with south-westerlies of about 8 knots. Then the wind moderates and veers anticlockwise until the land breeze slowly builds up again. Although not as clearly pronounced this structure can also be observed in the complementary pattern based on Δ^c. In particular the land breeze, which is a distinct feature of the ideal pattern based on Δ, is almost eliminated.

For summer 1971/72 we obtain $|\Delta| = 66$ and $|\Delta^c| = 25$, that is on 66 of the 91 days we were able to establish an ideal sea breeze circulation. Figure 4.8 presents the histograms of the *distance* parameter $e_k = \| \mathbf{s}_k - \overline{\mathbf{d}} \|$ calculated for each day in the season. Note the much wider spread of e_k in winter, suggesting that the summer pattern is more uniform. In fact, there are hardly any days in summer with $e_k > 0.5\,\overline{v}_{\mathbf{d}}$, which means that the land and sea breeze circulation can almost always be detected, even though it may be weaker on some days than on others. The similarity between the three summer patterns presented in Figure 4.7 confirms this conclusion.

Conducting the same analysis for winter 1971 yields an ideal circulation structure which is almost identical with that obtained in summer, but on a smaller velocity scale. As shown in Figure 4.9 (a) the land breeze is characterized by north-easterlies of about 3.5 knots, while the sea breeze blows with about 4.5 knots from a south-westerly direction. On 29 of the 92 days in winter 1971 we were able to detect an ideal circulation according to $|\Delta| = 29$, and $|\Delta^c| = 63$. This result is remarkable insofar as it is usually assumed that a distinct land and sea breeze cycle does not exist during winter. On the other hand, it is noted that the complementary winter pattern depicted in Figure 4.9 (b) is quite different from the patterns discussed so far. Even though the wind direction continues to describe a full anticlockwise circle, wind speed is almost uniform and no longer peaks twice a day.

Figure 4.7. Land and sea breeze circulation, summer 1971/72.

(a) Ideal pattern, based on Δ. *(b) Complementary pattern, based on* Δ^c. *(c) Average pattern, based on* $\Delta \cup \Delta^c$.

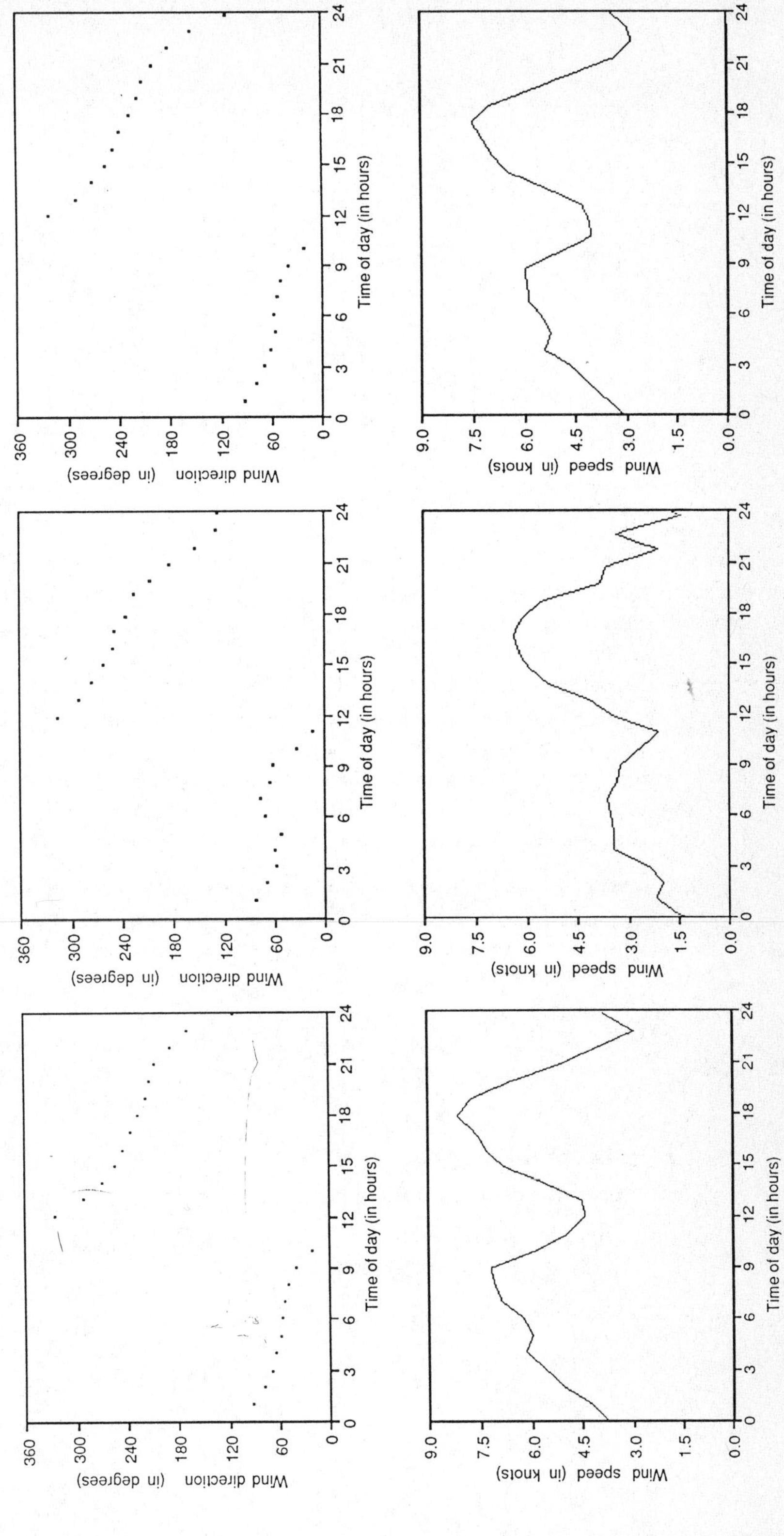

Figure 4.8. Histogram of the distance $e_k = \| \mathbf{s}_k - \bar{\mathbf{d}} \|$.

(*a*) *Summer 1971/72.* (*b*) *Winter 1971.*

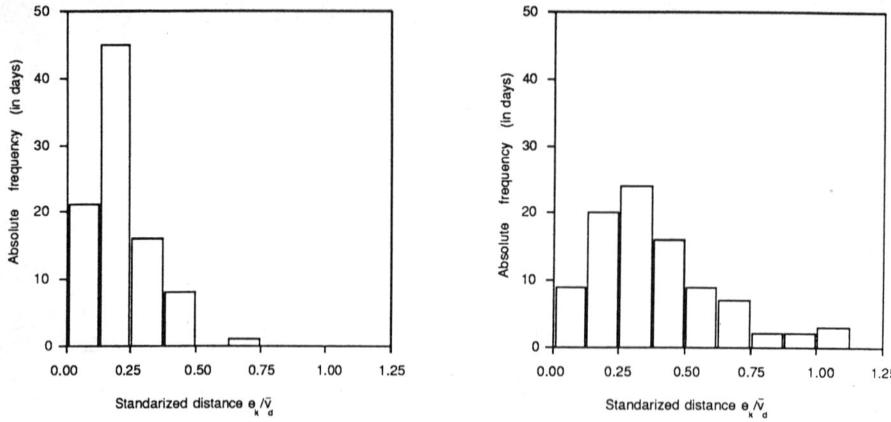

The general circulation pattern is also reflected in the frequency of daytimes at which certain extreme values occur. Figure 4.10 presents the distribution of daily maxima and minima of wind speed, speed change and of directional change. For example, diagram (h) shows that the minimum wind speed over a day will usually be recorded at midnight. In general it can be said that the diagrams (a) to (f) describing the situation in winter 1971 are similar to the diagrams (g) to (l) obtained for summer 1971/72. However, given that no distinction is made between days belonging to Δ and to Δ^c, the characteristics are more pronounced in summer. From Figures 4.10 (j) and (l) it can be seen that the daily cycle is most noticeable in the plot of daytime at which speed and direction changes are minimal.

To investigate the variation of the sea breeze patterns over the years, let $\mathbf{D}_{ghjk} = (\mathbf{d}_{24k+m})_{m=0,\ldots,23}$ be the vector comprising all wind records of day k which contribute to the sea breeze pattern h in season g of year j. Associated with this pattern is a land and sea breeze vector $\mathbf{B}_{ghj} = (\mathbf{b}_m)_{m=0,\ldots,23}$. The average pattern h in season g across the years is defined as $\mathbf{B}_{gh}$, while $\mathbf{B}_g$ and $\mathbf{B}$ correspond to the pattern that is season-specific only, and to the general land and sea breeze pattern, respectively. Also needed are

(i)	the intra-year sum of squares	s^2_{ghj}	$= \sum_k \| \mathbf{D}_{ghjk} - \mathbf{B}_{ghj} \|^2$,
(ii)	the inter-year sum of squares	s^2_{gh}	$= \sum_{jk} \| \mathbf{B}_{ghj} - \mathbf{B}_{gh} \|^2$,
(iii)	the inter-pattern sum of squares	s^2_g	$= \sum_{hjk} \| \mathbf{B}_{gh} - \mathbf{B}_g \|^2$,
(iv)	the inter-season sum of squares	s^2	$= \sum_{ghjk} \| \mathbf{B}_g - \mathbf{B} \|^2$, and
(v)	the total sum of squares	s^2_0	$= \sum_{ghjk} \| \mathbf{D}_{ghjk} - \mathbf{B} \|^2$.

Figure 4.9. Land and sea breeze circulation, winter 1971.

(a) Ideal pattern, based on Δ. *(b) Complementary pattern, based on* Δ^c. *(c) Average pattern, based on* $\Delta \cup \Delta^c$.

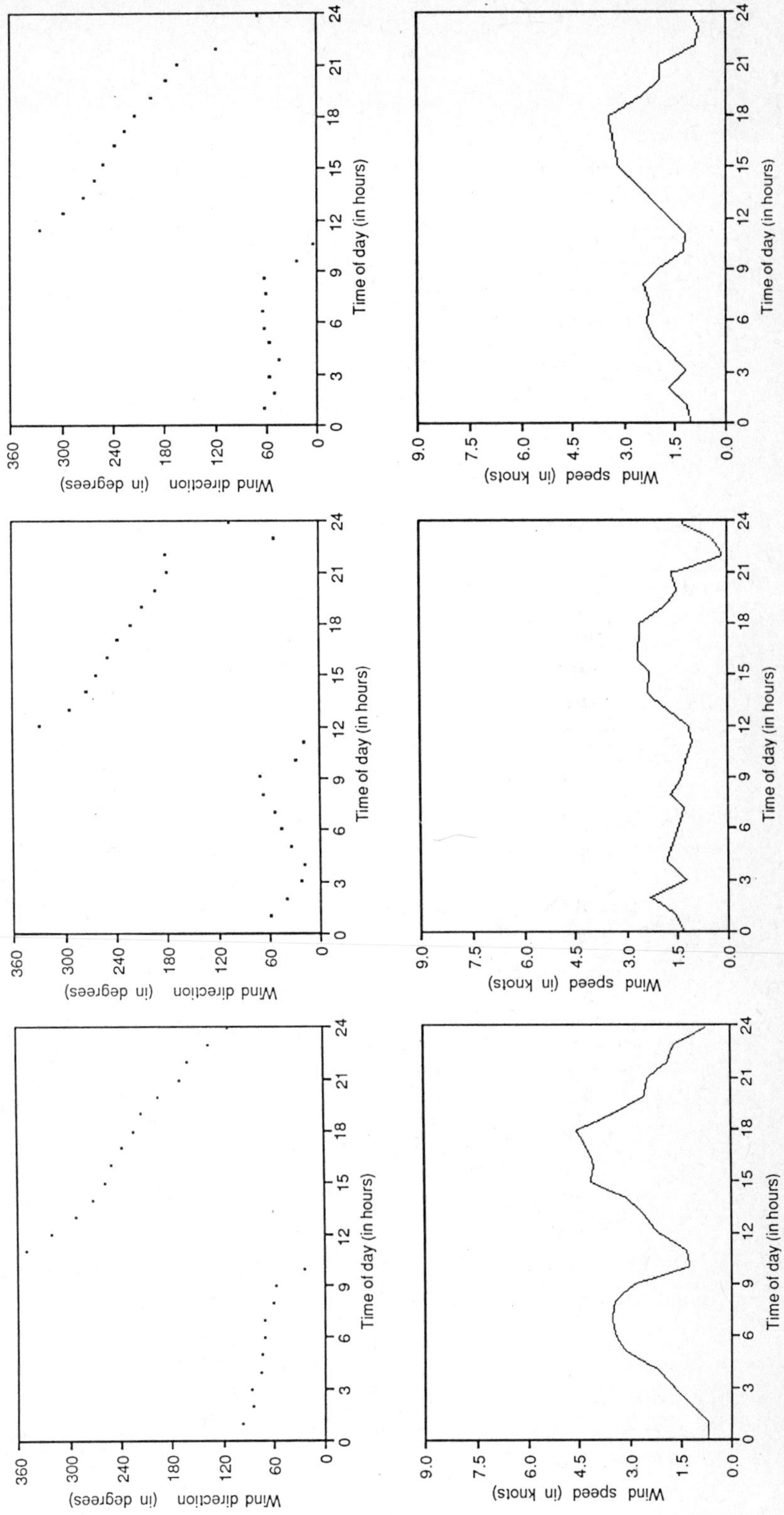

Figure 4.10. Distribution of daily maxima and minima of particular wind characteristics.

Summer 1971/72

(a) Maximum wind speed, W71.

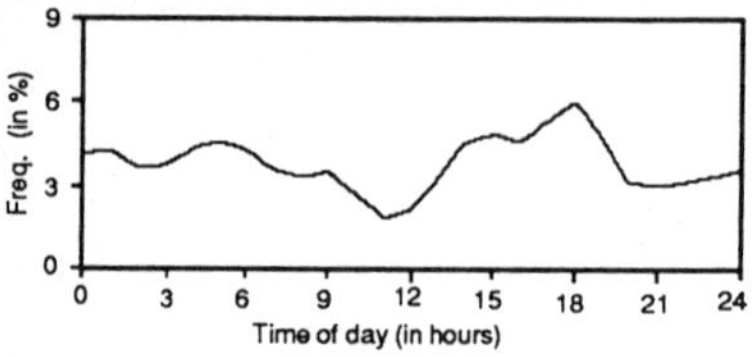

(b) Minimum wind speed, W71.

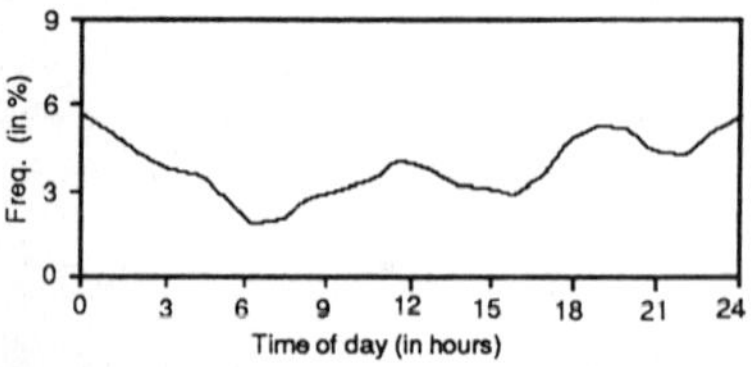

(c) Max. speed change, W71.

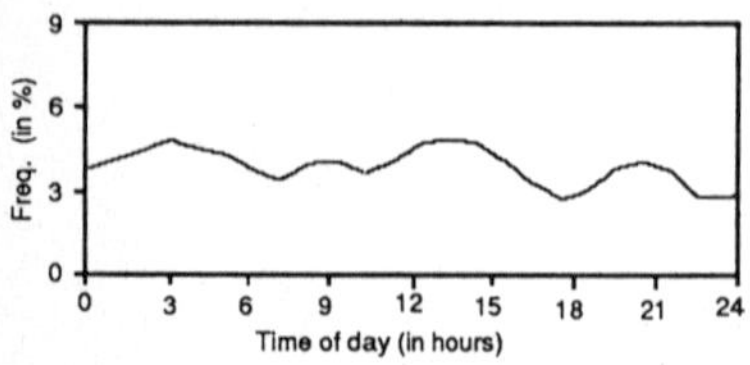

(d) Min. speed change, W71.

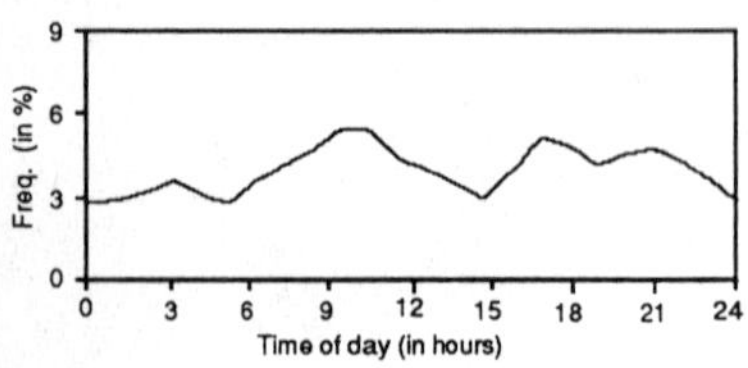

(e) Max. directional change, W71.

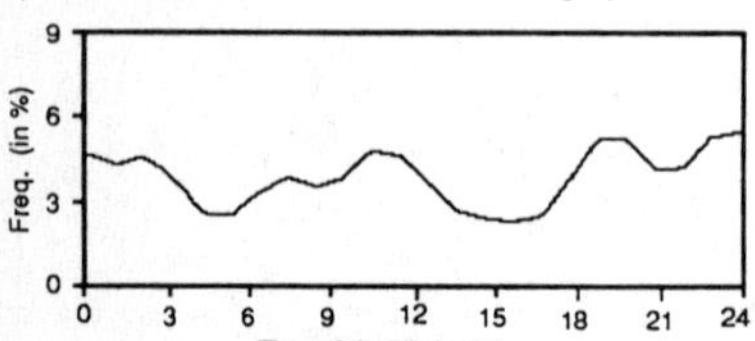

(f) Min. directional change, W71.

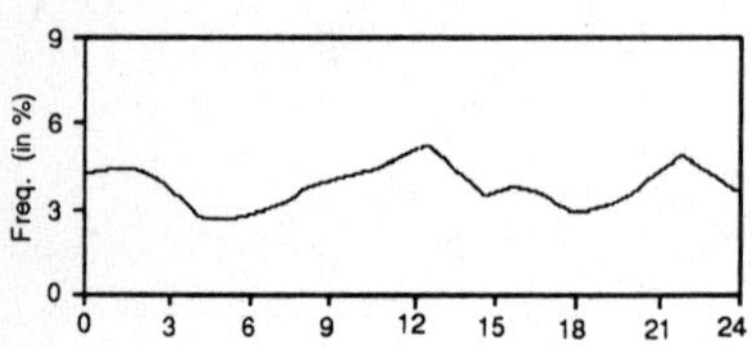

Winter 1971

(g) Maximum wind speed, S71/72.

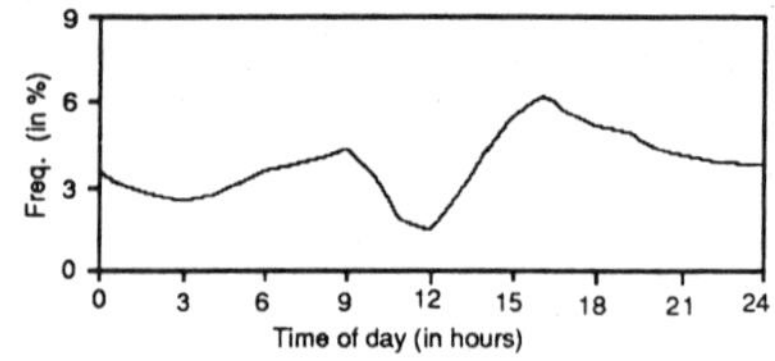

(h) Minimum wind speed, S71/72.

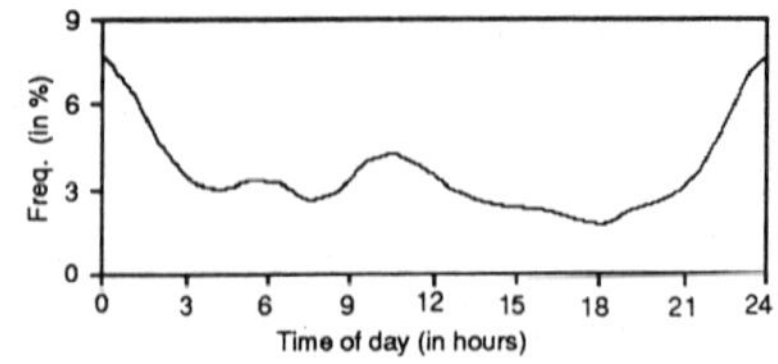

(i) Max. speed change, S71/72.

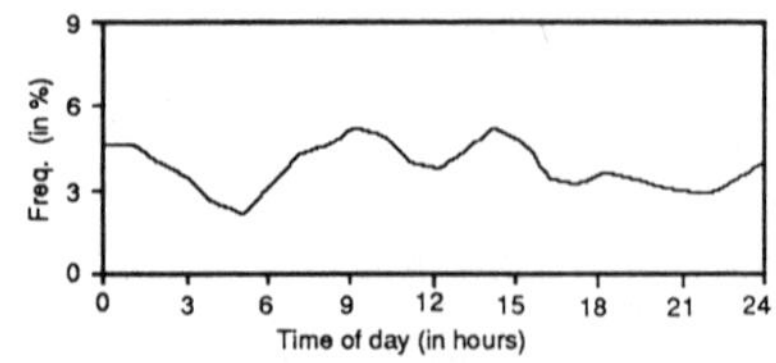

(j) Min. speed change, S71/72.

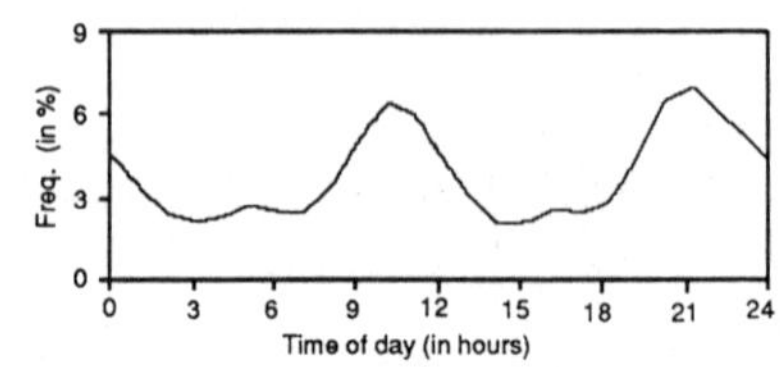

(k) Max. directional change, S71/72.

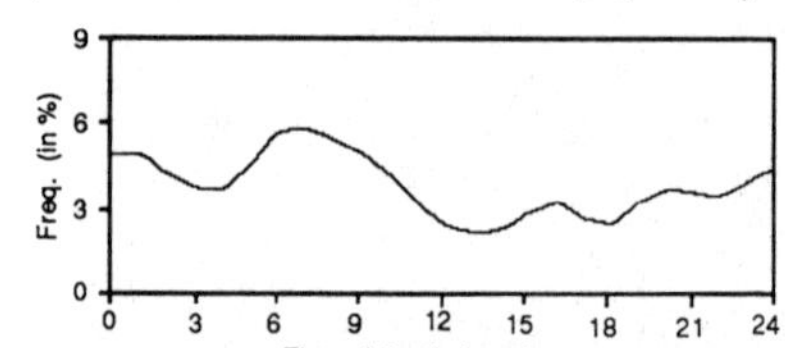

(l) Min. directional change, S71/72.

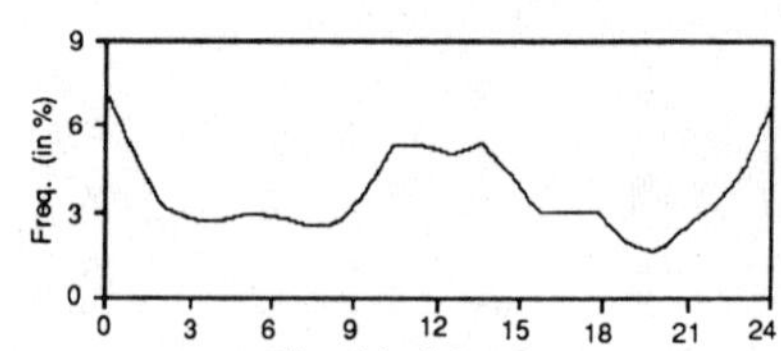

Table 4.2. Relative amount of variation within and across patterns (in %) (based on observations of the years 1971/72, 1973/74, 1975/76 and 1977/78).

Sum of squares	All seasons all patterns all years	All patterns all years winter	summer	All years winter Δ	Δ^c	summer Δ	Δ^c
Intra-year s^2_{+++}	95.0	53.5	44.7	5.4	48.4	26.8	18.5
Inter-year s^2_{++}	0.8	0.3	0.6	0.2	0.1	0.3	0.3
Inter-pattern s^2_{+}	0.8	0.3	0.6	-	-	-	-
Inter-season s^2	3.4	-	-	-	-	-	-
Total	100	100		100			

Defining $s^2_{+++} = \sum_{ghj} s^2_{ghj}$, $s^2_{++} = \sum_{gh} s^2_{gh}$ and $s^2_{+} = \sum_g s^2_g$ then yields

$$s_0^2 = s^2_{+++} + s^2_{++} + s^2_{+} + s^2 . \tag{2}$$

If equation (2) is divided by s_0^2 the terms on the right-hand side can be interpreted as the relative amount of variation within years, between years, between patterns and between seasons, respectively.

Values of all these ratios are presented in Table 4.2. Note that the figures in the first, second and third block are divided by s_0^2, $s_0^2 - s^2$ and $s_0^2 - s^2 - s^2_{+}$, respectively. It follows immediately that the fluctuation of the daily wind record about the seasonal sea breeze pattern accounts for most of the variation in $(\mathbf{d}_t)$. Across years as well as across patterns there is almost no variation which implies that the sea breeze patterns are very similar from year to year. That is, even though the sea breeze circle may vary from day to day as shown in Figure 4.1 (b), its average characteristics remain virtually unchanged. Compare, for example, the ideal patterns for 1971/72 and 1975/76 presented in Figures 4.7, 4.9 and 4.11. Even between the average summer and winter pattern is comparatively small variation which supports the earlier finding that their main difference is simply in the velocity scale.

However, the intra-year sums of squares for the winter Δ-pattern and Δ^c-pattern are markedly different, even though there are about twice as many days contributing to the complementary pattern as there are contributing to the ideal pattern. This suggests that the winter Δ^c-pattern reflects a synoptic configuration which is principally different from that associated with the other patterns. The high value of 48.4 per cent for the intra-year variation further indicates that the winter Δ^c-pattern is somewhat

Figure 4.11. Ideal land and sea breeze pattern, 1975/76.

(a) Winter 1975. *(b) Summer 1975/76.*

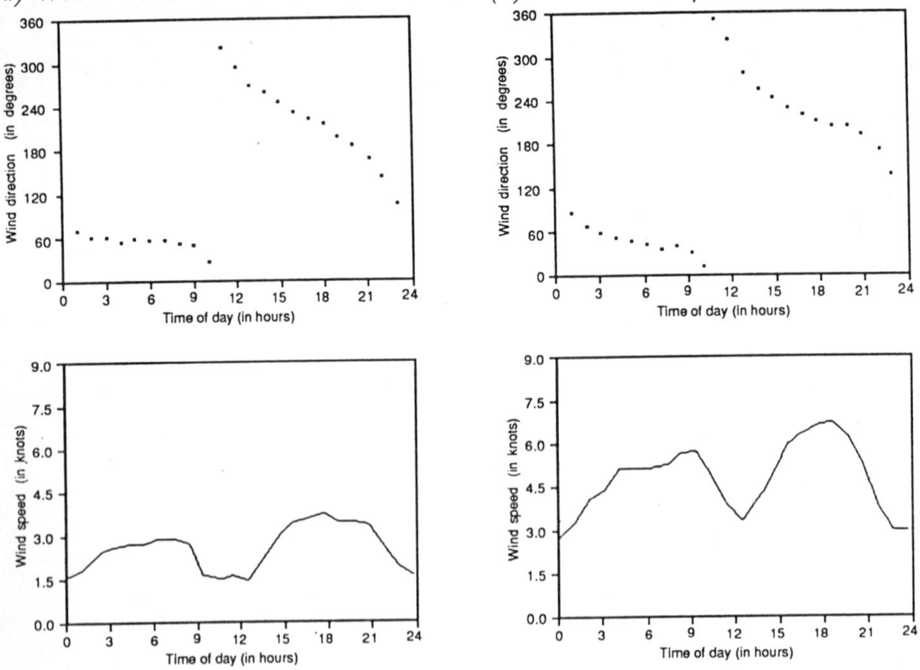

less distinct. Finally it is noted that the sea breeze patterns are not sensitive to the initial decomposition.

4.3.2 Removal of the sea breeze

Let $\mathbf{B} = (\mathbf{b}_m)_{m=0,\ldots,23}$ denote the sea breeze vector and $\mathbf{D}_k = (\mathbf{d}_{24k+m})_{m=0,\ldots,23}$ be the vector comprising all records of day k in the daily component $(\mathbf{d}_t)$. The mean sum of squares of the sea breeze vector and the mean sum of products with $\mathbf{d}_k$ are then defined as $s_{\mathbf{B}}^2 = \|\mathbf{B}\|^2/24 = \sum_{m=0}^{23} \|\mathbf{b}_m\|^2/24$ and $s_{\mathbf{B}k} = \mathbf{B}'\mathbf{D}_k/24 = \sum_{m=0}^{23} \mathbf{b}'_m \mathbf{d}_{24k+m}/24$, respectively. The statistic $s_{\mathbf{B}}^2$ gives an indication of the intensity of the circulation. From Table 4.3, listing the values of $s_{\mathbf{B}}^2$ for the ideal, complementary and average patterns, it can be seen that the land and sea breeze cycle is much stronger in summer than in winter, and further that the Δ-pattern is more pronounced than the Δ^c-pattern.

Now, calculate the degree $\gamma_k = s_{\mathbf{B}k}/s_{\mathbf{B}}^2$ of sea breeze circulation in $\mathbf{D}_k$ and define the series $(\mathbf{r}_t)$ of residual winds as illustrated in Figure 4.12. That is,

$$\mathbf{r}_t = \mathbf{d}_t - \max\{0, \gamma_k\}\, \mathbf{b}_m$$

Table 4.3. Mean sum of squares for the land and sea breeze circulation.

	$s^2_{\mathbf{B}}(\Delta)$	$s^2_{\mathbf{B}}(\Delta^c)$	$s^2_{\mathbf{B}}(\Delta \cup \Delta^c)$
Summer 1971/72	35.9	16.4	29.3
Winter 1971	8.2	3.1	4.8

where $t = 24k + m$, $\mathbf{b}_m$ is the sea breeze component associated with day k and time-of-day m, and $\mathbf{d}_t$ is a record of wind speed and direction in the daily component $(\mathbf{d}_t)$. Measuring the agreement between $\mathbf{D}_k$ and $\mathbf{B}$, γ_k can be interpreted as the strength of the sea breeze circulation on day k.

Figure 4.12. Definition of the residual wind series $(\mathbf{r}_t)$.

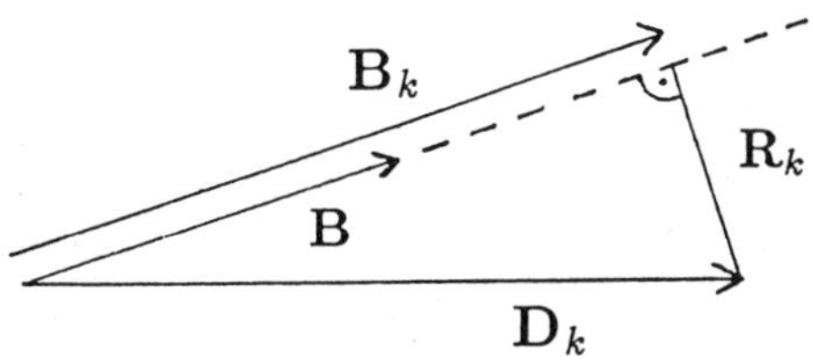

$$\mathbf{B}_k = (\mathbf{B}'\mathbf{D}_k) \,/\, \|\mathbf{B}\| \cdot \mathbf{B} \,/\, \|\mathbf{B}\| = \gamma_k \cdot \mathbf{B} \quad \text{and} \quad \mathbf{R}_k = \mathbf{D}_k - \mathbf{B}_k$$

To illustrate this procedure we again refer to the data of 1 to 8 December 1971. The residual component $(\mathbf{r}_t)$ for this period is plotted in Figure 4.13. With wind direction scattered over the full circle and wind speed peaking several times a day at irregular intervals, there is no more evidence of the pattern present in Figure 4.3. We further note that the average wind speed has been reduced from about 8 knots in Figure 4.3 to about 4 knots in Figure 4.13. The peak at 2300 h on 5 December is a reflection of comparatively high wind speeds at a time-of-day that is usually associated with moderate winds.

A number of summary values for 1 to 8 December 1971 are given in Table 4.4. In summer 1971/72 the distance $e_k = \|\, \mathbf{s}_k - \overline{\mathbf{d}} \,\|$ had to be compared with the critical distance $\epsilon = 3.1$, that is all days represented in Table 4.4 except for the 6 December contribute to the ideal circulation pattern. From a comparison of the mean sum of squares $s^2_{\mathbf{d}}$ of the daily component $(\mathbf{d}_t)$ with $s^2_{\mathbf{r}}$ of the residual component $(\mathbf{r}_t)$ it is evident that the land and sea breeze cycle constitutes the major component in $(\mathbf{d}_t)$. Further, the ratio $s^2_{\mathbf{r}}/s^2_{\mathbf{d}}$ gives an indication of the daily fluctuation about the sea breeze circulation. For example, on 7 and 8 December, when the land and sea breeze cycle was less evident in the data of Figure 4.3, this ratio is noticeably larger.

Figure 4.13 (a). Wind direction in ($\mathbf{r}_t$) *from 1 to 8 December 1971.*

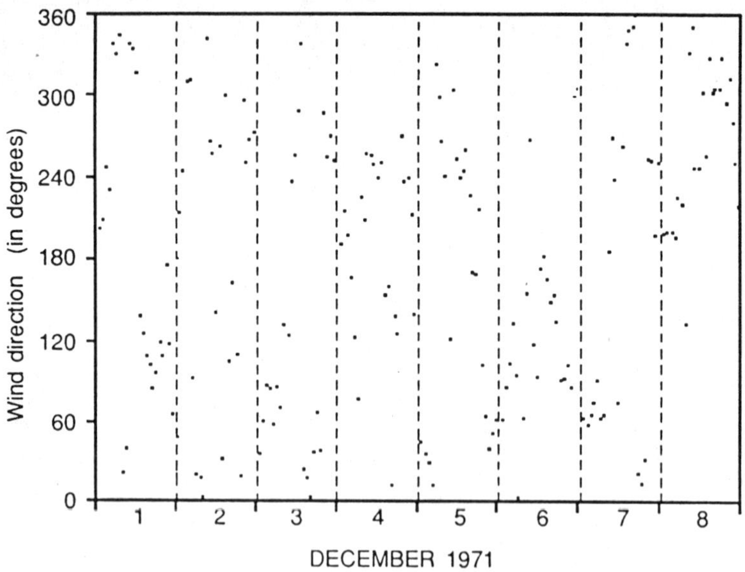

Figure 4.13 (b). Wind speed in ($\mathbf{r}_t$) *from 1 to 8 December 1971.*

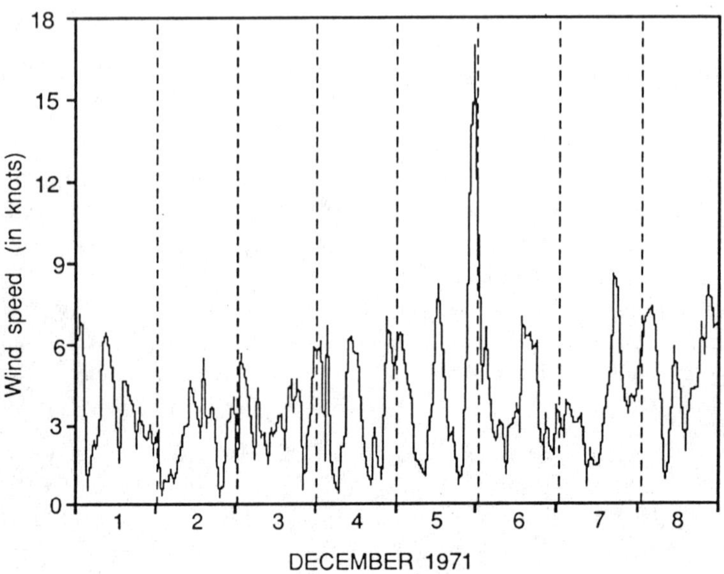

Table 4.4. Summary values for the period of 1 to 8 December 1971.

	1.12.	2.12.	3.12.	4.12.	5.12.	6.12.	7.12.	8.12.
Distance $e_k = \|\| \mathbf{s}_k - \overline{\mathbf{d}} \|\|$	0.76	0.87	1.38	2.40	1.82	3.99	2.05	2.86
Mean sum of squares $s^2_{\mathbf{d}}$	62.8	42.0	69.7	69.8	86.0	122.8	27.8	29.6
Mean sum of squares $s^2_{\mathbf{r}}$	16.2	7.9	12.0	16.5	37.0	19.5	15.6	28.8
Degree γ_k of circulation	1.2	1.0	1.3	1.3	1.2	2.1	0.6	0.2
Circulation pattern	Δ	Δ	Δ	Δ	Δ	Δ^c	Δ	Δ

The γ_k-values shown in Figure 4.14 provide an indication of the spread of the sea breeze intensities, and can be interpreted in terms of the circulation strength as follows: nil if $\gamma_k < 0.5$, normal if $0.5 \leq \gamma_k < 1.5$, and strong if $1.5 \leq \gamma_k$ (cf. Table 4.5). The histograms in Figure 4.14 and the numbers in Table 4.5 both support the earlier conclusion that there is a distinct and unique summer pattern, whereas the wider spread

Figure 4.14. Degree of sea breeze circulation in the daily component $(\mathbf{d}_t)$.

(a) Summer 1971/72. *(b) Winter 1971.*

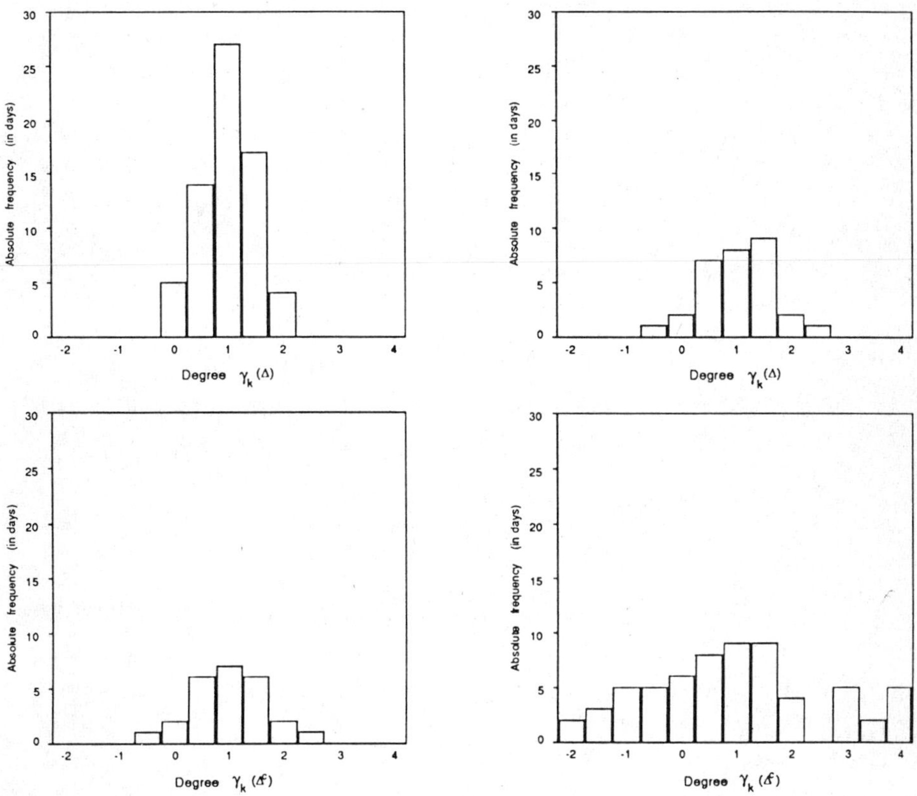

Table 4.5. Relative frequencies of sea breeze patterns.

	Winter 1971			Summer 1971/72		
	none	normal	strong	none	normal	strong
Ideal pattern	0.03	0.26	0.03	0.06	0.63	0.04
Complementary pattern	0.23	0.28	0.17	0.03	0.21	0.03

of γ_k for the winter Δ^c-pattern suggests that there exist a variety of weather types in winter. As will become evident in the following section this result can be related to the alternating influence of low and high pressure systems in winter, and also explains the difference between the Δ- and Δ^c-pattern depicted in Figures 4.9 (a) and (b).

4.3.3 Interaction with the geostrophic component

In this final section the parameter γ_k is related to the geostrophic component ($\mathbf{g}_t$) and the synoptic states introduced in the last chapter. Table 4.6 lists the average strength of the various circulation patterns conditional on the direction $\theta_\mathbf{g}$ of the geostrophic wind. In summer both the Δ- and Δ^c-pattern are slightly more pronounced if south-easterlies prevail. This type of association becomes even more obvious in the winter Δ-pattern where we have a distinct sea breeze for $30° \leq \theta_\mathbf{g} \leq 180°$, and none for the westerly sector. The Δ^c-pattern, on the other hand, shows completely different characteristics.

Table 4.6. Strength of the daily circulation conditional upon the geostrophic wind direction.

Summer 1971/72	30°	60°	90°	120°	150°	180°	210°	240°	270°	300°	330°	360°
Δ-pattern	-	-	1.0	1.1	1.2	1.0	1.0	0.9	-	-	-	-
Δ^c-pattern	-	-	1.4	2.0	1.5	0.8	0.6	0.5	-	-	-	-
Winter 1971	**30°**	**60°**	**90°**	**120°**	**150°**	**180°**	**210°**	**240°**	**270°**	**300°**	**330°**	**360°**
Δ-pattern	1.3	1.9	2.1	2.4	2.3	1.4	0.5	0.1	0.0	0.0	-0.1	0.2
Δ^c-pattern	1.3	0.9	0.6	0.4	0.8	1.2	1.6	1.2	0.3	0.4	1.9	2.6

It is most distinct if the geostrophic wind comes from an either northerly or southwesterly direction, suggesting that there are at least two different patterns in winter, whereas the similarity between the Δ- and Δ^c-pattern in summer indicates that the weather during this time of the year is of a more homogeneous type.

Further evidence that the winter Δ^c-pattern is different from the others is provided by Table 4.7 which presents the joint frequency distribution of the two winter patterns and the geostrophic wind direction. Just as in summer the ideal circulation is most pronounced if the geostrophic wind comes from an easterly direction, that is when Fremantle is under the influence of the high pressure belt. Conversely, if westerlies, which are generally associated with cold fronts moving across the region, prevail, the complementary Δ^c-pattern is almost always found.

Table 4.7. Relative frequency of the joint event of sea breeze pattern and geostrophic wind direction (in %) (winter 1971).

	30°	60°	90°	120°	150°	180°	210°	240°	270°	300°	330°	360°
Δ-pattern	2.1	6.8	4.9	5.8	5.0	3.6	0.7	0.6	0.6	0.6	0.8	0.8
Δ^c-pattern	3.1	4.8	2.3	2.6	3.9	6.4	3.8	8.3	12.0	9.9	7.2	3.4

In meteorological terms this can be explained as follows: In case of a dominant offshore flow, warm continental air is moved towards the sea producing a strong horizontal temperature gradient within a thick layer of the atmosphere, and thus a strong sea breeze. If the geostrophic wind blows onshore, however, the exchange of cold air inhibits the rise of the temperature over the land so that the horizontal temperature gradient and therefore the circulation are only minimal.

Generally it is found that a strong circulation is associated with light geostrophic winds, and *vice versa*. Given that the geostrophic wind is strongest from a westerly direction, it follows that the winter Δ^c-pattern is somewhat less intense since this pattern is most likely to occur if westerlies prevail. Compare, for example, the mean sum of squares for the winter Δ- and Δ^c-pattern given in Table 4.3.

Table 4.8 relates the strength of the daily circulation to the synoptic states introduced in chapter 3. The reader is reminded that *LN*, *SN*, *LE*, *etc.* denote *light* and *strong* winds from the *north, east, south* and *west*, respectively. Although light winds are generally associated with a stronger circulation than strong winds from the same direction, this relationship is particularly pronounced for the complementary Δ^c-patterns. For example, consider the case of geostrophic southerlies in winter (*LS* and *SS*). Whereas the *complementary* circulation is markedly different depending on the strength of the synoptic scale wind, the *ideal* circulation is hardly influenced by the speed of the geostrophic wind. The ideal circulation in summer, in fact, is little affected

Table 4.8. Strength of the daily circulation conditional upon the synoptic state.

Summer 1971/72	*LN*	*SN*	*LE*	*SE*	*LS*	*SS*	*LW*	*SW*	*0*
Δ-pattern	-	-	1.2	0.8	1.2	1.0	-	-	-
Δ^c-pattern	-	-	2.2	1.4	1.3	0.8	-	-	-
Winter 1971	*LN*	*SN*	*LE*	*SE*	*LS*	*SS*	*LW*	*SW*	*0*
Δ-pattern	0.9	0.4	2.4	1.8	1.7	1.7	0.0	0.0	0.8
Δ^c-pattern	2.6	1.0	0.9	0.4	1.6	0.4	1.0	0.2	1.0

by both, speed and direction of the geostrophic wind, while the strength of the complementary circulation varies between 0.8 and 2.2, thus exhibiting a much wider spread. The results presented in Table 4.8 also confirm the earlier findings, namely that an ideal circulation is normally triggered off by geostrophic winds from the south-easterly sector, and that the winter Δ^c-pattern is quite different from the others.

CHAPTER 5

SHORT-TERM EVENTS

In this chapter the analysis of the daily component $(\mathbf{d}_t)$ *is continued. Having removed the land and sea breeze component* $(\mathbf{b}_t)$ *from* $(\mathbf{d}_t)$*, the aim of this chapter is to detect and study particular events in the residual series* $(\mathbf{r}_t)$*.*

Section 5.1 is expository and describes the most common wind phenomena. In section 5.2 suitable definitions of the meteorological events are presented in terms of wind speed and direction. In particular, we are interested in calms, storms and oscillating winds. Given the definition of these events, calms are expected to be associated with stable atmospheric conditions, whereas storms are evidence of a synoptic pressure field that is highly asymmetric or changing rapidly. The series $(\mathbf{r}_t)$ *is then studied in section 5.3 to detect these events and to determine their seasonal characteristics, such as their distribution over the year and their day-time of onset. By relating the events to the geostrophic and sea breeze components it is found that calms occur almost exclusively in winter and are associated with the high pressure belt extending to the local region. It is further shown that storms are related to extratropical cyclones in winter and troughs developing from a depression over the north-west of the continent in summer. The removal of the storm component* $(\mathbf{e}_t)$ *from* $(\mathbf{r}_t)$ *is explained in appendix 5.4.*

5.1 Meteorological Patterns

So far synoptic pressure systems have been regarded as circular vortices, although they rarely resemble those in the real atmosphere. Usually they are highly asymmetric with the strongest winds and largest temperature gradients concentrated along frontal systems. Part of this complexity is due to the fact that synoptic systems are embedded in a global jet stream which is highly turbulent, as any disturbance introduced into the flow tends to amplify. Indeed, most synoptic scale systems in midlatitudes develop as a result of an atmospheric instability caused primarily by the vertical shear in the jet stream. If preexisting temperature gradients intensify these instabilities may even generate frontal systems which usually are spawned by the interaction of different air masses. Near frontal systems the isobars tend to be denser, and thus be associated with large pressure gradients and a highly asymmetric pressure field. However, both the geostrophic and sea breeze component, which were subtracted from the wind series ($\mathbf{w}_t$) in the previous chapters, reflect synoptic and mesoscale configurations that display a high degree of regularity and symmetry. Any abrupt deviation from a smooth flow will therefore be evident in the residual series ($\mathbf{r}_t$).

Referring to Figures 4.6 and 4.13 we note that the former clearly exhibits the storm which was initially added to the simulated wind series, whilst the latter shows an isolated storm which had been masked previously by the geostrophic and sea breeze component. In fact, much of our discussion in earlier chapters was devoted to eliminating the influence of distortions such as storms generated by frontal systems. The aim in this chapter is to detect these events, which are associated with local asymmetries in the synoptic pressure field, and to determine their characteristics. The storms are finally removed from ($\mathbf{r}_t$), leaving a residual series ($\mathbf{f}_t$) of short-term fluctuations.

Let us briefly summarise the most important short-term events and related synoptic patterns. Throughout the year the south-western corner of the continent is subject to a succession of eastward travelling low pressure cells, commonly referred to as extratropical cyclones. At the approach of the associated cold fronts the wind suddenly strengthens with speeds of up to 40 knots and instantaneous gusts of more than 60 knots. While the front passes through the area the wind swings anticlockwise from a predominantly north-westerly to a southerly direction. Extratropical cyclones are more common in winter than in summer since the associated low pressure cells pass south of the anticyclonic belt mentioned in chapter 1. The first three charts in Figure 3.11 show a typical synoptic pattern leading to extreme wind conditions. The intensity of these storms is related to the temperature difference across the cold front, which on occasions may cause severe wind conditions for more than 24 hours.

The second most important pattern producing a sudden change of wind speed and direction is associated with the high pressure belt lying across the Fremantle region in summer and a low pressure system over the north-west of the continent. As a trough begins to develop from the low in a southerly direction moderate north-easterlies prevail. However, when the trough moves across the region, splitting the high pressure ridge into a western and an eastern cell, the wind will suddenly strengthen and swing around to a southerly direction. This pattern is captured in Figure 3.15 (d) and can be associated with the change from NE to S in the schematic diagram of Figure 3.16.

Thunderstorms are produced by huge bands of cumulonimbus clouds, which are affected by intense solar heating, and often are initiated by cold fronts approaching masses of hot and moist air. Although there are about 20 thunderstorms around Fremantle per year they hardly appear on the wind record since they are usually confined to small areas and of only short duration. Occasionally a dissipating tropical cyclone occurs in the south-west mainly during summer when they originate in the eastern Indian Ocean off the Australian north-west coast. Extreme wind conditions are also associated with tornados, waterspouts, whirlwinds and squalls, but since there are not enough data available on all of the latter events, including thunderstorms, a detailed statistical analysis of their frequencies and intensities is impossible. For example, only 3 cyclones passed through the south-west between 1975 and 1983 with the most severe reaching speeds of 54 knots at Fremantle.

Following from this discussion, it can be expected that extreme wind conditions are more likely in winter than in summer and are associated with geostrophic westerlies in winter and geostrophic easterlies in summer. However, we are also interested in events which are characterized by low wind velocities and are denoted as calms. They are commonly related to the large high pressure system over southern Australia in winter, as this ensures small pressure gradients at a time when also continental heating does not have a strong influence on the weather and thus stable atmospheric conditions prevail. Calms therefore occur more often in winter between May and September than during the summer months.

5.2 Wind Classification

In this chapter we are not concerned with storms blowing over a period of time from more or less the same direction. Since they usually reflect a synoptic pressure field which is globally characterized by strong gradients, they have been removed from the data together with the geostrophic component. Instead, we will focus on short-term events and turbulences associated with sudden wind changes. It follows from the discussion in the last section that most storms detected in winter are related to extratropical cyclones, with a few others caused by thunderstorms. On the other hand, storms identified in summer will generally be associated with a trough moving across the Fremantle region.

Storms as defined in this chapter are to be interpreted physically as abrupt deviations from the general flow as determined by the geostrophic wind and sea breeze circulation. Similarly, calms signify a good agreement between the actual wind and its prediction from the geostrophic and sea breeze components. Whereas they are triggered off by stable atmospheric conditions, storms are evidence of turbulent air.

Earlier attempts to classify mesoscale (10 to 100 km) and synoptic scale (100 to 1000 km) wind fields are all based on raw wind observations. In order to classify mesoscale wind patterns and to quantify their temporal and spatial features Hardy & Walton (1978) generalized the concept of principal component analysis and applied it to a series of spatial wind fields observed at the west coast of California. For the Fremantle area Steedman & Craig (1979) identified three major wind categories, namely extratropical cyclones, the sea breeze, and the anticyclonic system in winter. However, the categories are not exhaustive, and since it is conceivable that the wind is actually influenced by various factors at the same time, a complete partition of the wind record as attempted in these studies could be difficult.

Let us now define the wind patterns of particular interest, namely calms, storms and oscillating winds. The data exhibit frequent fluctuations of wind speed and direction within small bands. Defining an event via a single reference speed could therefore result in recording a sequence of events which, in fact, represent one event interrupted for short intervals. As observed by Steedman & Craig (1979), for instance, calms are often disturbed by briefly freshening winds. It appears that the recording station is passed by patches of turbulent air which seem to have well defined boundaries and extend for about 10 to 15 km.

To eliminate this effect we take short-term moving averages of wind speed and introduce certain thresholds when defining the three events. Given the speed component (v_t) of the residual series $(\mathbf{r}_t)$ define (y_t) as follows:

$$y_t = \tfrac{1}{9}\, v_{t-2} + \tfrac{2}{9}\, v_{t-1} + \tfrac{3}{9}\, v_t + \tfrac{2}{9}\, v_{t+1} + \tfrac{1}{9}\, v_{t+2}\ .$$

On the basis of (y_t) each hour will now be classified as being part of a calm, storm, an oscillating wind or none of these events.

A *storm* is defined in terms of the four parameters s_a, s_b, s_c and s_d. Whereas the first three are reference speeds with $s_a \geq s_b \geq s_c$, the latter denotes a duration. The onset of a storm is defined as the time when (y_t) first exceeds s_a. As long as (y_t) stays above s_b the event is assumed to continue. However, if (y_t) drops below s_c, or below s_b for an extended period of time which is longer than s_d, the storm will end. This definition is best explained by reference to Figure 5.1 which plots (y_t) as a function of time.

Figure 5.1. Definition of a storm pattern.

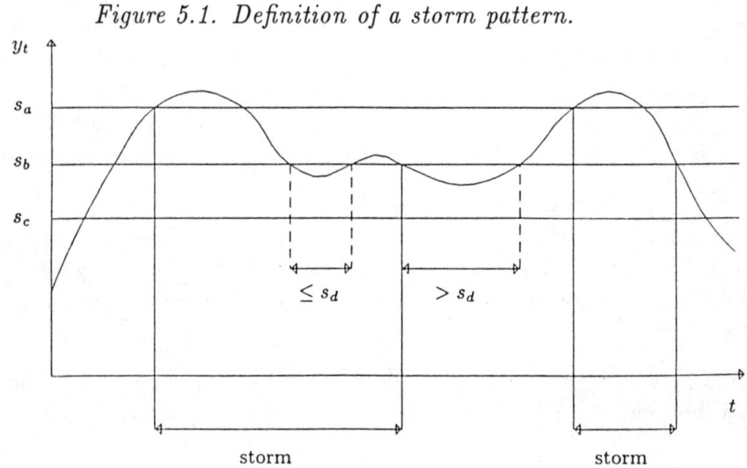

As soon as (y_t) increases above s_a we begin to register a storm. Eventually (y_t) drops below s_b. However, as this is only for a period shorter than s_d and since (y_t) does not drop below s_c, we continue to record a storm. The second time (y_t) falls below s_b it is for longer than s_d so that the storm ends the moment (y_t) hits s_b. A second storm begins when (y_t) again exceeds s_a. This time it ends because (y_t) drops below s_c. For the analysis in this chapter we chose $s_a = 10$ knots, $s_b = 9.5$ knots, $s_c = 9$ knots and $s_d = 3$ hours.

Using the four parameters c_a, c_b, c_c and c_d *calms* are specified completely analogously $(c_a \leq c_b \leq c_c)$. The beginning of a calm is defined as the moment when (y_t) first drops below c_a, and it ends when (y_t) either rises above c_c, or above c_b for a period longer than c_d. For the subsequent analysis we assigned the values $c_a = 1$ knot, $c_b = 2$ knots, $c_c = 3$ knots and $c_d = 3$ hours. The size of these parameters, which are used to define storms and calms respectively, was chosen by reference to Steedman & Craig (1979).

Finally, *oscillating winds* are defined in terms of the two parameters z_a and z_b, both integers smaller than or equal to 5. Whenever the number of changes within a 6-hour period in $(\mathbf{r}_t)$ from a clockwise to an anticlockwise rotating wind or vice versa exceeds z_a, we begin to register an oscillating wind. This event then continues until the number of changes within any subsequent 6-hour interval falls below z_b. The two parameters z_a and z_b were set to 5 and 4, respectively. Denoting the directional component of $(\mathbf{r}_t)$ by (θ_t) the following diagram may help to illustrate this definition (C=clockwise and A=anticlockwise).

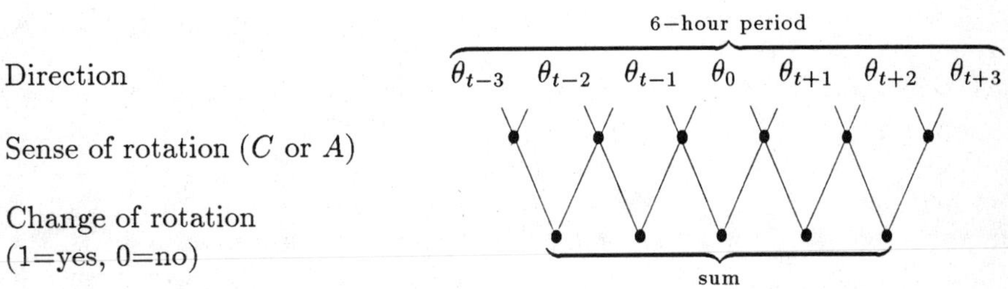

First, given the period from $t-3$ to $t+3$, the sense of rotation is established for each pair $(\theta_{t+s}, \theta_{t+s+1})$ of consecutive directions, $s = -3, \ldots, 2$. This yields a series of length 6 containing C's and A's. The number of changing runs, which cannot be greater than 5, is then compared with z_a and z_b, respectively.

It is noted that the events defined in this section are fairly elementary, and a more sophisticated classification may be required for a more detailed analysis. The definition of a storm, for example, involves wind speeds only, and it is obvious that more complex patterns could be defined by reference to both wind speed and direction. However, most studies of wind data have so far ignored the directional aspect and focussed on speeds only.

5.3 Characteristics of Short-Term Events

Having defined calms, storms and oscillating winds in the last section, the residual series ($\mathbf{r}_t$) will now be investigated under the aspect of these events. Of particular interest are the following characteristics:

(i) the frequency and spread of the events over the years,
(ii) their time of onset,
(iii) their duration,
(iv) any pattern in their occurrence, and
(v) their relationship with the geostrophic component.

5.3.1 Calms

Based on observations of the years 1971/72, 1973/74, 1975/76 and 1977/78 Table 5.1 shows the mean number of calms per season, with standard deviations over the 4 years given in brackets. Even though the numbers are not significantly different it appears that calms tend to occur more often in winter than during other seasons. This fact had been observed before (e.g. Bureau of Meteorology 1973, Steedman & Craig 1979), although the definition of a calm in these references is based on *raw* wind data and thus different from ours. However, from studying corresponding weather charts it becomes obvious that calms as defined here are generally associated with an extensive high pressure system and thus atmospheric conditions which are characterized by small gradients. In this sense calms in this study reflect the same synoptic scale configuration as they did in previous ones. Consequently, one could expect more calms in summer than in winter. Solar heating, giving rise to a sea breeze circulation and atmospheric turbulence, however, reduces the number of calms during the warmer season. The chance $\mathcal{P}(C)$ of a calm at any instant of time turns out to be $\mathcal{P}(C) = 0.027$ in winter and $\mathcal{P}(C) = 0.012$ in summer. That is, on average one records a calm every 37 hours in winter and every 83 hours in summer. Note, however, that calms usually last longer than only one hour and therefore that the inter-event periods are much longer.

Table 5.1. Distribution of calms over the year (based on 4 years' observations).

	Autumn	Winter	Spring	Summer
Mean of calms	5.0 (1.4)	7.8 (2.0)	4.3 (1.9)	5.5 (1.7)

The numbers in brackets are standard deviations over the 4 years.

Table 5.2. Duration of calms by season (based on 4 years' observations).

Season	Duration				Total
	≤ 4 hours	5-8 hours	9-12 hours	≥ 13 hours	
Autumn	6	7	3	4	20
Winter	14	7	7	3	31
Spring	8	8	1	0	17
Summer	15	7	0	0	22
Total	43	29	11	7	90

It is obvious that the number of calms is highly dependent on the parameters $c_a, \ldots, c_d$. However, the pattern of registering more calms in winter than in summer remains unchanged. Further, neither the initial decomposition of $(\mathbf{w}_t)$ nor the definition of the sea breeze cycle seem to have a major influence on the results presented in this section. This was established by applying different criteria when deriving the residual series $(\mathbf{r}_t)$ and by comparing the results for 1971/72 obtained for the different series.

An examination of the calm durations suggests the existence of two different populations. As shown in Table 5.2 the first type with up to 8 hours duration is present throughout the year, while the second type with durations of more than 8 hours is confined almost completely to the winter months. Especially in summer, calms tend to have a shorter duration, reflecting the comparatively unstable air flow during this time of the year.

Figure 5.2. Distribution of calm onsets over the day (over all seasons).

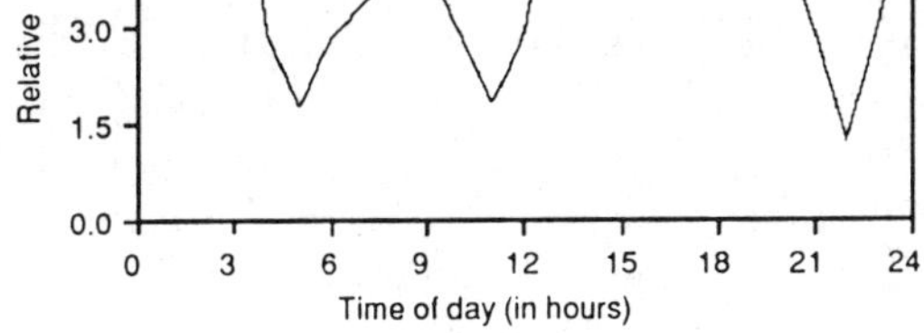

Table 5.3. Chance of a calm in winter (rescaled) conditional upon θ_g.

θ_g	30°	60°	90°	120°	150°	180°	210°	240°	270°	300°	330°	360°
$\mathcal{P}(C \mid \theta_g) / \mathcal{P}(C)$	1.30	1.81	2.56	1.38	0.30	0.26	0.08	0.32	0.44	0.44	0.97	1.66

As can be seen from Figure 5.2 the chance of a calm onset varies considerably over the day. They tend to begin either between 0100 h and 0300 h in the morning or between 1400 h and 1600 h in the afternoon, that is just about when the land and sea breezes start to intensify. Given the definition of a calm this indicates that the geostrophic and sea breeze components are a much better predictor of the actual wind when the land or sea breeze is strengthening than during any other time of the day.

An investigation of the relationship with the geostrophic component provides ample evidence that in winter calms are triggered off by geostrophic winds of about 4 to 8 knots from the easterly and northerly sectors, and are unlikely to occur when westerlies prevail. The chance of a calm conditional on the direction θ_g of the geostrophic wind, denoted by $\mathcal{P}(C \mid \theta_g)$, is listed in Table 5.3. Note that all entries are divided by $\mathcal{P}(C) = 0.027$. Since $\mathcal{P}(C \mid \theta_g) / \mathcal{P}(C) = \mathcal{P}(\theta_g \cdot C) / [\mathcal{P}(\theta_g) \cdot \mathcal{P}(C)]$, the numbers in Table 5.3 can be interpreted as departures from the case of independent events; in the sense that one would expect the numbers to be 1 if the calms were independent of the geostrophic wind direction.

Table 5.4. Chance of a calm (rescaled) conditional upon the synoptic state.

Synoptic state	*LN*	*SN*	*LE*	*SE*	*LS*	*SS*	*LW*	*SW*	*0*
Winter $\mathcal{P}(C \mid \text{syn. state}) / \mathcal{P}(C)$	1.44	0.91	1.95	1.76	0.33	0.00	0.54	0.33	1.24
Summer $\mathcal{P}(C \mid \text{syn. state}) / \mathcal{P}(C)$	-	-	1.06	0.77	2.05	0.97	-	-	-

In Table 5.4 the chance of a calm is related to the synoptic states introduced in chapter 3. The figures clearly show that in winter calms are associated with geostrophic winds from the north-easterly sector (*LE*, *SE* and *LN*), and also indicate that the strength of the synoptic scale wind has little influence on the chance of a calm in winter. In summer, however, the situation is quite different, as calms are more likely to occur if light winds and in particular, light southerlies (*LS*) prevail. Hence, given the synoptic pattern illustrated in Figure 3.18, calms tend to be recorded just when the trough, which extends from the low over the north-west of the continent, has passed through the region, and just before the wind again strengthens (*SS*). Especially in summer and to some extent in winter, calms therefore seem to be associated with an atmospheric pressure field that is characterized by small gradients.

Referring to the data in Figure 4.13, we note that at 0300 h on 2 December 1971 a calm of 4 hours duration is recorded. It is the only calm between 1 and 8 December 1971, even though the wind speed (v_t) of the residual component ($\mathbf{r}_t$) drops as low as 0.2 knots at 2000 h the same day. This is due to taking moving averages before checking ($\mathbf{r}_t$) for calms. From the weather charts presented in Figure 1.3 it can be seen that Fremantle was under the influence of an extensive high pressure system on 2 December 1971.

5.3.2 Storms

It is evident from Table 5.5 that most of the storms occur in winter, with an average frequency of 24.9 per season, in contrast with an average frequency of 15.6 in summer. Further, the numbers for autumn and summer are very close, which again indicates that these two seasons are similar in many respects.

Table 5.5. Distribution of storms over the year (based on 4 years' observations).

	Autumn	Winter	Spring	Summer
Mean of storms	16.0 (0.9)	24.9 (1.2)	22.5 (4.6)	15.6 (2.6)

The numbers in brackets are standard deviations over the 4 years.

Given the relatively small standard deviations, it is obvious that the differences are highly significant. Furthermore, as a study of synoptic charts reveals, the nature of the storms changes over the year. Whereas in winter they are related to extratropical cyclones approaching from the south-west (cf. Figure 3.12 (b)), they are mainly associated with an inland moving trough in summer (cf. Figure 3.15 (d)). As the trough develops southward from the low over the north-west, splitting the high pressure ridge into two systems, stronger gradients start to emerge. From Table 5.6, which presents the chance of a storm conditional on the direction $\theta_{\mathbf{g}}$ of the geostrophic component, it is evident that storms are triggered off by north-westerlies in winter and by easterlies

Table 5.6. Chance of a storm (rescaled) conditional upon $\theta_\mathbf{g}$.

θ_g	30°	60°	90°	120°	150°	180°	210°	240°	270°	300°	330°	360°
Winter $\mathcal{P}(S \mid \theta_g)/\mathcal{P}(S)$	0.99	0.53	0.33	0.27	0.22	0.28	0.47	0.75	1.35	2.21	2.64	1.88
Summer $\mathcal{P}(S \mid \theta_g)/\mathcal{P}(S)$	-	-	1.87	1.28	0.92	0.75	0.66	0.64	-	-	-	-

in summer. Here $\mathcal{P}(S)$ denotes the chance of a storm and $\mathcal{P}(S \mid \theta_\mathbf{g})$ the chance of a storm conditional on the geostrophic wind direction. It is noted that $\mathcal{P}(S)$ equals 0.129 in winter and 0.040 in summer.

The difference between the two types of storms is reflected also in their duration which, on average, is much longer in winter than in summer. Looking at Figure 5.3 we notice a separate peak in winter at 15 to 16 hours duration and a much longer tail than in summer. Apart from the fairly even tail up to 18 hours duration, most of the storms in summer last only a few hours and hence, can be attributed to short-term deviations from the mean flow as determined by the geostrophic and sea breeze components.

Figure 5.3. Distribution of storm duration.

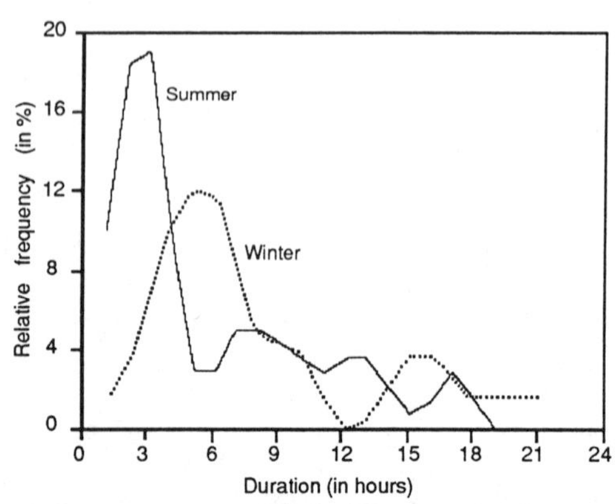

Table 5.7 shows the distribution of the resultant wind direction $\theta_\mathbf{e}$ which is defined as the direction of the vector obtained by summing the vectors $\mathbf{r}_t$ in the residual series $(\mathbf{r}_t)$ over the period of the storm. Despite the seasonal differences this direction tends to be westerly throughout the year. This confirms that any asymmetry in the synoptic pressure field in summer as well as in winter can be associated with a strengthening of the westerly component.

Table 5.7. Distribution of the resultant wind direction θ_e.

θ_e	30°	60°	90°	120°	150°	180°	210°	240°	270°	300°	330°	360°
$10^2 \times \mathcal{P}(\theta_e)$	9.2	7.3	4.1	2.2	5.2	10.3	12.7	12.2	10.1	7.6	8.7	10.4

Similar to Figure 5.2 above, Figure 5.4 shows the distribution of storm onsets over the day. It provides strong evidence that storms begin at times when the sea breeze pattern is associated with light winds, that is either at noon or at midnight.

Figure 5.4. Distribution of storm onsets over the day (over all seasons).

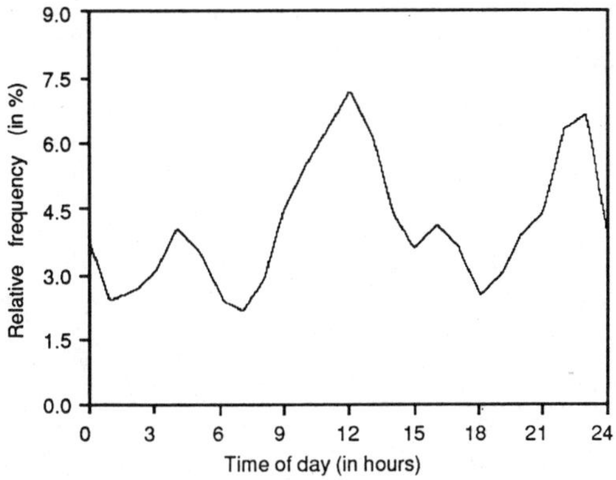

It may be interesting to draw together the main features of Figures 4.7 (a), 4.10 (a), 5.2 and 5.4, which exhibit the land and sea breeze cycle in summer and in winter, and the distribution of calm and storm onsets, respectively. The shaded areas in Figure 5.5 indicate the time of day when the land or sea breeze is peaking, and when calms and storms are more likely to begin. Observe that the day can almost be partitioned as calms tend to be triggered off by the developing breeze whereas storms seem to be associated with a weakening breeze and hence, faster rotating winds. This certainly is part of the reason why the wind during this time of the day is less predictable in terms of the geostrophic and sea breeze components. Also note that the sea breeze develops much faster than the land breeze, which is reflected in the diagram by the shaded areas being closer together during the day than at night.

An investigation of the inter-event periods suggests a tendency of storms to cluster. It is noted that just 67 per cent of all storms began on a day when no other storm was recorded. Indeed, let $(s_t)_{t=1,\dots,T}$ denote the *storm indicator* series with $s_t = 1$ if at time t a storm was recorded and $s_t = 0$ otherwise. The prospect of a second storm within 48 hours of the first event then becomes also apparent in the autocorrelation

Figure 5.5. Time of day when the land and sea breeze, and calm and storm onsets are peaking.

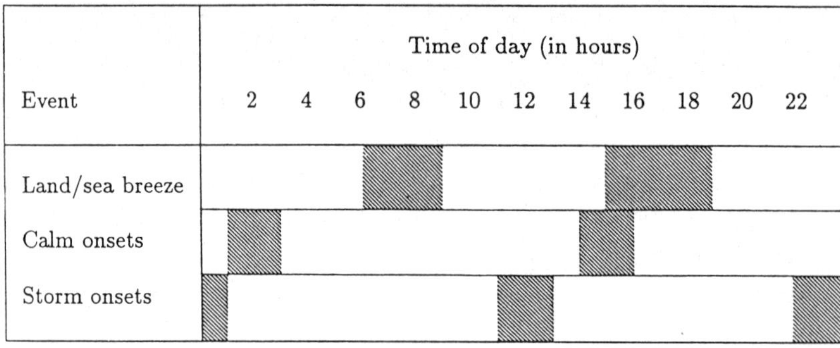

function $C_s(\tau)$ of (s_t) presented in Figure 5.6. This function has been defined in exactly the same way as for ordinary time series in $\mathcal{R}$. That is,

$$C_s(\tau) = \frac{1}{T-\tau} \sum_{t=\tau+1}^{T} s_t \, s_{t-\tau} - \left(\frac{1}{T} \sum_{t=1}^{T} s_t \right)^2 .$$

Figure 5.6 also indicates that there is a 7-day cycle which would, in fact, correspond to the travelling time of synoptic scale systems, even though the evidence does not seem particularly strong.

Figure 5.6. Autocorrelation function $C_s(\tau)$ of the storm indicator series (s_t).

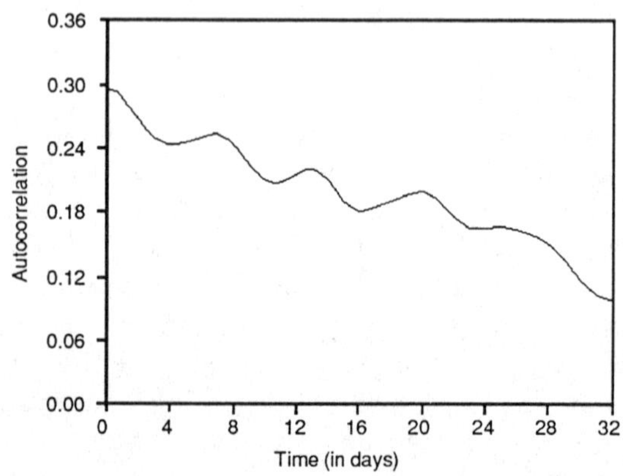

The storm which is apparent in Figure 4.13 was recorded as starting at 2200 h on 5 December 1971. It lasted for 3 hours, had a mean speed of 13.2 knots, which was 3.8 times as high as during the adjacent time intervals, and was associated with geostrophic easterlies. From the weather charts presented in Figure 1.3 it can be seen that a trough moved across the Fremantle area between 6 and 7 December 1971.

As a result of subtracting the sea breeze from the daily component $(\mathbf{d}_t)$ the wind speed in the residual series $(\mathbf{r}_t)$ may be higher at occasions than in $(\mathbf{d}_t)$, that is a few additional storms could have been generated in $(\mathbf{r}_t)$. It might therefore be interesting to examine both $(\mathbf{r}_t)$ and $(\mathbf{d}_t)$ simultaneously. Similarly, one could study $(\mathbf{r}_t)$ in conjunction with the series $(\mathbf{h}_t)$ of estimated geostrophic winds as defined in chapter 3.

5.3.3 Oscillating Winds

They are fairly evenly distributed over the year. Denoting the chance of an oscillating wind by $\mathcal{P}(Z)$ we have $\mathcal{P}(Z) = 0.031$ in winter and $\mathcal{P}(Z) = 0.040$ in summer. Oscillating winds are associated with moderate geostrophic winds from the south to north-westerly sector, and less likely to occur if north-easterlies prevail.

Although one would expect light winds to be more susceptible to directional fluctuations, oscillating winds have surprisingly high mean speeds, usually varying between 4 and 8 knots. On average, they last about 4 hours and tend to occur during the day. This can be explained by solar heating giving increased rise to turbulent air. It is also noted that storms and oscillating winds often follow each other closely, whereas calms are unlikely to occur within a few days of a storm. Whereas the synoptic configurations associated with storms and oscillating winds are somewhat similar they are completely different from those associated with calms. An oscillating wind of 7 hours duration was recorded between 1400 h and 2000 h on 4 December 1971 (cf. Figure 4.13).

Finally, it is noted that the results presented in sections 5.3.2 and 5.3.3 would be completely different if the same kind of analysis had been applied to $(\mathbf{w}_t)$ instead of $(\mathbf{r}_t)$. Since the average wind speed was much higher during the summer months, we would have recorded 36 per cent of all storms in summer and only 14 per cent in winter (18 per cent in autumn, 32 per cent in spring). This is mainly due to the sea breeze circulation which is much stronger in summer and which is present in $(\mathbf{w}_t)$ but not in $(\mathbf{r}_t)$. It can therefore be assumed that high wind speeds are attributable to a number of factors at the same time.

5.4 Appendix: Removal of Storms

Having investigated the three types of events and related them to the geostrophic component, the storms are now removed from $(\mathbf{r}_t)$. First, we determine the mean wind speeds $v_{\mathbf{e}}$ and v_0 during the storm, and during the adjacent pre- and post-storm periods, respectively. Given the speed component (v_t) of $(\mathbf{r}_t)$ and a storm which lasts from t_1 to t_2, one has $v_{\mathbf{e}} = [t_2 - t_1 + 1]^{-1} \sum_{t=t_1}^{t_2} v_t$ and $v_0 = 12^{-1} \sum_{i=1}^{6} (v_{t_1-i} + v_{t_2+i})$.

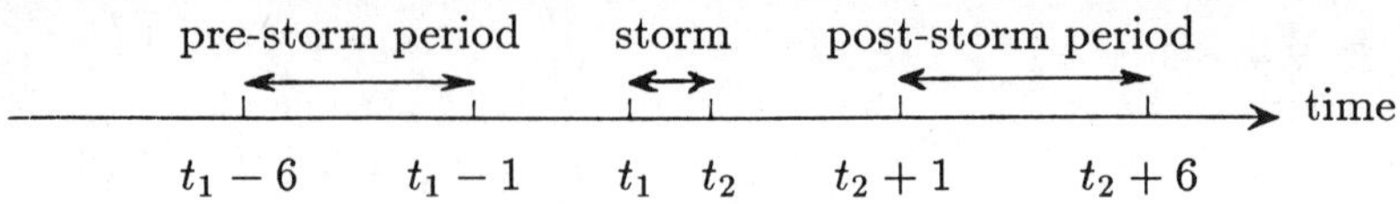

Then we take the moving average

$$\mathbf{y}_t = \tfrac{1}{9}\,\mathbf{r}_{t-2} + \tfrac{2}{9}\,\mathbf{r}_{t-1} + \tfrac{3}{9}\,\mathbf{r}_t + \tfrac{2}{9}\,\mathbf{r}_{t+1} + \tfrac{1}{9}\,\mathbf{r}_{t+2}$$

and subtract $\mathbf{e}_t = (1 - v_0/v_{\mathbf{e}})\,\mathbf{y}_t$ from $\mathbf{r}_t$ if $t_1 \le t \le t_2$. That is, the residual component $(\mathbf{f}_t)$ of the wind series $(\mathbf{w}_t)$ is defined as

$$\text{(1)} \qquad \mathbf{f}_t = \begin{cases} \mathbf{r}_t - \mathbf{e}_t & \text{if there is a storm at time } t \\ \mathbf{r}_t & \text{otherwise .} \end{cases}$$

Hence, the wind speeds during storm periods are adjusted so that their mean is comparable to that for the adjacent pre- and post-storm intervals. Figure 5.7 shows the same wind record as Figure 4.13 after the removal of the storm of 5-6 December 1971. Further, there does not seem to be any pattern in the residual component $(\mathbf{f}_t)$. Wind speed peaks at different times each day and at irregular intervals, while wind direction turns clockwise and anticlockwise, and at the same time covers the full circle.

Values for the mean sum of squares $s_{\mathbf{f}}^2$ of the residual series $(\mathbf{f}_t)$ are presented in the last row of Table 5.8. The corresponding values for the daily component $(\mathbf{d}_t)$ and the residual component $(\mathbf{r}_t)$ are copied from Table 4.4 and presented here for completeness.

Table 5.8. Summary values for the period of 1 to 8 December 1971.

Mean sum of squares	1.12.	2.12.	3.12.	4.12.	5.12.	6.12.	7.12.	8.12.
$s_{\mathbf{d}}^2$	62.8	42.0	69.7	69.8	86.0	122.8	27.8	29.6
$s_{\mathbf{r}}^2$	16.2	7.9	12.0	16.5	37.0	19.5	15.6	28.8
$s_{\mathbf{f}}^2$	16.2	7.9	12.0	16.5	21.8	15.9	15.6	28.8

Since $(\mathbf{f}_t)$ is identical with $(\mathbf{r}_t)$ for the period of 1 to 8 December 1971 except for the storm on 5-6 December, $s_{\mathbf{f}}^2$ differs from $s_{\mathbf{r}}^2$ only on these two days. The storm in $(\mathbf{r}_t)$ on 5-6 December is also reflected in the high values for $s_{\mathbf{r}}^2$. Note that the removal of the storm has resulted in bringing the values of $s_{\mathbf{f}}^2$ for these two days into the range of the other $s_{\mathbf{f}}^2$-values, even though they are still amongst the highest.

Figure 5.7 (a) Wind direction in $(\mathbf{f}_t)$ *from 1 to 8 December 1971.*

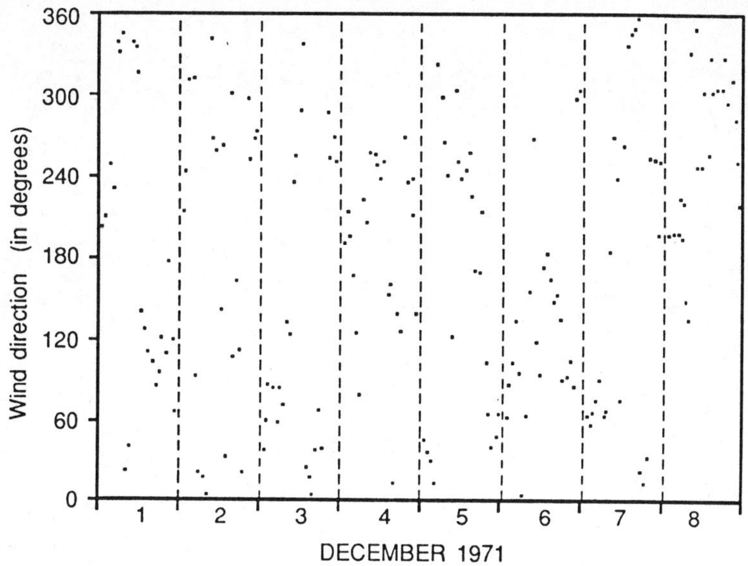

Figure 5.7 (b) Wind speed in $(\mathbf{f}_t)$ *from 1 to 8 December 1971.*

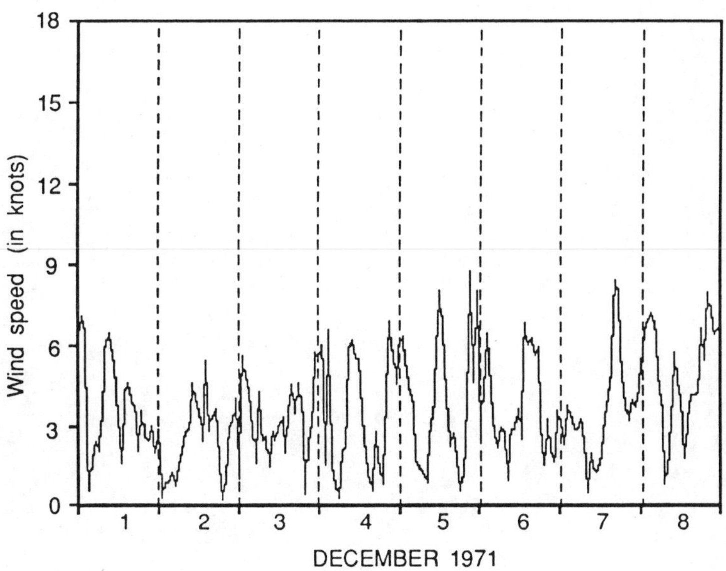

PART II

TIME SERIES OF DIRECTIONAL DATA

CHAPTER 6

TIME SERIES MODELS FOR DIRECTIONAL DATA

The residual series $(\mathbf{f}_t)$ obtained after the removal of the storms can in general not be related to any physical or meteorological phenomenon. However, the data are still highly autocorrelated. In an attempt to describe the series $(\mathbf{f}_t)$ by an autoregressive model, we will concentrate on the directional aspect. Although Part II of this monograph is motivated by the actual data analysis, it is mainly concerned with mathematical aspects and as such is self-contained.

In section 6.1 the results of directional distribution theory relevant to this study are briefly summarised. The reader is reminded that the von Mises and wrapped normal distributions play a similarly central role for directional variables as the normal distribution for real random variables, with each of the two distributions having some but not all of the desirable properties. Accordingly, two autoregressive models for directional time series are introduced in sections 6.2 and 6.3, which are related to the von Mises and wrapped normal distribution, and which are called von Mises and wrapped autoregressive process, respectively. As for the two distributions, it is shown that the two time series models can be approximated by one another so that one may choose whatever model is more appropriate under the given circumstances. The main difference between the two models is that the von Mises process will generally involve a certain kind of non-stationarity, while the wrapped autoregressive process will be assumed to be stationary.

Most of the discussion in Part II is devoted to fitting these two models to directional time series. Various methods of estimating the parameters are developed and compared. The application to the directional component of $(\mathbf{f}_t)$ is used to illustrate this approach.

6.1 Circular Variables

The objective of this and the following chapters is to describe the series $(\mathbf{f}_t)$ of short term fluctuations as an autoregressive process. Given that each record $\mathbf{f}_t$ in $(\mathbf{f}_t)$ can be represented as an element in $\mathcal{R}^2$ a model of the following form seems to be a natural choice,

$$\mathbf{f}_t = \sum_{j=1}^{p} \alpha_j \mathbf{f}_{t-j} + \alpha_0 \mathbf{U}_t \tag{1}$$

where $(\mathbf{U}_t)$ is a series of independent and identically distributed variables, $p \in \mathcal{N}$ is the order of the process and $\{\alpha_0, \ldots, \alpha_p\}$ is a set of appropriately chosen parameters in $\mathcal{R}$. Further, since wind data are usually expressed in terms of speed and direction a polar representation appears to be more appropriate than a cartesian representation, that is

$$\mathbf{f}_t = f_t \begin{pmatrix} \cos\theta_t \\ \sin\theta_t \end{pmatrix}$$

where f_t and θ_t denote speed and direction, respectively. It would therefore be desirable to specify a model that simplifies in a sensible and tractable way for the directional component (θ_t) of $(\mathbf{f}_t)$.

For example, specialising (1) to the directional component would yield a model of the following form

$$\begin{aligned} \theta_t &= \arg(\mathbf{f}_t) \\ \mathbf{f}_t &= \sum_{j=1}^{p} \alpha_j \begin{pmatrix} \cos\theta_{t-j} \\ \sin\theta_{t-j} \end{pmatrix} + \alpha_0 \begin{pmatrix} \cos\zeta_t \\ \sin\zeta_t \end{pmatrix} \end{aligned} \tag{2}$$

where $\arg(\mathbf{f}_t)$ is the direction of $\mathbf{f}_t$ and (ζ_t) is an innovation process of independent and identically distributed circular variables taking values between $-\pi$ and π. However, statistical inference based on this model is complicated as it involves scaling the weighted sum of unit vectors down to unit length.

There is an enormous literature on fitting autoregressive models to time series in $\mathcal{R}^k$ $(k \geq 1)$. Comparatively little work has been done, however, if the observations lie on a circle or on a torus, that is if the data involve a kind of periodicity. In this study we will therefore emphasise the directional aspect and solely concentrate on models for directional time series. As can be seen from Figure 5.7 the directions in $(\mathbf{f}_t)$ tend to oscillate within bands determined by the preceding observations. It is therefore suggested that a circular analogue of an ordinary, real valued autoregressive process be introduced.

Whereas there is a well developed theory based on the normal distribution for real valued variables, there is no single distribution which plays an equally central role for circular variables. However, in some repects the *von Mises* and *wrapped normal* distributions can be regarded as important contenders (Mardia 1972, p.68). Given a circular variable θ taking values between $-\pi$ and π the density of the former is defined as

$$f(\theta) = [2\pi I_0(\kappa)]^{-1} \exp\{\kappa \cos(\theta - \mu)\}$$

where $\mu \in (-\pi, \pi)$ is the mean direction, $\kappa \in \mathcal{R}_0^+$ is a concentration parameter and $I_n(\cdot)$ is the modified Bessel function of the first kind and order n ($n \in \mathcal{N}_0$). To denote that θ follows a von Mises distribution with mean direction μ and concentration κ we will also write $\theta \sim vM(\mu, \kappa)$. The vector $\kappa(\cos\mu, \sin\mu)'$ will be called the *concentration vector* of θ. Further, the angle θ will be identified with the random unit vector $\mathbf{U}_\theta$ pointing in the direction of θ. It is well known that

$$\mathcal{E}\begin{bmatrix} \cos\theta \\ \sin\theta \end{bmatrix} = A_1(\kappa)\begin{pmatrix} \cos\mu \\ \sin\mu \end{pmatrix}$$

where $A_n(\kappa) = I_n(\kappa)/I_0(\kappa)$ ($n \in \mathcal{N}$). This implies that the function

$$\rho(\mathbf{z}) = \mathcal{E}\left[\left\| \begin{pmatrix} \cos\theta \\ \sin\theta \end{pmatrix} - \mathbf{z} \right\|^2\right], \qquad \mathbf{z} \in \mathcal{R}^2 , \tag{3}$$

attains a minimum for $\mathbf{z} = A_1(\kappa)(\cos\mu, \sin\mu)'$. On the other hand, if θ_1 and θ_2 both follow a von Mises distribution, and if $\mathbf{U}_1$ and $\mathbf{U}_2$ denote the unit vectors associated with θ_1 and θ_2, then the angle $\arg(\mathbf{U}_1 + \mathbf{U}_2)$ will usually have a distribution of a different type.

Indeed, let $(\mathbf{U}_j)_{j=1,\ldots,p}$ ($p \in \mathcal{N}$) be a sample of independent and identically distributed unit vectors $\mathbf{U}_j \sim vM(\mu, \kappa)$. The distribution of the resultant vector

$$R\begin{pmatrix} \cos\theta \\ \sin\theta \end{pmatrix} = \sum_{j=1}^{p} \alpha_j \mathbf{U}_j , \qquad \alpha_j \in \mathcal{R} ,$$

is then obtained as

$$f(\theta, R) = \left[2\pi \prod_{j=1}^{p} I_0(\kappa\alpha_j)\right]^{-1} R\, h_p(R) \exp\{\kappa R \cos(\theta - \mu)\} \tag{4}$$

where

$$h_p(R) = \int_0^\infty J_0(rR) \prod_{j=1}^{p} J_0(r\alpha_j) r\, dr ,$$

with $J_0(\cdot)$ denoting the Bessel function of the first kind and order 0. This is a slightly more general result than an earlier one by Mardia (1972, p.97) and can be proved in a similar way. Hence, the direction θ of the resultant vector conditional on its length R has a von Mises distribution. However, the marginal distribution of θ has an almost intractable density involving integrals over Bessel functions. This illustrates the fact that inference for the model (2) can be extremely complicated.

The other distribution which plays a central role for circular variables taking values between $-\pi$ and π is given by the wrapped normal density defined as follows

$$f(\eta) = [2\pi\sigma^2]^{-1/2} \sum_{k=-\infty}^{\infty} \exp\left\{-\frac{(\eta + 2\pi k - \mu)^2}{2\sigma^2}\right\}.$$

A variable following this distribution can be conceived as the result of wrapping a normally distributed variable $Y \sim N(\mu, \sigma^2)$ around the circle, that is $\eta = Y \pmod{2\pi}$. To denote that η follows a wrapped normal distribution we will write $\eta \sim WN(\mu, \sigma^2)$ where μ and σ^2 are the parameters of the associated unwrapped variable. In contrast with the von Mises distribution the following property holds here. If η_1 and η_2 are two circular variables following a wrapped normal distribution, then any linear combination $\alpha_1\eta_1 + \alpha_2\eta_2$ with $\alpha_1, \alpha_2 \in \mathcal{R}$ is again wrapped normally distributed. However, the likelihood function is fairly complicated, so that statistical inference from this distribution may often be difficult.

It is noted, however, that the von Mises and wrapped normal distributions can be well approximated by one another (Kent 1978b, Watson 1982). In practice it can therefore be assumed that any property of any one distribution also holds approximately for the other. Consequently, one can choose whichever distribution is more appropriate or tractable under the given circumstances. Whereas the von Mises distribution leads to a simpler likelihood function and may be preferred in the context of hypothesis testing, the wrapped normal distribution is more suitable when circular variables are to be added. As for the approximation of a von Mises variable $\theta \sim vM(0, \kappa)$ by a wrapped normal variable $\eta \sim WN(0, \sigma^2)$ it is usual to equate the *first circular moments*. That is

$$A_1(\kappa) = \exp\{-\sigma^2/2\}. \tag{5}$$

This approximation works particularly well for large values of κ and for κ close to zero. In the first case the wrapped normal distribution practically reduces to a normal distribution as σ becomes so small in comparison to 2π that the periodicity of the points on the circle and hence the wrapping plays virtually no role. The von Mises distribution, on the other hand, also tends towards a normal distribution as κ tends towards infinity. In the second case where κ is small, both the von Mises and the wrapped normal distribution tend towards the uniform distribution with density $f(\theta) = [2\pi]^{-1}$.

Two time series models will now be introduced which are related to the von Mises and the wrapped normal distribution, respectively.

6.2 The von Mises Process

The familiar form of an autoregressive process (Y_t) in $\mathcal{R}$ is given by

$$Y_t = \sum_{j=1}^{p} \alpha_j Y_{t-j} + U_t \,, \qquad t \in \mathcal{Z} \,,$$

where (U_t) is a series of independent and identically distributed variables with $U_t \sim N(0, \sigma^2)$. The process (Y_t) is *weakly stationary* if and only if all the zeros of the associated polynomial

$$Q_p(x) = \sum_{j=0}^{p} \alpha_j x^j \quad \text{with} \quad \alpha_0 = -1 \tag{6}$$

lie outside the unit circle. The process (Y_t) can also be expressed in terms of conditional densities, that is

$$f(y_t \mid y_{t-1}, \ldots, y_{t-p}) = [2\pi\sigma^2]^{-1/2} \exp\left\{ -\frac{1}{2\sigma^2} \left(y_t - \sum_{j=1}^{p} \alpha_j y_{t-j} \right)^2 \right\} . \tag{7}$$

Regarding the von Mises distribution as the circular analogue of the normal distribution, a natural specification of a time series model for directional data seems to be as follows.

Definition: Let (θ_t) be a process of circular variables and let $\kappa_0, \ldots, \kappa_p \in \mathcal{R}_0^+$. If θ_t conditional on $(\theta_{t-1}, \ldots, \theta_{t-p})$ follows a von Mises distribution with concentration vector

$$\mathbf{v}_t = \sum_{j=1}^{p} \kappa_j \begin{pmatrix} \cos\theta_{t-j} \\ \sin\theta_{t-j} \end{pmatrix} + \begin{pmatrix} \kappa_0 \\ 0 \end{pmatrix} \tag{8}$$

then (θ_t) will be called a von Mises process.

The conditional density of a von Mises process can be obtained in the same way as (7) was derived. That is,

$$\begin{aligned} & f(\theta_t \mid \theta_{t-1}, \ldots, \theta_{t-p}) \\ & \quad = [2\pi I_0(v_t)]^{-1} \exp\left\{ \sum_{j=1}^{p} \kappa_j \cos(\theta_t - \theta_{t-j}) + \kappa_0 \cos\theta_t \right\} \end{aligned} \tag{9}$$

where $v_t = \|\mathbf{v}_t\|$ denotes the length of the vector $\mathbf{v}_t$.

Even though the densities (7) and (9) have a similar form there is an important difference which ought to be mentioned. The values of both, Y_t and θ_t, are largely determined by the preceding p observations. However, whereas in the case of an autoregressive process the variance σ^2 is constant, the concentration v_t in a von Mises process generally varies with time. The process (θ_t) will therefore also be called von Mises process with *changing* concentration.

If $\theta_{t-1}, \ldots, \theta_{t-p}$ are pointing in more or less the same direction the length of the concentration vector $\mathbf{v}_t$ will be relatively large and the chance of θ_t pointing in the same direction will also be large. On the other hand, if $\theta_{t-1}, \ldots, \theta_{t-p}$ are fairly different the concentration v_t will be small and there will be no particular direction which could be expected for θ_t. A von Mises process will therefore exhibit clusters of high as well as low concentration. Hence, if a series of wind directions could be described by a von Mises model of the form (9) then periods with a prevailing wind direction would be followed by periods when directions change more rapidly, and vice versa. This will be illustrated in chapter 10 where a von Mises process is fitted to the directional component (θ_t) of $(\mathbf{f}_t)$.

Of course, one could also define a von Mises process with *constant* concentration. Put $\mu_t = \arg(\mathbf{v}_t)$ with $\mathbf{v}_t$ as in (8), and consider

$$f(\theta_t \mid \theta_{t-1}, \ldots, \theta_{t-p}) = [2\pi I_0(\kappa)]^{-1} \exp\{\kappa \cos(\theta_t - \mu_t)\} \tag{10}$$

where $\kappa \in \mathcal{R}_0^+$ is a fixed parameter. In spite of the more simplistic density (10), inference is at least as hard to draw for this model as in the case of (9), since determination of μ_t implicitly requires that $\mathbf{v}_t$ is scaled to unit length. We are thus confronted with a similar problem to that in (2).

Example 6.1: Let $p = 1$ and $\kappa_0 = 0$. Then density (9), with $\kappa = \kappa_1$, reduces to

$$f(\theta_t \mid \theta_{t-1}) = [2\pi I_0(\kappa)]^{-1} \exp\{\kappa \cos(\theta_t - \theta_{t-1})\} . \tag{11}$$

Therefore (θ_t) reduces to a process with constant concentration. Further, (θ_t) constitutes a Markov chain with a uniform marginal distribution, that is $f(\theta_t) = [2\pi]^{-1}$. Since the state space of this Markov chain is irreducible and compact the uniform distribution is invariant and unique. Hence, the joint density of θ_t and θ_{t-1} is given by

$$f(\theta_t, \theta_{t-1}) = [4\pi^2 I_0(\kappa)]^{-1} \exp\{\kappa \cos(\theta_t - \theta_{t-1})\} . \tag{12}$$

The maximum likelihood estimate $\hat{\kappa}$ of κ, based on the observations $\theta_0, \ldots, \theta_T$, is therefore easily determined as

$$\hat{\kappa} = A_1^{-1}\left(\frac{1}{T}\sum_{t=1}^{T} \cos(\theta_t - \theta_{t-1})\right) . \tag{13}$$

□

Apart from this particular case estimation of the parameters $\kappa_0, \ldots, \kappa_p$ is a lot more complicated. First, we look at maximum likelihood estimation and then propose a least squares type estimator.

Given the order p of the von Mises process and the sample $\{\theta_0, \ldots, \theta_T\}$, the likelihood function conditional on $\{\theta_0, \ldots, \theta_{p-1}\}$ is obtained as

$$\begin{aligned} L &= L(\kappa_p, \ldots, \kappa_0; \theta_T, \ldots, \theta_p \mid \theta_{p-1}, \ldots, \theta_0) \\ &= \prod_{t=p}^{T} \left[2\pi I_0(v_t)\right]^{-1} \exp\left\{ \sum_{j=1}^{p} \kappa_j \cos(\theta_t - \theta_{t-j}) + \kappa_0 \cos\theta_t \right\}, \end{aligned}$$

with $v_t = \|\mathbf{v}_t\|$ and $\mathbf{v}_t$ as in (8). Let

$$\gamma_t = \begin{pmatrix} \cos\theta_t \\ \cos(\theta_t - \theta_{t-1}) \\ \vdots \\ \cos(\theta_t - \theta_{t-p}) \end{pmatrix} \quad \text{and} \quad \mathbf{\Gamma}_t = \begin{pmatrix} 1 & 0 \\ \cos\theta_{t-1} & \sin\theta_{t-1} \\ \vdots & \vdots \\ \cos\theta_{t-p} & \sin\theta_{t-p} \end{pmatrix},$$

and define $\kappa = (\kappa_0, \ldots, \kappa_p)'$. The concentration vector $\mathbf{v}_t$ of the variable θ_t conditional on $(\theta_{t-1}, \ldots, \theta_{t-p})$ can then be written as $\mathbf{v}_t = \mathbf{\Gamma}_t'\kappa$, and the estimating equation for κ takes the form

$$\frac{d \ln L}{d\kappa} = \sum_{t=p}^{T} \left\{ -\frac{A_1(v_t)}{v_t} \mathbf{\Gamma}_t \mathbf{\Gamma}_t' \kappa + \gamma_t \right\} = \mathbf{0}\,. \tag{14}$$

Since v_t is a function of κ an iterative procedure will generally be required to determine κ. Note that in case of example 6.1 $v_t = \kappa = \kappa_1$ for all t and hence, that equation (14) reduces to (13).

Alternatively, one could devise a least squares type estimator along the following lines. First, note that the vector which minimises the cost function (3), given a von Mises distribution with concentration vector $\kappa(\cos\mu, \sin\mu)'$, is equal to $A_1(\kappa)(\cos\mu, \sin\mu)'$. Second, consider the case of fitting a von Mises process to the time series (θ_t). Let $\mathbf{v}_t = \mathbf{\Gamma}_t'\kappa$ be the concentration vector of the von Mises variable θ_t, $\mathbf{w}_t$ be the vector which minimises the corresponding cost function and $\langle \mathbf{\Gamma}_t \rangle$ be the subspace spanned by the rows of $\mathbf{\Gamma}_t$. It follows that $\mathbf{w}_t$ is an element of $\langle \mathbf{\Gamma}_t \rangle$. Now assume that $\mathbf{w}_t$ can be expressed in the form $\mathbf{w}_t = \mathbf{\Gamma}_t'\beta$ with $\beta = (\beta_0, \ldots, \beta_p)' \in \mathcal{R}^{p+1}$ independent of t. That is, $(\cos\theta_t, \sin\theta_t)'$ is projected onto $\langle \mathbf{\Gamma}_t \rangle$.

Substituting the expression for $\mathbf{w}_t$ into the cost function yields

$$\rho(\beta) = \sum_{t=p}^{T} \left\| \begin{pmatrix} \cos\theta_t \\ \sin\theta_t \end{pmatrix} - \mathbf{\Gamma}_t'\beta \right\|^2.$$

This leads to the following estimating equation for β

$$\sum_{t=p}^{T} \left\{ \mathbf{\Gamma}_t \begin{pmatrix} \cos\theta_t \\ \sin\theta_t \end{pmatrix} - \mathbf{\Gamma}_t \mathbf{\Gamma}_t' \beta \right\} = \mathbf{0}\,. \tag{15}$$

Given the relationship between the concentration vector on the one hand and the vector minimising the cost function (3) on the other hand, it follows that $w_t = \|\mathbf{w}_t\| = A_1(v_t)$ and $\arg(\mathbf{w}_t) = \arg(\mathbf{v}_t)$. Hence,

$$\mathbf{\Gamma}'_t\beta = \frac{\mathbf{w}_t}{w_t}\|\mathbf{w}_t\| = \frac{\mathbf{v}_t}{v_t}A_1(v_t) = \frac{A_1(v_t)}{v_t}\mathbf{\Gamma}'_t\kappa \ .$$

By substituting this expression into (15) one obtains an estimating equation for κ which is identical to (14) above. That is, maximum likelihood and least squares estimation lead to the same result. It should be noted, however, that the least squares argument is based on the vectors $(\cos\theta_t, \sin\theta_t)'$ rather than the angles θ_t $(t = 0, \ldots, T)$. As for a numerical example see chapter 10 where a first order von Mises model is fitted to the series of residual wind directions obtained in chapter 5.

6.3 The Wrapped Autoregressive Process

In this section a process that is related to the wrapped normal distribution will be introduced. Just like the wrapped normal variable it can be conceived as the result of wrapping an autoregressive process (Y_t) around the circle. That is

$$\eta_t = Y_t \ (\mathrm{mod}\ 2\pi) \ ,$$

where $Y_t = \sum_{j=1}^{p} \alpha_j Y_{t-j} + U_t$ with U_t independently and identically distributed as $N(0, \sigma^2)$. Although only the circular process (η_t) is observed, fitting a *wrapped autoregressive* (WAR) model requires estimation of the parameters $\alpha_1, \ldots, \alpha_p$ and σ^2 which describe the *unobserved* series (y_t). The parameters of a wrapped autoregressive process are nevertheless easier to estimate than those of a von Mises process, as the former is intimately related to a normal autoregressive process in $\mathcal{R}$. Clearly, the density of Y_t given $(Y_{t-1}, \ldots, Y_{t-p})$ is of the form (7).

Central to fitting an AR model to a time series in $\mathcal{R}$ is the autocovariance function of the underlying process. However, before a similar concept can be developed for time series of directional data a measure of angular correlation needs to be specified. In chapter 7 various proposals will be studied and compared. We then come back in chapters 8 and 9 to fitting a WAR model to a series of angular data. In the remainder of this section it will be shown how the wrapped autoregressive process is related to the von Mises process introduced in section 6.2.

Suppose that $\theta_0 \sim vM(0, \kappa_0)$ and $(\theta_t - \theta_{t-1}) \sim vM(0, \kappa)$ for $t = 1, 2, \ldots$ are independent circular variables $(\kappa_0, \kappa \in \mathcal{R}^+)$. This means that (θ_t) forms a random walk on the circle. Given that the von Mises and wrapped normal distributions can be well approximated by one another, one has

$$(\theta_t - \theta_{t-1}) \overset{\text{approx.}}{\sim} WN(0, \sigma^2) \quad \text{with} \quad A_1(\kappa) = \exp\{-\sigma^2/2\}$$

and

$$\theta_0 \overset{\text{approx.}}{\sim} WN(0,\sigma_0^2) \quad \text{with} \quad A_1(\kappa_0) = \exp\{-\sigma_0^2/2\}$$

where $\overset{\text{approx.}}{\sim}$ means *approximately distributed as.* Hence,

$$\theta_t \overset{\text{approx.}}{\sim} WN(0,\sigma_0^2 + t\sigma^2)\,.$$

Now, approximating the wrapped normal by a von Mises distribution yields

$$\theta_t \overset{\text{approx.}}{\sim} vM(0,\kappa_t)$$

with $A_1(\kappa_t) = \exp\{-(\sigma_0^2 + t\sigma^2)/2\} = A_1(\kappa_0)\, A_1^t(\kappa)$. Consequently, $\kappa_t \to 0$ as $t \to \infty$, that is θ_t is asymptotically uniform. This example shows that the approximation above could be a useful device in examining the properties of the von Mises process. As in this example, it can be shown in general that wrapping any non-stationary autoregressive process around the circle leads to an asymptotically uniform distribution.

Let $k = \min\{t,p\}$ and consider the von Mises process $(\theta_t)_{t=0,1,\ldots}$ assuming that θ_0 as well as θ_t given $(\theta_{t-1},\ldots,\theta_{t-k})$ follow von Mises distributions. Denote the mean direction and concentration of θ_t by μ_t and v_t, respectively. Obviously, μ_t and v_t are defined by the previous k observations, that is

$$v_t \begin{pmatrix} \cos\mu_t \\ \sin\mu_t \end{pmatrix} = \sum_{j=1}^{\min(t,p)} \kappa_j \begin{pmatrix} \cos\theta_{t-j} \\ \sin\theta_{t-j} \end{pmatrix} + \kappa_0 \begin{pmatrix} \cos\mu_0 \\ \sin\mu_0 \end{pmatrix}.$$

Since any von Mises variable θ can be well approximated by a wrapped normal variable η, θ_0 can be approximated by $\eta_0 \sim WN(\mu_0,\sigma_0^2)$ with $\sigma_0^2 = -2\ln A_1(\kappa_0)$. Similarly, θ_t given $(\theta_{t-1},\ldots,\theta_0)$ can be approximated by $\eta_t \sim WN(\mu_t,\sigma_t^2)$ with $\sigma_t^2 = -2\ln v_t$ for all $t \geq 1$. This means that the conditional distribution of θ_t given $(\theta_{t-1},\ldots,\theta_0)$ is approximately the same as the conditional distribution of η_t given $(\theta_{t-1},\ldots,\theta_0)$ and thus the conditional distribution of η_t given $(\eta_{t-1},\ldots,\eta_0)$.

More specifically, consider the density

$$\begin{aligned} &f(\theta_t \mid \theta_{t-1},\ldots,\theta_{t-p}) \\ &= [2\pi I_0(v_t)]^{-1} \exp\Big\{\sum_{j=1}^{p} \kappa_j \cos(\theta_t - \theta_{t-j}) + \kappa_0 \cos(\theta_t - \mu_0)\Big\}\,. \end{aligned} \tag{16}$$

Each factor $\exp\{\kappa_j \cos(\theta_t - \theta_{t-j})\}$ in (16) can be approximated by const.$\sum_{k=-\infty}^{\infty} \exp\{-(\theta_t - \theta_{t-j} + 2\pi k)^2/2\sigma_j^2\}$ where $\sigma_j^2 = -2\ln A_1(\kappa_j)$. It follows that the conditional distribution of θ_t given $(\theta_{t-1},\ldots,\theta_{t-p})$ is approximately the same as that of η_t given $(\eta_{t-1},\ldots,\eta_{t-p})$ which has the following density

$$\begin{aligned} &f(\eta_t \mid \eta_{t-1},\ldots,\eta_{t-p}) \\ &= \text{const.} \sum_{k_0\ldots k_p} \exp\Big\{-\sum_{j=1}^{p} (\eta_t - \eta_{t-j} + 2\pi k_j)^2/2\sigma_j^2 \\ &\qquad - (\eta_t - \mu_0 + 2\pi k_0)^2/2\sigma_0^2\Big\} \end{aligned}$$

$$= [2\pi\sigma^2]^{-1/2} \sum_{k=-\infty}^{\infty} \exp\left\{-\frac{1}{2\sigma^2}\left(\eta_t - \sum_{j=1}^{p} \alpha_j \eta_{t-j} - 2\pi k\right)^2\right\} \tag{17}$$

where $1/\sigma^2 = \sum_{j=0}^{p} 1/\sigma_j^2$ and $\alpha_j = \sigma^2/\sigma_j^2$ $(j = 1, \ldots, p)$. This, however, is the density of a wrapped normal variable with mean $\sum_{j=1}^{p} \alpha_j \eta_{t-j}$ and variance σ^2. Denoting the corresponding unwrapped variables by $Y_t, \ldots, Y_{t-p}$ this implies that η_t given $(\eta_{t-1}, \ldots, \eta_{t-p})$ can be obtained by wrapping Y_t given $(Y_{t-1}, \ldots, Y_{t-p})$ around the circle, where

$$f(y_t \mid y_{t-1}, \ldots, y_{t-p}) = [2\pi\sigma^2]^{-1/2} \exp\left\{-\frac{1}{2\sigma^2}\left(y_t - \sum_{j=1}^{p} \alpha_j y_{t-j}\right)^2\right\}. \tag{18}$$

This means that (Y_t) forms an autoregressive process and hence, that (η_t) is a wrapped autoregressive process. Note that the parameters $\alpha_1, \ldots, \alpha_p$ satisfy the following constraints:

$$\alpha_j \geq 0 \qquad \text{for } j = 1, \ldots, p \tag{19a}$$

$$\sum_{j=1}^{p} \alpha_j \leq 1 \tag{19b}$$

with strict inequality in (19b) to ensure stationarity of (Y_t). If $\sum_{j=1}^{p} \alpha_j = 1$ then one of the zeros of the associated polynomial (6) is equal to 1, and the wrapped process (η_t) tends towards a marginal uniform distribution.

Let us summarise this result. The von Mises process (θ_t) can be approximated by a WAR process (η_t) obtained by wrapping an AR process (Y_t) with only positive coefficients around the circle. This constraint merely reflects the fact that the coefficients $\kappa_0, \ldots, \kappa_p$ in the von Mises process were assumed to be non-negative.

Concerning this approximation there is one difference between (θ_t) and (η_t) which should be pointed out. Whereas the concentration vector of θ_t given $(\theta_{t-1}, \ldots, \theta_{t-p})$ has a variable length, the variable η_t conditional on $(\eta_{t-1}, \ldots, \eta_{t-p})$ has a constant variance. Approximating (θ_t) by (η_t) thus leads to a different result compared with the case where each variable θ_t given $(\theta_{t-1}, \ldots, \theta_{t-p})$ is approximated individually by a wrapped normal variable. However, these wrapped variables would generally be related to each other in a form that is a lot more complicated than a WAR process. Hence, by smoothing the variation in the concentration vectors of (θ_t), the WAR approximation may not be as good as a wrapped normal approximation whose characteristics are determined every time, but it certainly is more tractable.

This discussion also helps to identify the main difference between a WAR process and a model of the form (2). Whereas in the former the angles are simply added, in the latter one forms sums of unit vectors. This essentially is the reason for the mathematical complication associated with (2), since the sum of unit vectors will not necessarily be a unit vector while the sum of angles will always be an angle.

Let us now return to example 6.1 and consider the case where the angular differences are approximated by a wrapped normal distribution.

Example 6.1 (cont.): Consider the directional process (θ_t) with density (11), that is $p = 1$ and $\kappa_0 = 0$, and let (Y_t) denote the process underlying the wrapped normal approximation (η_t). The conditional density of Y_t given Y_{t-1} is then obtained as

$$f(y_t \mid y_{t-1}) = [2\pi\sigma^2]^{-1/2} \exp\{-(y_t - y_{t-1})^2/2\sigma^2\} ,$$

where $\sigma^2 = -2\ln A_1(\kappa)$. If in addition Y_{t-1} follows a normal distribution then $\mathrm{var}\,(Y_t) = \mathrm{var}\,(Y_{t-1}) + \sigma^2$. This implies that (η_t) tends towards a uniform circular variable with limiting joint density

$$f(\eta_t, \eta_{t-1}) = \left[2\pi\sqrt{2\pi\sigma^2}\,\right]^{-1} \sum_{k=-\infty}^{\infty} \exp\{-(\eta_t - \eta_{t-1} + 2\pi k)^2/2\sigma^2\} .$$

Furthermore, the relationship

$$\begin{aligned} c(1) &= \mathcal{E}[\cos(\eta_t - \eta_{t-1})] - \mathcal{E}[\cos\eta_t]\; \mathcal{E}[\cos\eta_{t-1}] - \mathcal{E}[\sin\eta_t]\; \mathcal{E}[\sin\eta_{t-1}] \\ &= \mathcal{E}\big[\exp\{i(\eta_t - \eta_{t-1})\}\big] \\ &= \exp\{-\sigma^2/2\} \end{aligned}$$

can be used to estimate σ^2 and κ, respectively. Because of equation (5) we arrive at the same estimator as (13) in example 6.1, but whereas there it was derived as the maximum likelihood estimator here it is obtained through equating the empirical and theoretical *covariance* $c(1)$. A detailed discussion of fitting a WAR process will follow in chapter 9. □

Example 6.2: If in (9) $p = 1$ and $\kappa_0 \neq 0$ then the wrapped normal approximation leads to a process (Y_t) with conditional density

$$f(y_t \mid y_{t-1}) = [2\pi\sigma^2]^{-1/2} \exp\{-(y_t - \alpha y_{t-1})^2/2\sigma^2\} , \tag{20}$$

where $\sigma^2 = -2\ \ln A_1(\kappa_0)\ \ln A_1(\kappa_1)\ /\ \ln\{A_1(\kappa_0)\ A_1(\kappa_1)\}$ and $\alpha = \ln A_1(\kappa_0)\ /\ \ln\{A_1(\kappa_0)\ A_1(\kappa_1)\} = [1 + \ln A_1(\kappa_1)/\ln A_1(\kappa_0)]^{-1}$. If κ_0 tends towards 0 then $\alpha \to 1$, that is the non-stationary boundary. In fact, for any von Mises process with density (9) one has $\kappa_0 = 0$ if and only if the unwrapped process (18) underlying the wrapped normal approximation is non-stationary. On the other hand, $\kappa_0 = 0$ is equivalent to a uniform marginal distribution. □

Given that the directional component (θ_t) of $(\mathbf{f}_t)$ has an almost uniform distribution it is suggested that the series $(\theta_t - \theta_{t-1})$ of differenced wind directions is modelled as a WAR process, and that (θ_t) is modelled as a von Mises process with a parameter κ_0 close to zero. Both these possibilities will be considered in chapter 10.

CHAPTER 7

MEASURES OF ANGULAR ASSOCIATION

In the last chapter two time series models for directional data were introduced. However, to define the autocovariance function of the circular process and to estimate the parameters of these models, a concept of angular dependence needs to be developed. In this chapter we explore various methods of measuring the association between two circular variables by a scalar and propose a new correlation coefficient for bivariate angular data.

In section 7.1 desirable properties for measures of angular association are specified. These measures are generally related to bivariate distribution functions through the maximum entropy characterization. In section 7.2 we therefore present a discussion of bivariate angular distribution functions. It is shown that one is confronted again with the problem that there are two important distributions for circular variables, both of which have some but not all of the desirable properties. Whereas the wrapped normal distribution has the disadvantage that it does not belong to the exponential family, the von Mises distribution has the disadvantage that it cannot be extended to a bivariate distribution such that its marginals are von Mises. A proof of this result is presented in the appendix 7.5.

The most important measures of angular association are reviewed in section 7.3. In section 7.4 an alternative correlation coefficient is proposed, which is particularly useful for directional time series and which can be extended to a measure of vector dependence. It is shown that this new proposal is related to the von Mises process through the maximum entropy characterization. When examining the properties of this measure and comparing it with those of the other measures it is found that each particular measure of association has some but not all of the desirable properties. This again reflects the dilemma described above.

7.1 Desirable Properties

In many situations one would like to describe the strength of dependence between two vector valued variables $\mathbf{X}$ and $\mathbf{Y} \in \mathcal{R}^k$ $(k \geq 1)$ by a numerical value, possibly in the range $[0,1] \subseteq \mathcal{R}$ with 0 and 1 corresponding to independence and functional dependence, respectively. In particular the problem of measuring the association between two angular variables has been the subject of much current research. Different proposals were made by Watson & Beran (1967), Downs (1974), Mardia (1975b), Thompson (discussion in Mardia 1975b), Johnson & Wehrly (1977), Mardia & Puri (1978), Stephens (1979), Jupp & Mardia (1980), Rivest (1982), Fisher & Lee (1982) and Fisher & Lee (1983). Most of the correlation coefficients suggested in that literature are readily generalised to measures of vector association. A brief summary of these proposals will be given in section 7.3. For tests of independence the reader is also referred to Rothman (1971), Mardia (1975b) and Puri & Rao (1977).

Rényi (1959) states seven postulates which any real valued measure $\delta(\mathbf{X},\mathbf{Y})$ of dependence should fulfill. These properties are based on a concept of general functional dependence, and are satisfied by the *maximal correlation coefficient* introduced by Gebelein (1941). This coefficient is defined as the supremum over a set of correlation coefficients obtained for Borel-measurable transformations of $\mathbf{X}$ and $\mathbf{Y}$. However, since these transformations are fairly general, the maximal correlation coefficient could be hard to calculate and may not always be practical.

When stating desirable properties for a measure of association between two angular variables, our attention will therefore be restricted to linear operations $\mathbf{H} \in \mathcal{R}^{2\times 2}$ satisfying $|\det \mathbf{H}| = 1$, that is rotations and reflections. Let $\mathbf{U}_\theta$ and $\mathbf{U}_\eta \in \mathcal{R}^2$ be the unit vectors which can be identified with the angular variables θ and η, respectively. Drawing on the analogy between the circular moments $\mathcal{E}[\cos p\theta]$ and $\mathcal{E}[\sin p\theta]$ $(p \geq 1)$ of an angular variable θ, and the moments $\mathcal{E}[Y^p]$ of a random variable Y in $\mathcal{R}$, $\mathbf{U}_\theta$ and $\mathbf{U}_\eta$ are called *linearly dependent* if and only if there exists an $\mathbf{H} \in \mathcal{R}^{2\times 2}$ with $|\det \mathbf{H}| = 1$ such that $\mathbf{H}\mathbf{U}_\theta = \mathbf{U}_\eta$. This relationship will also be written as $\mathbf{H}(\theta) = \eta$.

Corresponding to the postulates formulated by Rényi we now state the properties which any real valued measure $\delta(\theta,\eta)$ of angular association should satisfy.

(i) Domain :
$\delta(\theta,\eta)$ is defined for all non-trivial variables θ and η.

(ii) Symmetry :
$\delta(\theta,\eta) = \delta(\eta,\theta)$.

(iii) Boundaries :
$0 \leq \delta(\theta,\eta) \leq 1$.

(iv) Independence :
if θ and η are independent then $\delta(\theta,\eta) = 0$.

(v) Dependence :
if θ and η are linearly dependent then $\delta(\theta,\eta) = 1$.

(vi) Invariance :
$\delta(\mathbf{H}_1(\theta),\, \mathbf{H}_2(\eta)) = \delta(\theta,\eta)$ for all $\mathbf{H}_1, \mathbf{H}_2 \in \mathcal{R}^{2\times 2}$ with $|\det \mathbf{H}_1| = |\det \mathbf{H}_2| = 1$.

These properties essentially coincide with those suggested by Jupp & Mardia (1980). However, these authors also demand that the sample correlation coefficient has an asymptotic distribution which does not depend on the marginals under the null hypothesis of independence. Although this property is useful in the context of hypothesis testing it is not regarded as being as fundamental as the ones mentioned above and has therefore been excluded from that list. The properties (i) to (vi) will help in assessing the various measures of association and to select the one most suitable for an analysis of angular time series.

All the correlation coefficients mentioned above satisfy properties (i) to (iii), and except for the proposal by Thompson (discussion in Mardia 1975b) they are all continuous functions of the data. Their differences will thus be reflected in the extent to which they satisfy (iv) to (vi) and a few properties that are useful in specific situations. Note that property (vi) implies that $\delta(\theta,\eta)$ is rotation invariant.

7.2 Bivariate Angular Distributions

The entropy (cf. Rao 1973, p.173) is a measure of the uncertainty or information contained in a random variable X and can be used for characterizing density functions. In particular, measures of association and bivariate probability distributions are intimately connected through the maximum entropy characterization. That is, if X is a random variable in $\mathcal{R}$ and $h_1,\ldots,h_p$ $(p \in \mathcal{N})$ are linearly independent real valued functions of X, then the density, which maximises the entropy subject to the constraints $\mathcal{E}[\,h_j(X)\,] = \text{const.}$ $(j = 1,\ldots,p)$, is of the form

$$(1) \qquad f(x) = \text{const.}\ \exp\left\{\sum_{j=1}^{p} c_j h_j(x)\right\} \quad \text{with} \quad c_1,\ldots,c_p \in \mathcal{R}$$

provided there exist parameters $c_1,\ldots,c_p$ such that the constraints are satisfied. For a proof the reader is referred to Kagan, Linnik & Rao (1973, p.409).

For example, consider the circular variable θ. In case no constraint is specified, the maximum entropy characterization would lead to the uniform distribution while fixing $\mathcal{E}[\mathbf{U}_\theta]$ would lead to the von Mises distribution. Indeed, given the constraint

$$\mathcal{E}\begin{bmatrix}\cos\theta\\ \sin\theta\end{bmatrix} = A_1(\kappa)\begin{pmatrix}\cos\mu\\ \sin\mu\end{pmatrix} \quad \text{with } \mu \in (-\pi,\pi] \text{ and } \kappa \in \mathcal{R}_0^+ \ ,$$

density (1) reduces to $f(\theta) = [2\pi I_0(\kappa)]^{-1}\exp\{\kappa\cos(\theta-\mu)\}$.

Any bivariate distribution belonging to the exponential family and linking two random variables can therefore be associated with a measure of association. Referring to the desirable properties of these measures, stated in the last section, it is noted that Rényi (1959) had included the additional postulate, that in case of a bivariate normal variable the measure of association should reduce to the ordinary Pearson correlation coefficient. The question thus arises whether there exists a bivariate circular distribuiton and an associated measure of dependence which play an analogous role for angular variables. This would imply that the measure associated with this distribution has all of the properties (i) to (vi) and is as important for circular variables as the Pearson correlation coefficient for random variables in $\mathcal{R}$.

In the last chapter it was mentioned that the von Mises and wrapped normal distributions play a central role for circular variables. Since the latter does not belong to the exponential family, however, it does not suggest any measure of angular association. In this section we will therefore concentrate on the von Mises distribution. A number of bivariate angular distributions will be introduced and their relationship with the von Mises distribution be established. First, let us recapitulate the following well known characterization of the von Mises distribution, namely that it can be constructed as a conditional bivariate normal distribution. If $\mathbf{U} = (X, Y) \sim N\big((1,0),\, \kappa^{-1}\mathbf{I}_2\big)$, where $\mathbf{I}_2$ denotes the identity matrix of order 2, then $\mathbf{U}$ given the length $R = \sqrt{X^2+Y^2} = r$ has a von Mises distribution with mean direction 0 and concentration $\kappa\, r$ (Mardia 1972, p.60). Equivalently, given a planar Brownian motion starting at the origin with direction 0 and drift $\kappa > 0$, the hitting density to the unit circle is $vM(0,\kappa)$ (Gordon & Hudson 1977).

This characterization suggests that an appropriate bivariate angular distribution can be derived from a 4-dimensional normal variable $(X_1, Y_1, X_2, Y_2)'$ in the following way. Assume

$$
\begin{aligned}
&(X_1, Y_1, X_2, Y_2) \\
&\sim N\left((1,0,1,0),\, \frac{1}{\kappa^2+2\kappa\iota}\begin{pmatrix} \kappa+\iota & 0 & \iota & 0 \\ 0 & \kappa+\iota & 0 & \iota \\ \iota & 0 & \kappa+\iota & 0 \\ 0 & \iota & 0 & \kappa+\iota \end{pmatrix}\right)
\end{aligned}
\tag{2}
$$

with $\kappa > 0$ and $\iota > -\kappa/2$. Further, let (R_j, θ_j) be the polar representation of $\mathbf{U}_j = (X_j, Y_j)'$ $(j = 1,2)$. Then the joint angular distribution conditional on the resultant lengths $R_1 = r_1$ and $R_2 = r_2$ has a density of the form

$$
\begin{aligned}
&f(\theta_1, \theta_2 \mid R_1, R_2) \\
&\quad = \text{const.}\ \exp\big\{\kappa r_1 \cos\theta_1 + \iota r_1 r_2 \cos(\theta_1 - \theta_2) + \kappa r_2 \cos\theta_2\big\}\ .
\end{aligned}
\tag{3}
$$

Proof: Let

$$
\mathbf{z} = \begin{pmatrix} x_1 - 1 \\ y_1 \\ x_2 - 1 \\ y_2 \end{pmatrix} \quad \text{and} \quad \mathbf{\Sigma}^{-1} = \begin{pmatrix} \kappa+\iota & 0 & -\iota & 0 \\ 0 & \kappa+\iota & 0 & -\iota \\ -\iota & 0 & \kappa+\iota & 0 \\ 0 & -\iota & 0 & \kappa+\iota \end{pmatrix}.
$$

Then

$$
f(\mathbf{u}_1, \mathbf{u}_2) = \frac{\kappa^2+2\kappa\iota}{4\pi^2} \exp\big\{-\tfrac{1}{2}\, \mathbf{z}'\mathbf{\Sigma}^{-1}\mathbf{z}\big\}\ .
$$

Transforming (x_1, y_1, x_2, y_2) into polar coordinates $(r_1, \theta_1, r_2, \theta_2)$ according to $x_j = r_j \cos\theta_j$ and $y_j = r_j \sin\theta_j$ $(j = 1, 2)$ yields

$$f(r_1, \theta_1, r_2, \theta_2) = c\ \exp\{\kappa(r_1\cos\theta_1 + r_2\cos\theta_2) + \iota r_1 r_2 \cos(\theta_1 - \theta_2)\}$$

where $c = [4\pi^2]^{-1}\,(\kappa^2 + 2\kappa\iota) r_1 r_2 \exp\{-\kappa - \frac{1}{2}(\kappa + \iota)(r_1^2 + r_2^2)\}$. □

The density (3) thus maximises the entropy subject to constraints on $\mathcal{E}[\mathbf{U}_1]$, $\mathcal{E}[\mathbf{U}_2]$, and $\mathcal{E}[\mathbf{U}_1'\,\mathbf{U}_2]$. Also note that both variables, θ_1 conditional on θ_2 and θ_2 conditional on θ_1 follow von Mises distributions; however, that the marginal distributions of θ_1 and θ_2 have rather complicated forms. Using the Neumann addition formula (Mardia 1972, formula (3.4.68) and Watson 1952, p.358) it is easy to show that the integration constant in (3) is

$$\left[4\pi^2 \sum_{n=-\infty}^{\infty} I_{|n|}(\kappa_1 r_1)\, I_{|n|}(\iota r_1 r_2)\, I_{|n|}(\kappa_2 r_2)\right]^{-1} .$$

With respect to the von Mises process (θ_t) introduced in the last chapter we note that both θ_t conditional on $(\theta_{t-1}, \ldots, \theta_{t-p})$ and θ_{t+1} conditional on $(\theta_t, \ldots, \theta_{t-p+1})$ have von Mises distributions with densities of the form (6.9). Further, the joint density of θ_{t+1} and θ_t given $(\theta_{t-1}, \ldots, \theta_p)$ differs from (3), apart from the integration constant, only in the factor $[I_0(v_{t+1})]^{-1}$ as the concentration v_{t+1} in (6.9) involves θ_t. The presence of such a factor also explains why the marginal distributions of θ_1 and θ_2, given the joint density (3), will in general not be of von Mises type. Given that in case of a bivariate wrapped normal distribution the conditional and marginal distributions are again wrapped normal it may be interesting to note that for the wrapped normal approximation to (θ_t) in section 6.3 the factor $[I_0(v_{t+1})]^{-1}$ was ignored.

Given the importance of the von Mises distribution, especially in the context of maximum likelihood estimation and hypothesis testing, it would be desirable to obtain a bivariate angular distribution where both variables have von Mises marginals. Following a proposal by Johnson & Wehrly (1978) let $f_1(\theta_1)$ and $f_2(\theta_2)$ be predefined marginals and put

$$F_j(\theta_j) = \int_{-\pi}^{\theta_j} f_j(\eta)\, d\eta \quad , \qquad j = 1, 2\ ,$$

where $\theta_j \in (-\pi, \pi]$. For any density $g(\theta)$ on the circle,

$$f(\theta_1, \theta_2) = 2\pi g\big(2\pi F_1(\theta_1) - 2\pi F_2(\theta_2)\big)\, f_1(\theta_1)\, f_2(\theta_2)$$

then defines a bivariate angular distribution with the specified marginals $f_1(\theta_1)$ and $f_2(\theta_2)$.

Proof (Johnson & Wehrly 1978): Putting $\eta = 2\pi F_1(\theta_1)$ yields

$$\int f(\theta_1, \theta_2)\, d\theta_1 = \int g\big(\eta - 2\pi F_2(\theta_2)\big)\, f_2(\theta_2)\, d\eta = f_2(\theta_2)\, .$$

See also Wehrly & Johnson (1979). □

However, if $g(\theta)$ is the uniform density then θ_1 and θ_2 will be independent. Hence, any kind of dependence between θ_1 and θ_2 has to be built into $g(\theta)$ but this will in general lead to complicated and intractable densities of a rather artificial form. In fact, it can be shown that if both θ_1 and θ_2 have non-uniform von Mises marginals and a joint density as follows,

$$f(\theta_1,\theta_2) = \text{const. } \exp\left\{ \sum_{j=0}^{p}\sum_{k=0}^{q}(\alpha_{jk}\cos j\theta_1 \cos k\theta_2 + \gamma_{jk}\cos j\theta_1 \sin k\theta_2 \right.$$

$$\left. + \delta_{jk}\sin j\theta_1 \cos k\theta_2 + \beta_{jk}\sin j\theta_1 \sin k\theta_2)\right\} \tag{4}$$

where $p, q \in \mathcal{N}_0$ and $\alpha_{jk}, \beta_{jk}, \gamma_{jk}, \delta_{jk} \in \mathcal{R}$, then θ_1 and θ_2 must be independent. Since the proof of this result is fairly tedious it has been relegated to appendix 7.5.

This result suggests that two statistically dependent circular variables cannot be related in a simple way and at the same time have von Mises marginals. The problem discussed in chapter 6, that is using the von Mises distribution in a time series context, is mainly due to this difficulty. For instance, given the weighted average of independent and identically von Mises distributed unit vectors, the density of the resultant direction will in general take a quite complicated form (formula (6.4)).

As far as bivariate angular distributions are concerned it seems that one is confronted with a situation similar to the univariate case, where the von Mises distribution has some but not all of the desirable properties and the wrapped normal distribution has some of the others. Similarly, the bivariate wrapped normal distribution has conditional and marginal distributions of the same type, while a distribution with density (3) may be more appropriate for maximum likelihood estimation and hypothesis testing even though it does not involve von Mises marginals. However, like the wrapped normal approximation to the von Mises distribution it is easy to show that the two bivariate distributions approximate each other closely. It can therefore be assumed that the two distributions approximately share the same properties and that one may select whichever distribution is more convenient under the circumstances. Using the method described in section 6.3 the bivariate wrapped normal approximation to density (3) with $r_1 = r_2 = 1$ and $\kappa, \iota > 0$ yields the correlation

$$\rho = [1 + \ln A_1(\iota)/\ln A_1(\kappa)]^{-1} \tag{5}$$

for the two unwrapped normal variables. Consequently, ρ tends towards 0 or 1 if ι/κ tends towards 0 or ∞, respectively, as can be expected from the covariance matrix in (2). Note that (5) is identical with the formula for the parameter α derived in example 6.2. Finally, given the entropy characterization (1), density (3) suggests the sum of products as a measure of association. Indeed, this aspect provides the rationale for the correlation coefficient which will be introduced in section 7.4.

7.3 Review of Measures of Association

Given that angles can be conceived as unit vectors it is noted that the concept of angular dependence can be embedded in a concept of vector dependence without difficulty. For example, the spectra depicted in Figure 2.1 are based on a measure of vector rather than angular association. In this section we will summarise the most important measures of vector/angular association and briefly discuss them in view of the properties stated in section 7.1.

Let $\mathbf{X}$ and $\mathbf{Y}$ be random vectors in $\mathcal{R}^k$ $(k > 1)$ with covariance matrix $\boldsymbol{\Sigma}$ partitioned as

$$\boldsymbol{\Sigma} = \begin{pmatrix} \boldsymbol{\Sigma}_{XX} & \boldsymbol{\Sigma}_{XY} \\ \boldsymbol{\Sigma}_{YX} & \boldsymbol{\Sigma}_{YY} \end{pmatrix} \tag{6}$$

and suppose that both $\boldsymbol{\Sigma}_{XX}$ and $\boldsymbol{\Sigma}_{YY}$ are nonsingular and of rank k. Also, put $\boldsymbol{\Pi}_{XY} = \mathcal{E}[\mathbf{XY}']$, and $\boldsymbol{\Pi}_{YX}$, $\boldsymbol{\Pi}_{XX}$ and $\boldsymbol{\Pi}_{YY}$ accordingly.

(a) Watson & Beran (1967) propose a statistic which is the sample analogue of

$$\rho_{WB} = \mathcal{E}[\mathbf{X}'\mathbf{Y}] = \text{tr}\ \boldsymbol{\Pi}_{XY}\ . \tag{7}$$

In order to test whether there is significant first order serial correlation in a time series $(\mathbf{x}_t)$ Watson & Beran suggest the statistic

$$r_{WB} = \frac{1}{T}\sum_{t=1}^{T} \mathbf{x}_{t-1}'\mathbf{x}_t$$

with large values of r_{WB} indicating significance. The attempt to determine a critical value was based initially on an argument involving all permutations of the index set $\{0, \ldots, T\}$. However, Epp, Tukey & Watson (1971) managed to derive feasible expressions for $\mathcal{E}[r_{WB}]$ and var(r_{WB}) under the null hypothesis of independence. For this case it will be shown in section 7.4 that r_{WB} is asymptotically normal, not necessarily, however, with mean 0. Properties (iv) to (vi) are generally not satisfied.

(b) Stephens (1979) discusses the sample analogue of

$$\rho_{St} = \text{tr}\,(\boldsymbol{\Pi}_{YX}\boldsymbol{\Pi}_{XY})^{1/2}\ . \tag{8}$$

It is shown that

$$\rho_{St} = \max\{\ \mathcal{E}[\mathbf{X}'\mathbf{HY}] \mid \mathbf{H} \in \mathcal{R}^{k\times k} \text{ orthogonal}\ \}\ ,$$

meaning that ρ_{St} measures how well the random vectors $\mathbf{X}$ and $\mathbf{Y}$ can be matched using orthogonal transformations. Hence, ρ_{St} is invariant under orthogonal transformations, and if $\|\mathbf{X}\| = \|\mathbf{Y}\| = 1$ one obviously has $0 \leq \rho_{St} \leq 1$, with $\rho_{St} = 1$ if $\mathbf{X}$ and $\mathbf{Y}$ are dependent as defined in section 7.1. However, independence of $\mathbf{X}$ and $\mathbf{Y}$ does not imply $\rho_{St} = 0$. The null distribution depends on the marginals of $\mathbf{X}$ and $\mathbf{Y}$, and is generally difficult to derive.

(c) The correlation coefficient ρ_{DM} introduced independently by *Downs* (1974) and *Mardia* (1975b) is similar to ρ_{St} except that the variables are centered and standardised. Downs proposes the measure

$$\rho_{DM}^2 = \frac{[\mathrm{tr}\,(\boldsymbol{\Sigma}_{XY}\boldsymbol{\Sigma}_{YX})^{1/2}]^2}{\mathrm{tr}\,\boldsymbol{\Sigma}_{XX} \cdot \mathrm{tr}\,\boldsymbol{\Sigma}_{YY}}\,, \tag{9}$$

with the sign of ρ_{DM} being determined by the determinant of the orthogonal transformation that achieves the best agreement between $\mathbf{X}$ and $\mathbf{Y}$. Mardia, on the other hand, suggests the following sample correlation coefficient for the angular variables θ and η. Let (θ_t) and (η_t) $(t = 1, \ldots, T)$ be two random samples with mean direction 0, and mean resultant lengths r_θ and r_η, respectively. Then

$$r_M^2 = \frac{\max\{D_+, D_-\}}{(1 - r_\theta)\,(1 - r_\eta)}$$

where $T^2 D_\pm = \left[\sum_{t=1}^T \cos(\theta_t \pm \eta_t) - T r_\theta r_\eta\right]^2 + \left[\sum_{t=1}^T \sin(\theta_t \pm \eta_t)\right]^2$. The equivalence of r_M and the sample analogue r_{DM} of ρ_{DM} was shown by Downs & Eifler (discussion in Mardia 1975b). Like ρ_{St}, ρ_{DM}^2 is invariant under orthogonal transformations and further, $0 \leq \rho_{DM}^2 \leq 1$ with $\rho_{DM}^2 = 1$ if and only if $\mathbf{X} - \mathcal{E}[\mathbf{X}] = \lambda\,\mathbf{H}(\mathbf{Y} - \mathcal{E}[\mathbf{Y}])$, $\lambda \in \mathcal{R}$, $\mathbf{H} \in \mathcal{R}^{k \times k}$ orthogonal. Independence of $\mathbf{X}$ and $\mathbf{Y}$ obviously implies $\rho_{DM} = 0$.

(d) Johnson & Wehrly (1977) propose the *dominant canonical correlation* between $\mathbf{X}$ and $\mathbf{Y}$ as a measure of association, that is

$$\rho_{JW} = \lambda^{1/2} \tag{10}$$

where λ is the largest eigenvalue of $\boldsymbol{\Sigma}_{XX}^{-1}\boldsymbol{\Sigma}_{XY}\boldsymbol{\Sigma}_{YY}^{-1}\boldsymbol{\Sigma}_{YX}$. In the case of two angular variables θ and η this reduces to

$$\rho_{JW} = \max\left\{\mathrm{corr}\left(\cos(\theta - \alpha), \cos(\eta - \beta)\right) \mid \alpha, \beta \in (-\pi, \pi]\right\}. \tag{11}$$

Since the eigenvalues are invariant under orthogonal transformations of $\mathbf{X}$ and $\mathbf{Y}$, ρ_{JW} meets all the requirements for measures of association listed in section 7.1. In particular, if $\|\mathbf{X}\| = \|\mathbf{Y}\| = 1$, then $\rho_{JW} = 1$ if and only if θ and η are linearly dependent. Johnson & Wehrly also show that for large sample sizes the sample analogue r_{JW} tends towards a normal distribution with mean ρ_{JW}.

Example 6.1 (cont.): Consider the density (6.12), that is

$$f(\theta, \eta) = [4\pi^2 I_0(\kappa)]^{-1} \exp\{\kappa \cos(\theta - \eta)\} \quad \text{with } \kappa > 0$$

for a bivariate angular variable (θ, η). Note that this distribution maximises the entropy subject to $\mathcal{E}[\cos(\theta - \eta)] = A_1(\kappa)$ and $\mathcal{E}[\sin(\theta - \eta)] = 0$. Further,

$$\operatorname{corr}\left(\cos(\theta - \alpha), \cos(\eta - \beta)\right) = A_1(\kappa)\cos(\alpha - \beta)\,, \quad \alpha, \beta \in (-\pi, \pi]\,,$$

which attains a maximum if $\alpha - \beta = 0 \pmod{2\pi}$. Hence, $\rho_{JW} = A_1(\kappa)$. □

Although in this example one of the parameters α and β can be chosen freely, generally both rotations are necessary to maximise the correlation between $\cos(\theta - \alpha)$ and $\cos(\eta - \beta)$. This is illustrated by the following example presented in Johnson & Wehrly (1977).

Example 7.1: Suppose that (θ, η) follows a wrapped bivariate normal distribution with mean (μ_1, μ_2) and covariance matrix $(\sigma_{jk})_{jk=1,2}$, and represent θ and η as unit vectors $\mathbf{X}$ and $\mathbf{Y}$, respectively. The eigenvalues of $\mathbf{\Sigma}_{XX}^{-1}\mathbf{\Sigma}_{XY}\mathbf{\Sigma}_{YY}^{-1}\mathbf{\Sigma}_{YX}$ are then found to be

$$\begin{aligned}\lambda_1 &= [\cosh\sigma_{12} - 1] \;/\; [2\,\sinh(\sigma_{11}^2/2)\,\sinh(\sigma_{22}^2/2)] \quad \text{and}\\ \lambda_2 &= |\sinh\sigma_{12}| \;/\; [\sinh\sigma_{11}^2\;\sinh\sigma_{22}^2]^{1/2}\,.\end{aligned}$$

Referring to formula (11) the canonical correlation λ_1 corresponds to a rotation of θ by μ_1 and of η by μ_2, while λ_2 is associated with a rotation of θ by $\mu_1 \pm \pi/2$ and of η by $\mu_2 \pm \pi/2$ depending on whether σ_{12} is positive or negative. Johnson & Wehrly refer to a numerical study which suggests that $\lambda_2 > \lambda_1$ if σ_{11} and σ_{22} are relatively close to each other, and $\lambda_1 > \lambda_2$ otherwise. □

The last example shows that the two rotations of θ and η, necessary to determine ρ_{JW}, are dependent on the parameters of the bivariate distribution.

(e) Jupp & Mardia (1980) propose

$$\rho_{JM}^2 = \operatorname{tr}\left(\mathbf{\Sigma}_{XX}^{-1}\mathbf{\Sigma}_{XY}\mathbf{\Sigma}_{YY}^{-1}\mathbf{\Sigma}_{YX}\right) \tag{12}$$

as a measure of association. Equivalently, that is

$$\rho_{JM}^2 = \sum_{j=1}^{k} \lambda_j^2$$

where $\lambda_1, \ldots, \lambda_k$ are the canonical correlations. In particular, $\rho_{JM}^2 = 0$ if and only if $\mathbf{\Sigma}_{XY} = \mathbf{0}$. Further, $\rho_{JM}^2 \le k$, with equality if and only if $\mathbf{X}$ and $\mathbf{Y}$ are related by a linear vector equation with matrix coefficients. The following property distinguishes this measure from the correlation coefficients considered so far. Under the null hypothesis, $Tr_{JM}^2 \xrightarrow{\text{d}} \chi_4^2$ where T is the sample size, r_{JM}^2 is the sample analogue of ρ_{JM}^2 and χ_4^2 is the χ^2-distribution with 4 degrees of freedom. Hence, r_{JM}^2 is asymptotically robust in the sense that its asymptotic null distribution does not depend on the distribution parameters. In the case of a bivariate angular variable ρ_{JM}^2 is invariant under orthogonal transformations, and if the marginals are also symmetric then ρ_{JM}^2 coincides with the measure suggested by Mardia & Puri (1978).

(f) Fisher & Lee (1983) define a correlation coefficient for two angular variates that is motivated by the extent to which one variable can be predicted from the other using orthogonal transformations. That is

$$\rho_{FL} = \frac{\det \mathbf{\Pi}_{XY}}{[\det \mathbf{\Pi}_{XX} \cdot \det \mathbf{\Pi}_{YY}]^{1/2}} . \tag{13}$$

Suppose that (θ_1, η_1) and (θ_2, η_2) are independently distributed as (θ, η). Then ρ_{FL} can also be written as

$$\rho_{FL} = \frac{\mathcal{E}[\sin(\theta_1 - \theta_2)\ \sin(\eta_1 - \eta_2)]}{\{\mathcal{E}[\sin^2(\theta_1 - \theta_2)]\ \mathcal{E}[\sin^2(\eta_1 - \eta_2)]\}^{1/2}}$$

analogous to the alternative form of the Pearson correlation coefficient for two linear variables U and V:

$$\rho(U, V) = \frac{\mathcal{E}[(U_1 - U_2)\,(V_1 - V_2)]}{\{\mathcal{E}[(U_1 - U_2)^2]\ \mathcal{E}[(V_1 - V_2)^2]\}^{1/2}} .$$

Hence, for highly concentrated distributions of θ and η, ρ_{FL} becomes similar to the ordinary Pearson coefficient. Given that in this case the circle virtually coincides with the real line, this is quite a desirable property. Like ρ_{WB} and ρ_{DM}, but in contrast with most other proposals, ρ_{FL} also is a signed measure taking values between -1 and 1, with $\rho_{FL} = \pm 1$ if and only if θ and η are linearly dependent, that is $\mathbf{H}(\theta) = \eta$ for some $\mathbf{H} \in \mathcal{R}^{2\times 2}$ with $\det \mathbf{H} = \pm 1$. If θ and η are independent then $\rho_{FL} = 0$. Property (vi) is satisfied by ρ_{FL}^2, but not by ρ_{FL} due to the distinction between positive and negative correlation.

(g) The coefficient suggested by *Rivest* (1982) for circular variates is based on the smaller of the two eigenvalues of $\mathbf{\Pi}_{XY}$. It is derived in a similar way to (13), but does not share all the properties of ρ_{FL}.

Other measures of angular association include a U-statistic defined by Fisher & Lee (1982) analogous to Kendall's tau, and a rank correlation coefficient proposed by Mardia (1975b) which has all the properties listed in section 7.1. A slight modification of this coefficient yields an analogue of Spearman's rho (Fisher & Lee 1982). A test for independence of angular data is constructed by Rothman (1971) using the empirical cumulative distribution function. Based on the number of observations falling into half-circles, Puri & Rao (1977) develop a test for independence of axial data. Asymptotic null distributions have been derived for all the test statistics mentioned in this paragraph.

As for a summary assume $\mathbf{\Pi}_{XY} = \mathbf{\Pi}_{YX}$ and $\mathbf{\Sigma}_{XY} = \mathbf{\Sigma}_{YX}$, and let the eigenvalues of $\mathbf{\Pi}_{XY}$, $\mathbf{\Sigma}_{XY}$ and $\mathbf{\Sigma}_{XX}^{-1}\mathbf{\Sigma}_{XY}\mathbf{\Sigma}_{YY}^{-1}\mathbf{\Sigma}_{YX}$ be denoted by λ_j^{Π}, λ_j^{Σ} and λ_j^2 $(j = 1, 2)$, respectively. Table 7.1 shows the functions, upon which the various measures of association are based. On the basis of these functions the correlation coefficients described in this section can be broadly classified as follows:

Table 7.1. Summary of correlation coefficients.

Category	Reference	Symbol	Equation number	Basic function
I	Watson & Beran (1967)	ρ_{WB}	(7)	$\lambda_1^\Pi + \lambda_2^\Pi$
	Stephens (1979)	ρ_{St}	(8)	$\lambda_1^\Pi + \lambda_2^\Pi$
	Downs (1974) and Mardia (1975b)	ρ_{DM}	(9)	$\lambda_1^\Sigma + \lambda_2^\Sigma$
II	Johnson & Wehrly (1977)	ρ_{JW}	(10)	$\max\{\lambda_1, \lambda_2\}$
	Jupp & Mardia (1980)	ρ_{JM}	(12)	$(\lambda_1^2 + \lambda_2^2)^{1/2}$
	Fisher & Lee (1983)	ρ_{FL}	(13)	$\lambda_1^\Pi \cdot \lambda_2^\Pi$
	Rivest (1982)			$\min\{\lambda_1^\Pi, \lambda_2^\Pi\}$

I. Coefficients like ρ_{WB}, ρ_{St} and ρ_{DM} that are based on *linear* functions of the eigenvalues of $\mathbf{\Pi}_{XY}$ or $\mathbf{\Sigma}_{XY}$.
II. Coefficients like ρ_{JW}, ρ_{JM} and ρ_{FL} that are based on *non-linear* functions of the canonical correlations or the eigenvalues of $\mathbf{\Pi}_{XY}$.
III. *Rank* correlation coefficients as presented in Mardia (1975b) and Fisher & Lee (1982).

Although the correlation coefficients reviewed in this section generally satisfy the requirements for measures of angular association, they are often too cumbersome to be practical in a time series context where the autocovariance function may have to be evaluated over a wide range of time lags. Part of this complexity is necessary to ensure that property (vi) is satisfied, that is $\delta\big(\mathbf{H}_1(\theta), \mathbf{H}_2(\eta)\big) = \delta(\theta, \eta)$ for orthogonal matrices $\mathbf{H}_1, \mathbf{H}_2 \in \mathcal{R}^{2\times 2}$. In a times series context, however, this property can be relaxed as it does not make sense to transform variables, which are part of the same times series, independently. Linearity of the correlation coefficient, on the other hand, would be a desirable property in such a situation.

In the following section an alternative measure of association is proposed, which is linear and easy to calculate. A numerical comparison with other correlation coefficients will be presented in chapter 8.

7.4 A Proposal for Vector Valued Time Series

Let θ and η be two random angular variables. Given the von Mises process (6.9) introduced in chapter 6 and the bivariate density (3) introduced in conjunction with the maximum entropy characterization (1), it seems natural to base the measure of association on $\mathcal{E}[\cos(\theta-\eta)]$ and to propose the following correlation coefficient

$$\text{(14)} \qquad \sigma_P(\theta,\eta) = \max\{\ \mathcal{E}[\cos(\mathbf{H}(\theta)-\eta)] - \mathcal{E}[\cos \mathbf{H}(\theta)]\ \mathcal{E}[\cos\eta] - \mathcal{E}[\sin \mathbf{H}(\theta)]\ \mathcal{E}[\sin\eta]\ \mid\ \mathbf{H}\in\mathcal{R}^{2\times 2} \text{ orthogonal}\ \}\,.$$

The matrix $\mathbf{H}$ has been included in the expression $\mathcal{E}[\cos(\theta-\eta)]$ so that the correlation coefficent is invariant under independent orthogonal transformations and the two angular variables are matched as closely as possible. Further, the introduction of the term $-\mathcal{E}[\cos \mathbf{H}(\theta)]\ \mathcal{E}[\cos\eta] - \mathcal{E}[\sin \mathbf{H}(\theta)]\ \mathcal{E}[\sin\eta]$ ensures that $\sigma_P(\theta,\eta)=0$ if θ and η are independent. Note that in vector notation the correlation coefficient can also be written as

$$\sigma_P(\theta,\eta) = \max\{\ \mathcal{E}[\ \mathbf{U}_\theta'\mathbf{H}\mathbf{U}_\eta\ ] - \mathcal{E}[\mathbf{U}_\theta]'\ \mathbf{H}\ \mathcal{E}[\mathbf{U}_\eta]\mid\ \mathbf{H}\in\mathcal{R}^{2\times 2} \text{ orthogonal}\ \}\,.$$

The standardised form

$$\text{(15)} \qquad \rho_P(\theta,\eta) = \frac{\sigma_P(\theta,\eta)}{[\sigma_P(\theta,\theta)\cdot\sigma_P(\eta,\eta)]^{1/2}}$$

essentially coincides with ρ_{DM} as defined in (9), the only difference being that ρ_{DM} is a signed measure while ρ_P will always be positive. Obviously, ρ_P satisfies all the properties (i) to (vi) listed in section 7.1 with $0\le\rho_P(\theta,\eta)\le 1$.

Property (vi) implies that θ can be transformed independently of η. However, consider the case of a *stationary* angular time series (θ_t) with autocovariance function

$$c_P(\tau) = \sigma_P(\theta_t,\theta_{t+\tau}) \qquad ,\tau\in\mathcal{N}_0\,.$$

Suppose that the maxima implicit in $c_P(\tau)$ are attained for the orthogonal transformation

$$\mathbf{H}^\tau = \begin{pmatrix} \cos(\alpha\tau) & \sin(\alpha\tau) \\ -\sin(\alpha\tau) & \cos(\alpha\tau)\end{pmatrix}$$

with $\alpha\neq 0 \pmod{2\pi}$. The time series (θ_t) could then be decomposed into a constant rotation (αt) and a term $(\theta_t-\alpha t)$ which fluctuates about a constant direction. Hence, if the time series (θ_t) was detrended yielding the series $(\theta_t-\alpha t)$, then it could be assumed that all the maxima are attained for $\mathbf{H}=\mathbf{I}_2$. In this case (14) reduces to

$$\text{(16)} \qquad \sigma_Q(\theta,\eta) = \mathcal{E}[\cos(\theta-\eta)] - \mathcal{E}[\cos\theta]\ \mathcal{E}[\cos\eta] - \mathcal{E}[\sin\theta]\ \mathcal{E}[\sin\eta]\,.$$

The corresponding modified correlation coefficient ρ_Q no longer satisfies properties (v) and (vi). However, in addition to (i) to (iv) the following still holds:

(v′) Dependence :

$$\text{if } \theta = \eta \text{ then } \rho_Q(\theta, \eta) = 1.$$

(vi′) Invariance :

$$\rho_Q(\mathbf{H}(\theta), \mathbf{H}(\eta)) = \rho_Q(\theta, \eta) \quad \text{for all } \mathbf{H} \in \mathcal{R}^{2\times 2} \text{ orthogonal.}$$

The generality of (v) and (vi), on the other hand, seems to be dispensable in a time series context since it is usually only simultaneous transformations of all the variables that are meaningful.

Even though σ_P and σ_Q have been introduced to measure the extent of angular association, they can be extended without difficulty to variables in $\mathcal{R}^k$ ($k > 1$). Suppose that $\mathbf{X}$ and $\mathbf{Y}$ are random vectors in $\mathcal{R}^k$ with covariance matrix (6) and non-singular submatrices $\mathbf{\Sigma}_{XX}$ and $\mathbf{\Sigma}_{YY}$. Then (14) takes the more general form

$$\sigma_P(\mathbf{X}, \mathbf{Y}) = \max\{\mathrm{tr}\,(\mathbf{H}\mathbf{\Sigma}_{XY}) \mid \mathbf{H} \in \mathcal{R}^{k\times k} \quad \text{orthogonal}\}\,. \tag{17}$$

For non-singular matrices $\mathbf{\Sigma}_{XY}$ it can be shown that the maximum in (17) is attained for $\mathbf{H}_0 = (\mathbf{\Sigma}_{YX}\mathbf{\Sigma}_{XY})^{1/2}\,\mathbf{\Sigma}_{XY}^{-1}$ (Mackenzie 1957).

Hence,

$$\sigma_P(\mathbf{X}, \mathbf{Y}) = \mathrm{tr}\,(\mathbf{\Sigma}_{YX}\mathbf{\Sigma}_{XY})^{1/2}\,. \tag{18}$$

Further, if $\mathbf{X}$ and $\mathbf{Y}$ are variables in a vector valued time series then $\mathbf{\Sigma}_{XY}$ will often be symmetric. Note that in this case $\mathbf{H}_0 = \mathbf{I}_k$, the identity matrix of order k, and that (18) reduces to

$$\begin{aligned} \sigma_Q(\mathbf{X}, \mathbf{Y}) &= \mathrm{tr}\ \mathbf{\Sigma}_{XY} \\ &= \mathcal{E}[\mathbf{X}'\mathbf{Y}] - \mathcal{E}[\mathbf{X}']\,\mathcal{E}[\mathbf{Y}]\,. \end{aligned} \tag{19}$$

In fact, it will subsequently be assumed that $\mathbf{\Sigma}_{XY} = \mathbf{\Sigma}_{YX}$. The correlation coefficient σ_Q can then be written in the alternative form

$$\sigma_Q(\mathbf{X}, \mathbf{Y}) = \sum_{j=1}^{k} \lambda_j^{\Sigma}$$

where $\lambda_1^{\Sigma}, \ldots, \lambda_k^{\Sigma}$ are the eigenvalues of $\mathbf{\Sigma}_{XY}$. That is, σ_Q is based on the same function as ρ_{DM} (cf. Table 7.1). Further, for random vectors $\mathbf{X}$, $\mathbf{Y}$ and $\mathbf{Z}$ in $\mathcal{R}^k$, and $\alpha, \beta \in \mathcal{R}$ it follows immediately from definition (19) that

$$\sigma_Q(\alpha\mathbf{X} + \beta\mathbf{Y}, \mathbf{Z}) = \alpha\,\sigma_Q(\mathbf{X}, \mathbf{Z}) + \beta\,\sigma_Q(\mathbf{Y}, \mathbf{Z})\,. \tag{20}$$

The linearity of σ_Q implies that standard spectral techniques can be based on this measure of association as will be shown at the end of this section. This measure will therefore facilitate the analysis of the *covariance structure* of vector valued time series. Without further restrictions none of the other measures of association satisfies equation (20), except for the statistic introduced by Watson & Beran (1967). Note, however, that ρ_{St} and ρ_{DM} reduce to ρ_{WB} and ρ_Q, respectively, if the cross covariance matrices are symmetric.

It is now shown that one can devise a test of independence on the basis of σ_Q. Suppose that $(\mathbf{X}_t, \mathbf{Y}_t)$ $(t = 1, \ldots, T)$ are independently distributed as $(\mathbf{X}, \mathbf{Y})$. Put $\mathbf{S}_{XX} = \overline{\mathbf{XX}'} - \overline{\mathbf{X}}\,\overline{\mathbf{X}}'$, $\mathbf{S}_{YY} = \overline{\mathbf{YY}'} - \overline{\mathbf{Y}}\,\overline{\mathbf{Y}}'$ and $\mathbf{S}_{XY} = \overline{\mathbf{XY}'} - \overline{\mathbf{X}}\,\overline{\mathbf{Y}}'$ where the bar denotes the ordinary mean, that is $\overline{\mathbf{XX}'} = T^{-1}\sum_{t=1}^{T}\mathbf{X}_t\mathbf{X}_t'$, *etc.* Further, let

$$z_Q = [\operatorname{tr}\mathbf{S}_{XY}] \;/\; [\operatorname{tr}\mathbf{S}_{XX}\mathbf{S}_{YY}]^{1/2} .$$

If $\mathbf{X}$ and $\mathbf{Y}$ are independent then it follows that

$$\sqrt{T}\, z_Q \xrightarrow{\mathrm{d}} N(0,1) . \tag{21}$$

Hence, under the null hypothesis z_Q is asymptotically independent of the marginal distributions of $\mathbf{X}$ and $\mathbf{Y}$, and thus asymptotically robust against concentration in the case of a bivariate circular variable.

Proof: Without loss of generality assume $\mathcal{E}[\mathbf{X}] = \mathcal{E}[\mathbf{Y}] = \mathbf{0}$. The weak law of large numbers applied to $\overline{\mathbf{Y}}$ ensures $\overline{\mathbf{Y}} \xrightarrow{\mathrm{pr}} \mathbf{0}$, and the central limit theorem applied to $\overline{\mathbf{X}}$ yields $\sqrt{T}\,\overline{\mathbf{X}} \xrightarrow{\mathrm{d}} \mathbf{0}$. From Slutsky's Theorem (Serfling 1980, p.19) it then follows that $\sqrt{T}\operatorname{tr}\mathbf{S}_{XY}$ is asymptotically distributed as $\sqrt{T}\operatorname{tr}\overline{\mathbf{XY}'}$, and hence

$$\sqrt{T}\operatorname{tr}\mathbf{S}_{XY} \xrightarrow{\mathrm{d}} N(0, \operatorname{tr}\boldsymbol{\Sigma}_{XX}\boldsymbol{\Sigma}_{YY}) .$$

Consequently,

$$[\sqrt{T}\operatorname{tr}\mathbf{S}_{XY}] \;/\; [\operatorname{tr}\boldsymbol{\Sigma}_{XX}\boldsymbol{\Sigma}_{YY}]^{1/2} \xrightarrow{\mathrm{d}} N(0,1) .$$

Using $\mathbf{S}_{XX} \xrightarrow{\mathrm{pr}} \boldsymbol{\Sigma}_{XX}$ and $\mathbf{S}_{YY} \xrightarrow{\mathrm{pr}} \boldsymbol{\Sigma}_{YY}$ finally gives

$$[\sqrt{T}\operatorname{tr}\mathbf{S}_{XY}] \;/\; [\operatorname{tr}\mathbf{S}_{XX}\mathbf{S}_{YY}]^{1/2} \xrightarrow{\mathrm{d}} N(0,1) .$$

□

Which standardisation of σ_Q should be preferred therefore depends on the objective. Whereas ρ_Q seems to be a better choice for a correlation coefficient, z_Q is more appropriate for hypothesis testing. Clearly, $z_Q^2 \le k$. The parameter σ_Q will now be related to the correlation coefficients reviewed in section 7.3. The main difference between ρ_{WB} and σ_Q as well as between ρ_{St} and σ_P is that ρ_{WB} and ρ_{St} are based on raw moments while σ_P and σ_Q are based on central moments. Further, as ρ_{St} reduces to ρ_{WB} if $\boldsymbol{\Pi}_{XY} = \boldsymbol{\Pi}_{YX}$, so does σ_P reduce to σ_Q if $\boldsymbol{\Sigma}_{XY} = \boldsymbol{\Sigma}_{YX}$. Note that both ρ_{WB} and σ_Q are proposed in the context of angular time series where the assumption of symmetry is quite reasonable. The relationship between σ_Q and the correlation coefficients in category II is more complicated since these coefficients are non-linear functions of certain eigenvalues. However, let us briefly comment on a comparison of σ_Q with ρ_{JW}.

Example 6.1 (cont.): Suppose that the bivariate circular variable (θ, η) has the following density

$$f(\theta,\eta) = [4\pi^2 I_0(\iota)]^{-1}\exp\{\iota\cos(\theta-\eta)\} , \qquad \iota \in \mathcal{R}_0^+ .$$

Writing $\mathbf{X} = (\cos\theta, \sin\theta)'$ and $\mathbf{Y} = (\cos\eta, \sin\eta)'$ the covariance matrix (6) becomes

$$\boldsymbol{\Sigma} = \tfrac{1}{2} \begin{pmatrix} \mathbf{I}_2 & A_1(\iota)\,\mathbf{I}_2 \\ A_1(\iota)\,\mathbf{I}_2 & \mathbf{I}_2 \end{pmatrix}$$

which implies $\rho_{JW}(\theta,\eta) = A_1(\iota)$ and $\rho_{JM} = \sqrt{2}\,A_1(\iota)$. On the other hand, $\rho_Q(\theta,\eta) = \sigma_Q(\theta,\eta) = \operatorname{tr}\boldsymbol{\Sigma}_{XY} = A_1(\iota)$. Hence, $\rho_{JW}(\theta,\eta) = \sigma_Q(\theta,\eta)$. □

However, this result does not hold for arbitrary values of κ and ι in (3). For example, assume that (θ,η) follows a distribution with density

$$f(\theta,\eta) = [4\pi^2 I_0(\iota) I_0(\kappa)]^{-1} \exp\{\kappa\cos\theta + \iota\cos(\theta-\eta)\} \quad , \ \kappa,\iota \in \mathcal{R}_0^+ \,.$$

Then $\sigma_Q(\theta,\eta) \leq \rho_{JW}(\theta,\eta)$, with equality if and only if $\kappa = 0$ or $\iota = 0$.

Proof: Define

$$\begin{aligned} a &= A_2(\kappa)/2 \,, \\ a_\iota &= A_2(\kappa) A_2(\iota)/2 \,, \\ b &= A_1^2(\kappa) \,, \\ \text{and} \quad b_\iota &= A_1^2(\kappa) A_1^2(\iota) \,. \end{aligned}$$

Then the covariance matrix (6) reduces to

$$\begin{aligned} \boldsymbol{\Sigma} &= (\sigma_{jk}) \\ &= \begin{pmatrix} \frac{1}{2} + a_\iota - b_\iota & 0 & A_1(\iota)(\frac{1}{2} + a - b) & 0 \\ 0 & \frac{1}{2} - a_\iota & 0 & A_1(\iota)(\frac{1}{2} - a) \\ A_1(\iota)(\frac{1}{2} + a - b) & 0 & \frac{1}{2} + a - b & 0 \\ 0 & A_1(\iota)(\frac{1}{2} - a) & 0 & \frac{1}{2} - a \end{pmatrix} \end{aligned}$$

and the eigenvalues λ_1^2 and λ_2^2 of $\boldsymbol{\Sigma}_{XX}^{-1}\boldsymbol{\Sigma}_{XY}\boldsymbol{\Sigma}_{YY}^{-1}\boldsymbol{\Sigma}_{YX}$ are obtained as $\lambda_1^2 = \sigma_{13}^2/[\sigma_{11}\sigma_{33}]$ and $\lambda_2^2 = \sigma_{24}^2/[\sigma_{22}\sigma_{44}]$. Hence,

$$\begin{aligned} \sigma_Q(\theta,\eta) &= \sigma_{13} + \sigma_{24} \\ &= [\lambda_1^2\sigma_{11}\sigma_{33}]^{1/2} + [\lambda_2^2\sigma_{22}\sigma_{44}]^{1/2} \\ &\leq \max(\lambda_1,\lambda_2)\, \{[\,(\tfrac{1}{2} + a_\iota)\,(\tfrac{1}{2} + a)\,]^{1/2} + [\,(\tfrac{1}{2} - a_\iota)\,(\tfrac{1}{2} - a)\,]^{1/2}\} \,. \end{aligned}$$

Now, for $x, y \in [-\frac{1}{2}, \frac{1}{2}] \subseteq \mathcal{R}$ it is easy to show that

$$[(\tfrac{1}{2} + x)\,(\tfrac{1}{2} + y)]^{1/2} + [(\tfrac{1}{2} - x)\,(\tfrac{1}{2} - y)]^{1/2} \leq 1$$

with equality if and only if $x = y$. Using this result one thus has

$$\sigma_Q(\theta,\eta) \leq \rho_{JW}(\theta,\eta)$$

with equality if and only if $\kappa = 0$ or $\iota = 0$. □

In chapter 8 the different measures of association will be compared numerically. Their performance will first be studied for pairs (θ,η) of directional data which are independent and identically distributed, and then for an angular time series (θ_t).

This section will now be completed with a brief comment on the role of σ_Q in a time series context. Let $(\mathbf{X}_t)_{t\in\mathcal{Z}}$ be a stochastic process of vector variables in $\mathcal{R}^k$ $(k \geq 1)$ and define

$$c(t, t+\tau) = \sigma_Q(\mathbf{X}_t, \mathbf{X}_{t+\tau}) \qquad t, \tau \in \mathcal{Z} . \tag{22}$$

The process $(\mathbf{X}_t)$ is then called *weakly stationary* if and only if $c(0,\tau) = c(t, t+\tau)$ for all $t, \tau \in \mathcal{Z}$. In this case we will simply write $c(\tau)$ and call $c(\cdot)$ the *autocovariance function* of the vector process $(\mathbf{X}_t)$.

Given a time series $(\mathbf{x}_1, \ldots, \mathbf{x}_T)$ in $\mathcal{R}^k$ the *empirical* autocovariance function is defined accordingly as

$$c(\tau) = \frac{1}{T} \sum_{t=\max\{-\tau+1,1\}}^{\min\{T-\tau,T\}} \mathbf{x}_t' \, \mathbf{x}_{t+\tau} - \frac{T-|\tau|}{T^3} \left\| \sum_{t=1}^{T} \mathbf{x}_t \right\|^2$$

$$\tau = -T+1, \ldots, T-1 .$$

Note that the two terms are estimators of $\mathcal{E}[\mathbf{X}_t' \, \mathbf{X}_{t+\tau}]$ and $\mathcal{E}[\mathbf{X}_t'] \cdot \mathcal{E}[\mathbf{X}_{t+\tau}]$ respectively. That is, $c(\tau)$ is an estimator of $\sigma_Q(\mathbf{X}_t, \mathbf{X}_{t+\tau})$ as given in equation (19). Let

$$\tilde{\mathbf{f}}(\omega) = \sum_{t=-T+1}^{T-1} \mathbf{f}(t) \exp\{i\omega t\} ,$$

$$\omega = -2\pi(T-1)/T, \ldots, 2\pi(T-1)/T ,$$

be the finite Fourier transform of a vector valued function $\mathbf{f}(t)$ $(0 \leq t \leq T-1)$. Note that $\tilde{\mathbf{f}}(-\omega) = \overline{\tilde{\mathbf{f}}(\omega)}$ and hence, that the *information* contained in $\tilde{\mathbf{f}}(\omega)$ is equivalent to that contained in $\mathbf{f}(t)$. For any vector series $(\mathbf{x}_t)_{t=1,\ldots,T}$ in $\mathcal{R}^k$ it then follows that

$$\tilde{c}(\omega) = \begin{cases} \|\tilde{\mathbf{x}}(\omega)\|^2/T & \text{if } \omega \neq 0 \\ 0 & \text{otherwise} . \end{cases} \tag{23}$$

Proof: Define

$$g(\tau) = \frac{1}{T} \sum_{t=\max\{-\tau+1,1\}}^{\min\{T-\tau,T\}} \mathbf{x}_t' \, \mathbf{x}_{t+\tau}$$

and take the Fourier transform

$$\begin{aligned} \tilde{g}(\omega) &= \frac{1}{T} \sum_{\tau=-T+1}^{T-1} \sum_{t=\max\{-\tau+1,1\}}^{\min\{T-\tau,T\}} \mathbf{x}_t' \, \mathbf{x}_{t+\tau} \exp\{i\omega\tau\} \\ &= \frac{1}{T} \sum_{t=1}^{T} \sum_{s=1}^{T} \mathbf{x}_s' \, \mathbf{x}_t \exp\{i\omega(s-t)\} \\ &= \frac{1}{T} \|\tilde{\mathbf{x}}(\omega)\|^2 . \end{aligned}$$

Further, let $h(\tau) = (T - |\tau|)\, T^{-3} \, \left\| \sum_{s=1}^{T} \mathbf{x}_s \right\|^2$. Then

$$\tilde{h}(\omega) = \sum_{\tau=-T+1}^{T-1} \frac{1}{T^3} \exp\{i\omega\tau\} \sum_{t=\max\{-\tau+1,1\}}^{\min\{T-\tau,T\}} \left\| \sum_{s=1}^{T} \mathbf{x}_s \right\|^2$$

$$= \begin{cases} \left\| \sum \mathbf{x}_s \right\|^2 / T & \text{if } \omega = 0 \\ 0 & \text{otherwise .} \end{cases}$$

Since $\tilde{c}(\omega) = \tilde{g}(\omega) - \tilde{h}(\omega)$ equation (23) follows. □

As for univariate time series in $\mathcal{R}$, $\tilde{c}(\cdot)$ will be called the *spectrum* of the vector series $(\mathbf{x}_t)$. Formula (23) provides a convenient method of calculating the spectrum for a vector series in $\mathcal{R}^k$ with σ_Q as the measure of association. Based on an autocovariance function the spectrum of a vector series can be interpreted in the same way as the spectra for univariate time series in $\mathcal{R}$. As shown in Figure 2.1, for example, it helps to identify dominant cycles in vector valued time series.

7.5 Appendix: Non-von Mises Marginals

It will now be shown that two circular variables θ and η with a joint density of the form (4) do not have non-uniform von Mises marginal distributions, unless θ and η are independent. That is, given the function

$$h(\theta, \eta) = \sum_{j=0}^{p} \sum_{k=0}^{q} (\cos j\theta, \sin j\theta) \begin{pmatrix} \alpha_{jk} & \gamma_{jk} \\ \delta_{jk} & \beta_{jk} \end{pmatrix} \begin{pmatrix} \cos k\eta \\ \sin k\eta \end{pmatrix}$$

with at least one non-zero cross term linking θ and η, one of the following constraints must be violated:

$$\text{(24a)} \qquad \int_{-\pi}^{\pi} \exp\{h(\theta, \eta)\}\, d\theta = \text{const.} \exp\{\kappa_\eta \cos(\eta - \mu_\eta)\} \quad \text{with } \kappa_\eta \neq 0$$

$$\text{(24b)} \qquad \int_{-\pi}^{\pi} \exp\{h(\theta, \eta)\}\, d\eta = \text{const.} \exp\{\kappa_\theta \cos(\theta - \mu_\theta)\} \quad \text{with } \kappa_\theta \neq 0\,.$$

First, let us prove the following result.

Lemma 7.1: Let $p \in \mathcal{N}$ and $\alpha_j, \beta_j \in \mathcal{R}$ $(j = 0, \ldots, p)$. Then, if $p > 1$,

$$I = \int_{-\pi}^{\pi} \exp\Big\{\sum_{j=0}^{p} \alpha_j \cos j\theta + \beta_j \sin j\theta\Big\}\, d\theta$$

$$(25) \qquad = 2\pi \exp\{\alpha_0\} \sum_{n_2,\ldots,n_p=-\infty}^{\infty} I_{|s|}\Big(\sqrt{\alpha_1^2+\beta_1^2}\,\Big) \times \prod_{j=2}^{p} \cos n_j(j\psi_1 - \psi_j)\, I_{|n_j|}\Big(\sqrt{\alpha_j^2+\beta_j^2}\,\Big)$$

where $s = \sum_{h=2}^{p} h\, n_h$ and $\psi_j = \arctan(\beta_j/\alpha_j)$ $(j = 1, \ldots, p)$. For $p = 1$ the right-hand side of (25) reduces to $2\pi \exp\{\alpha_0\}\, I_0(\sqrt{\alpha_1^2+\beta_1^2}\,)$.

Proof: A Taylor series expansion of the exponential function yields

$$\begin{aligned} I &= \int_{-\pi}^{\pi} \exp\Big\{\sum_{j=0}^{p}(\alpha_j \cos j\theta + \beta_j \sin j\theta)\Big\}\, d\theta \\ &= \sum_{k,l=0}^{\infty} \frac{1}{k!\, l!} \int_{-\pi}^{\pi} \Big(\sum_{j=0}^{p} \alpha_j \cos j\theta\Big)^k \Big(\sum_{j=0}^{p} \beta_j \sin j\theta\Big)^l\, d\theta \\ &= \sum_{k,l=0}^{\infty} \frac{1}{k!\, l!} \int_{-\pi}^{\pi} \Big(\sum_{j=0}^{p} \tfrac{1}{2}\alpha_j(\exp\{-ij\theta\} + \exp\{ij\theta\})\Big)^k \\ &\qquad \times \Big(\sum_{j=0}^{p} \tfrac{1}{2} i\beta_j(\exp\{-ij\theta\} - \exp\{ij\theta\})\Big)^l\, d\theta\,. \end{aligned}$$

Put $a_0 = \alpha_0$, $a_{-j} = a_j = \frac{1}{2}\alpha_j$, $b_0 = 0$, $b_{-j} = \frac{1}{2}\beta_j$ and $b_j = -\frac{1}{2}\beta_j$ $(j = 1, \ldots, p)$. Then

$$\begin{aligned} I &= \sum_{k,l=0}^{\infty} \frac{1}{k!\, l!} \int_{-\pi}^{\pi} \Big(\sum_{j=-p}^{p} a_j \exp\{ij\theta\}\Big)^k \Big(\sum_{j=-p}^{p} ib_j \exp\{ij\theta\}\Big)^l\, d\theta \\ &= \sum_{k,l=0}^{\infty} \frac{1}{k!\, l!} \int_{-\pi}^{\pi} \Big[\sum_{k_{-p},\ldots,k_p=0}^{k} \frac{k!}{k_{-p}! \cdots k_p!}\, a_{-p}^{k_{-p}} \cdots a_p^{k_p} \exp\Big\{i\theta \sum_{j=-p}^{p} jk_j\Big\}\Big] \\ &\qquad \times \Big[\sum_{l_{-p},\ldots,l_p=0}^{l} \frac{l!}{l_{-p}! \cdots l_p!}\, (ib_{-p})^{l_{-p}} \cdots (ib_p)^{l_p} \exp\Big\{i\theta \sum_{j=-p}^{p} jl_j\Big\}\Big]\, d\theta \end{aligned}$$

where the numbers $k_{-p}, \ldots, k_p$ and $l_{-p}, \ldots, l_p$ have to satisfy the constraints $k_{-p} + \cdots + k_p = k$ and $l_{-p} + \cdots + l_p = l$, respectively. Now, integration with respect to θ yields

$$I = 2\pi \sum_{k,l=0}^{\infty} \sum_{k_{-p},\ldots,k_p=0}^{k} \sum_{l_{-p},\ldots,l_p=0}^{l} \frac{a_{-p}^{k_{-p}} \cdots a_p^{k_p}}{k_{-p}! \cdots k_p!} \cdot \frac{(ib_{-p})^{l_{-p}} \cdots (ib_p)^{l_p}}{l_{-p}! \cdots l_p!}$$

with $k_{-p}+\cdots+k_p=k$, $l_{-p}+\cdots+l_p=l$, and $\sum_{h=-p}^{p} h(k_h+l_h)=0$. From rearranging the summation we have

$$I = 2\pi \sum_{\substack{k_{-p},\ldots,k_p=0\\ l_{-p},\ldots,l_p=0}}^{\infty} \frac{a_{-p}^{k_{-p}}\cdots a_p^{k_p}}{k_{-p}!\cdots k_p!} \cdot \frac{(ib_{-p})^{l_{-p}}\cdots(ib_p)^{l_p}}{l_{-p}!\cdots l_p!} \tag{26}$$

with the constraint $\sum_{h=-p}^{p} h(k_h+l_h)=0$.

In the absence of this constraint the function I would be equal to $2\pi\exp\{\sum_{j=-p}^{p}(a_j+ib_j)\}$, and von Mises marginals could be attained for a density of the form (4). Due to the presence of this constraint, however, I is more similar to a modified Bessel function. In fact, if $p=1$ then $I=2\pi\exp\{\alpha_0\}\ I_0(\sqrt{\alpha_1^2+\beta_1^2}\ )$. As p increases, on the other hand, the constraint becomes less dominant and I behaves more and more like an exponential function.

For given k_{-h}, k_h, l_{-h} and l_h $(2\le h\le p)$ let $s=\sum_{h=2}^{p} h(k_h+l_h-k_{-h}-l_{-h})$ and

$$P=\prod_{j=2}^{p} \frac{a_{-j}^{k_{-j}}\,a_j^{k_j}}{k_{-j}!\ k_j!}\cdot\frac{(ib_{-j})^{l_{-j}}(ib_j)^{l_j}}{l_{-j}!\ l_j!}\ .$$

Then $l_{-1}=k_1+l_1+s-k_{-1}$ and

$$\begin{aligned}
I &= 2\pi\exp\{a_0\} \sum_{\substack{k_{-p},\ldots,k_{-2},k_1,\ldots,k_p=0\\ l_{-p},\ldots,l_{-2},l_1,\ldots,l_p=0}}^{\infty}\ \sum_{k_{-1}=0}^{k_1+l_1+s} P \\
&\quad\times \frac{a_1^{k_1}a_{-1}^{k_{-1}}(ib_1)^{l_1}(ib_{-1})^{k_1+l_1+s-k_{-1}}}{k_1!\ k_{-1}!\ l_1!\ (k_1+l_1+s-k_{-1})!}\cdot\frac{(k_1+l_1+s)!}{(k_1+l_1+s)!} \\
&= 2\pi\exp\{a_0\} \sum_{\substack{k_{-p},\ldots,k_{-2},k_1,\ldots,k_p=0\\ l_{-p},\ldots,l_{-2},l_1,\ldots,l_p=0}}^{\infty} P\cdot\frac{a_1^{k_1}(ib_1)^{l_1}(a_{-1}+ib_{-1})^{k_1+l_1+s}}{k_1!\ l_1!\ (k_1+l_1+s)!}
\end{aligned}$$

where $k_1+l_1+s\ge 0$. Putting $m=k_1+l_1$ and changing the summation yields

$$\begin{aligned}
I &= 2\pi\exp\{a_0\} \sum_{\substack{k_{-p},\ldots,k_{-2},k_2,\ldots,k_p=0\\ l_{-p},\ldots,l_{-2},l_2,\ldots,l_p=0}}^{\infty} P \\
&\quad\times \sum_{m=\max(0,-s)}^{\infty}\ \sum_{k_1=0}^{m} \frac{a_1^{k_1}(ib_1)^{m-k_1}(a_{-1}+ib_{-1})^{m+s}}{k_1!\ (m-k_1)!\ (m+s)!} \\
&= 2\pi\exp\{a_0\} \sum_{\substack{k_{-p},\ldots,k_{-2},k_2,\ldots,k_p=0\\ l_{-p},\ldots,l_{-2},l_2,\ldots,l_p=0}}^{\infty} P \\
&\quad\times \sum_{m=\max(0,-s)}^{\infty} \frac{(a_1+ib_1)^m(a_{-1}+ib_{-1})^{m+s}}{m!\ (m+s)!}\ .
\end{aligned}$$

Defining $\psi_j = \arctan \beta_j/\alpha_j$ $(j = 1, \ldots, p)$ and using the identity

$$(a_1 + ib_1)^m (a_{-1} + ib_{-1})^{m+s}$$
$$= \left((a_1 + ib_1)^{1/2}(a_{-1} + ib_{-1})^{1/2})\right)^{2m+s} \left(\frac{a_{-1} + ib_{-1}}{a_1 + ib_1}\right)^{s/2}$$
$$= \left(\tfrac{1}{2}(\alpha_1^2 + \beta_1^2)^{1/2}\right)^{2m+s} \left(\frac{\alpha_1 + i\beta_1}{(\alpha_1^2 + \beta_1^2)^{1/2}}\right)^s$$

gives

$$I = 2\pi \exp\{a_0\} \sum_{\substack{k_{-p},\ldots,k_{-2},k_2,\ldots,k_p=0 \\ l_{-p},\ldots,l_{-2},l_2,\ldots,l_p=0}}^{\infty} P \exp\{i\psi_1 s\}\, I_{|s|}\left(\sqrt{\alpha_1^2 + \beta_1^2}\,\right)$$

$$= 2\pi \exp\{a_0\} \sum_{\substack{m_{-p},\ldots,m_{-2}=0 \\ m_2,\ldots,m_p=0}}^{\infty} \exp\{i\psi_1 s\}\, I_{|s|}\left(\sqrt{\alpha_1^2 + \beta_1^2}\,\right)$$
$$\times \sum_{k_{-p}=0}^{m_{-p}} \cdots \sum_{k_{-2}=0}^{m_{-2}} \sum_{k_2=0}^{m_2} \cdots \sum_{k_r=0}^{m_p}$$
$$\prod_{j=2}^{p} \frac{a_{-j}^{k_{-j}} (ib_{-j})^{m_{-j}-k_{-j}}\ m_{-j}!\ a_j^{k_j} (ib_j)^{m_j-k_j}\ m_j!}{k_{-j}!\ (m_{-j} - k_{-j})!\ m_{-j}!\ k_j!\ (m_j - k_j)!\ m_j!}$$

$$= 2\pi \exp\{a_0\} \sum_{\substack{m_{-p},\ldots,m_{-2}=0 \\ m_2,\ldots,m_p=0}}^{\infty} \exp\{i\psi_1 s\}\, I_{|s|}\left(\sqrt{\alpha_1^2 + \beta_1^2}\,\right) \prod_{j\in J} \frac{(a_j + ib_j)^{m_j}}{m_j!}$$

where $J = \{-p, \ldots, -2, 2, \ldots, p\}$ and $s = \sum_{h=2}^{p} h(m_h - m_{-h})$. Let $n_j = m_j - m_{-j}$ and $p_j = m_j + m_{-j} = |n_j| + 2q_j$ with $q_j \in \mathcal{N}_0$. Then $m_j = \frac{1}{2}(n_j + p_j) = \frac{1}{2}(n_j + |n_j|) + q_j$ and $m_{-j} = \frac{1}{2}(-n_j + p_j) = \frac{1}{2}(-n_j + |n_j|) + q_j$. Further, $s = \sum_{h=2}^{p} h\, n_h$. Now, substituting the identity

$$(a_{-j} + ib_{-j})^{m_{-j}} (a_j + ib_j)^{m_j}$$
$$= \left((a_{-j} + ib_{-j})^{1/2}(a_j + ib_j)^{1/2}\right)^{2q_j + |n_j|} \left(\frac{a_{-j} + ib_{-j}}{a_j + ib_j}\right)^{-n_j/2}$$

into the last expression gives the final result,

$$I = 2\pi \exp\{a_0\} \sum_{n_2,\ldots,n_p=-\infty}^{\infty} \exp\{i\psi_1 s\}\, I_{|s|}\left(\sqrt{\alpha_1^2 + \beta_1^2}\,\right)$$
$$\times \sum_{q_2,\ldots,q_p=0}^{\infty} \prod_{j=2}^{p} \exp\{-i\psi_j n_j\} \frac{\left(\frac{1}{2}(\alpha_j^2 + \beta_j^2)^{1/2}\right)^{2q_j + |n_j|}}{q_j!\ (q_j + |n_j|)!}.$$

That is

$$I = 2\pi \exp\{\alpha_0\} \sum_{n_2,\dots,n_p=-\infty}^{\infty} I_{|s|}\left(\sqrt{\alpha_1^2+\beta_1^2}\right) \times \prod_{j=2}^{p} \cos n_j(j\psi_1 - \psi_j)\, I_{|n_j|}\left(\sqrt{\alpha_j^2+\beta_j^2}\right). \tag{27}$$

□

To prove that $I = \int_{-\pi}^{\pi} \exp\{\sum_{j=0}^{p} \alpha_j \cos j\theta + \beta_j \sin j\theta\}\, d\theta$ is not of the form const. $\exp\{\kappa_\eta \cos(\eta - \mu_\eta)\}$ with $\kappa_\eta \neq 0$, the following asymptotic results are required.

Lemma 7.2: The function I as in (25) is asympotically equivalent to

$$I_{1,\infty} = \left(\frac{2\pi}{\sqrt{\alpha_1^2+\beta_1^2}}\right)^{1/2} \times \exp\left\{\sum_{j=0}^{p}\sum_{k=0}^{q} (\cos j\psi_1, \sin j\psi_1) \begin{pmatrix} \alpha_{jk} & \gamma_{jk} \\ \delta_{jk} & \beta_{jk} \end{pmatrix} \begin{pmatrix} \cos k\eta \\ \sin k\eta \end{pmatrix}\right\}, \tag{28}$$

where $\psi_1 = \beta_1/\alpha_1$.

Proof: It is well known that for large $|z_1|$ ($z_1 \in \mathcal{R}$) the function $I_{|s|}(z_1)$ behaves like $[2\pi z_1]^{-1/2} \exp\{z_1\}$ (Abramowitz & Stegun 1966, formula (9.7.1)). Further, for $z \in \mathcal{R}$ and $\eta \in (-\pi, \pi]$, $\sum_{n=-\infty}^{\infty} \cos(n\eta)\, I_{|n|}(z) = \exp\{z \cos\eta\}$ (Abramowitz & Stegun 1966, formula (9.6.34)). From Lemma 7.1 it therefore follows that the function $I = \int_{-\pi}^{\pi} \exp\{\sum_{j=0}^{p} \alpha_j \cos j\theta + \beta_j \sin j\theta\}\, d\theta$ is asymptotically equivalent to

$$I_{1,\infty} = \left(\frac{2\pi}{\sqrt{\alpha_1^2+\beta_1^2}}\right)^{1/2} \exp\left\{\alpha_0 + \sum_{j=1}^{p} \sqrt{\alpha_j^2+\beta_j^2}\, \cos(j\psi_1 - \psi_j)\right\}. \tag{29}$$

For $j = 1, \dots, p$ one has $\tan\psi_j = \beta_j/\alpha_j$ and hence,

$$\cos\psi_j = \frac{1}{[1+\tan^2\psi_j]^{1/2}} = \frac{\alpha_j}{[\alpha_j^2+\beta_j^2]^{1/2}} \tag{30a}$$

and

$$\sin\psi_j = \frac{\tan\psi_j}{[1+\tan^2\psi_j]^{1/2}} = \frac{\beta_j}{[\alpha_j^2+\beta_j^2]^{1/2}}. \tag{30b}$$

Substituting these expressions into (29) yields

$$I_{1,\infty} = \left(\frac{2\pi}{\sqrt{\alpha_1^2+\beta_1^2}}\right)^{1/2} \exp\left\{\alpha_0 + \sum_{j=1}^{p} (\alpha_j \cos j\psi_1 + \beta_j \sin j\psi_1)\right\}.$$

Finally, since

$$\alpha_j = \sum_{k=0}^{q} (\alpha_{jk} \cos k\eta + \gamma_{jk} \sin k\eta) \tag{31a}$$

and

$$\beta_j = \sum_{k=0}^{q} (\delta_{jk} \cos k\eta + \beta_{jk} \sin k\eta) \tag{31b}$$

the result follows. □

For $I_{1,\infty}$ to be of the form const. $\exp\{\kappa_\eta \cos(\eta - \mu_\eta)\}$ with $\kappa_\eta \neq 0$, $\sqrt{\alpha_1^2 + \beta_1^2}$ must be independent of η, that is $\alpha_1 = c_1 \cos k_1(\eta - \mu_1)$ and $\beta_1 = c_1 \sin k_1(\eta - \mu_1)$ with $c_1 \in \mathcal{R}$, $k_1 \in \mathcal{Z}$ and $\mu_1 \in (-\pi, \pi]$. Without loss of generality it can be assumed that $\mu_1 = 0$. Hence, $\psi_1 = k_1\eta$. For $p = 1$ the function $I_{1,\infty}$ is obviously not of the required form. It will therefore be assumed that $p \geq 2$.

Notice that the asymptotic approximation $I_{1,\infty}$ leads to a condition which the term $\sqrt{\alpha_1^2 + \beta_1^2}$ has to satisfy. By rearranging the summation in (25), a similar constraint on $\sqrt{\alpha_j^2 + \beta_j^2}$ can also be derived for $j = 2, \ldots, p$.

Lemma 7.3: Let $h \in \{1, \ldots, p\}$, $\nu \in \{1, \ldots, h\}$ and $\phi_{\nu,h} = [\psi_h - 2\pi\nu]/h$. The function I defined as in (25) is then asympotically equivalent to

$$\begin{aligned} I_{h,\infty} = &\left(\frac{2\pi}{\sqrt{\alpha_h^2 + \beta_h^2}}\right)^{1/2} \frac{1}{h} \sum_{\nu=1}^{h} \\ &\exp\left\{\sum_{j=0}^{p} \sum_{k=0}^{q} (\cos j\phi_{\nu,h}, \sin j\phi_{\nu,h}) \begin{pmatrix} \alpha_{jk} & \gamma_{jk} \\ \delta_{jk} & \beta_{jk} \end{pmatrix} \begin{pmatrix} \cos k\eta \\ \sin k\eta \end{pmatrix}\right\}. \end{aligned} \tag{32}$$

Proof: Referring to equation (27) note that

$$\begin{aligned} I = 2\pi \exp\{\alpha_0\} &\sum_{n_2, \ldots, n_p = -\infty}^{\infty} \exp\{i\psi_1 s\}\, I_{|s|}\left(\sqrt{\alpha_1^2 + \beta_1^2}\,\right) \\ &\times \prod_{j=2}^{p} \exp\{-i\psi_j n_j\}\, I_{|n_j|}\left(\sqrt{\alpha_j^2 + \beta_j^2}\,\right) \end{aligned}$$

where $s = \sum_{j=2}^{p} j\, n_j$. Now, define $m = h\, n_h$ and $r = s - m$. Since

$$I_{|n_h|}\left(\sqrt{\alpha_h^2 + \beta_h^2}\,\right) \overset{\text{asymp.}}{\approx} \left[2\pi\sqrt{\alpha_h^2 + \beta_h^2}\,\right]^{-1/2} \exp\left\{\sqrt{\alpha_h^2 + \beta_h^2}\,\right\}$$

it follows that I is asympotically equivalent to

$$
\begin{aligned}
I_{h,\infty} &= \left(\frac{2\pi}{\sqrt{\alpha_h^2+\beta_h^2}}\right)^{1/2} \exp\left\{\alpha_0 + \sqrt{\alpha_h^2+\beta_h^2}\,\right\} \\
&\times \sum_{n_2,\ldots,n_p=-\infty}^{\infty} \exp\{i\psi_1(r+m) - i\psi_h n_h\}\, I_{|r+m|}\left(\sqrt{\alpha_1^2+\beta_1^2}\,\right) \\
&\times \prod_{\substack{j=2\\ j\neq h}}^{p} \exp\{-i\psi_j n_j\}\, I_{|n_j|}\left(\sqrt{\alpha_j^2+\beta_j^2}\,\right).
\end{aligned}
\tag{33}
$$

Subsequently use will be made of the following identity

$$
\frac{1}{h}\sum_{\nu=1}^{h} \exp\{2\pi i\nu m/h\} = \begin{cases} 1 & \text{if } m \in h\mathcal{Z} \\ 0 & \text{otherwise}\,. \end{cases}
$$

This formula provides a convenient method to replace the summation index n_h in (33) by $m = h\,n_h$. Using this fact and changing the order of summation then leads to

$$
\begin{aligned}
I_{h,\infty} &= \left(\frac{2\pi}{\sqrt{\alpha_h^2+\beta_h^2}}\right)^{1/2} \exp\left\{\alpha_0 + \sqrt{\alpha_h^2+\beta_h^2}\,\right\} \\
&\times \frac{1}{h}\sum_{\nu=1}^{h} \sum_{\substack{n_2,\ldots,n_{h-1},m,\\ n_{h+1},\ldots,n_p=-\infty}}^{\infty} \exp\{i(-\phi_{\nu,h}+\psi_1)(r+m)\}\, I_{|r+m|}\left(\sqrt{\alpha_1^2+\beta_1^2}\,\right) \\
&\times \prod_{\substack{j=2\\ j\neq h}}^{p} \exp\{i(j\phi_{\nu,h}-\psi_j)n_j\}\, I_{|n_j|}\left(\sqrt{\alpha_j^2+\beta_j^2}\,\right).
\end{aligned}
$$

By applying the identity $\sum_{n=-\infty}^{\infty}\cos(n\eta)\, I_{|n|}(z) = \exp\{z\cos\eta\}$ this expression reduces to

$$
\begin{aligned}
I_{h,\infty} &= \left(\frac{2\pi}{\sqrt{\alpha_h^2+\beta_h^2}}\right)^{1/2} \exp\left\{\alpha_0 + \sqrt{\alpha_h^2+\beta_h^2}\,\right\} \\
&\times \frac{1}{h}\sum_{\nu=1}^{h} \exp\left\{\sqrt{\alpha_1^2+\beta_1^2}\,\cos(-\phi_{\nu,h}+\psi_1)\right\} \\
&\times \prod_{\substack{j=2\\ j\neq h}}^{p} \exp\left\{\sqrt{\alpha_j^2+\beta_j^2}\,\cos(j\phi_{\nu,h}-\psi_j)\right\},
\end{aligned}
$$

which is the same as

$$
(34)\qquad \begin{aligned} I_{h,\infty} &= \left(\frac{2\pi}{\sqrt{\alpha_h^2+\beta_h^2}}\right)^{1/2} \\ &\times \frac{1}{h}\sum_{\nu=1}^{h} \exp\left\{\alpha_0 + \sum_{j=1}^{p}\sqrt{\alpha_j^2+\beta_j^2}\,\cos\left(j\phi_{\nu,h}-\psi_j\right)\right\}. \end{aligned}
$$

Substituting (30) into (34) and applying the same steps as in the proof of Lemma 7.2 yields the result. □

Note that formula (32) reduces to (28) if $h = 1$. It follows that none of the expressions $\sqrt{\alpha_j^2+\beta_j^2}$ must be dependent on η ($j = 1,\ldots,p$). Hence, $\alpha_j = c_j \cos k_j(\eta - \mu_j)$ and $\beta_j = c_j \sin k_j(\eta-\mu_j)$ with $c_j \in \mathcal{R}$, k_j integer between $-p$ and p, and $\mu_j \in (-\pi,\pi]$. Further, $|k_h| \neq |k_j|$ for $h \neq j$. However, referring to the relationship (31) this implies that

$$
\begin{aligned}
\alpha_{jk} &= c_j \cos\, k_j\mu_j \\
\gamma_{jk} &= c_j \sin\, k_j\mu_j \\
\beta_{jk} &= c_j \cos\, k_j\mu_j \\
\delta_{jk} &= -c_j \sin\, k_j\mu_j
\end{aligned}
$$

if $k = k_j$, and $\alpha_{jk} = \gamma_{jk} = \beta_{jk} = \delta_{jk} = 0$ otherwise. For all $j \in \{1,\ldots,p\}$ it follows that

$$
\sum_{k=0}^{q} \begin{pmatrix} \alpha_{jk} & \gamma_{jk} \\ \delta_{jk} & \beta_{jk} \end{pmatrix} \begin{pmatrix} \cos k\eta \\ \sin k\eta \end{pmatrix} = c_j \begin{pmatrix} \cos(k_j\eta - k_j\mu_j) \\ \sin(k_j\eta - k_j\mu_j) \end{pmatrix}.
$$

By substituting this expression into (28) we obtain

$$
I_{1,\infty} = \left[2\pi/c_1\right]^{1/2} \exp\left\{\alpha_0 + \sum_{j=1}^{p} c_j\, \cos\left(jk_1\eta - k_j\eta + k_j\mu_j\right)\right\}.
$$

For condition (24a) to hold it is therefore necessary that $\cos(jk_1 - k_j)\eta$ is proportional to $\cos\eta$ and hence, that $jk_1 - k_j = 1$ for all $j \in \{2,\ldots,p\}$ with $k_j \neq 0$. Similarly, given the symmetry between θ and η, $hm_1 - m_h = 1$ for all $h \in \{2,\ldots,p\}$ with $m_h \neq 0$, and $-p \leq m_1, m_h \leq p$. Further, $m_{|k_j|} = j \operatorname{sgn} k_j$ and $k_{|m_h|} = h \operatorname{sgn} m_h$ for all $j,h \in \{2,\ldots,p\}$ with $k_j \neq 0$ and $m_h \neq 0$, respectively. However, it can easily be shown that there are no numbers k_j and m_h such that these conditions are satisfied. Hence, we are now in the position to state the main result.

Unless the circular variables θ and η are independent, there is no bivariate density of the form

$$
\begin{aligned}
f(\theta,\eta) = \text{const.}\ \exp\Big\{ \sum_{j=0}^{p}\sum_{k=0}^{q} (\alpha_{jk}\cos j\theta\ \cos k\eta + \gamma_{jk}\cos j\theta\ \sin k\eta \\
+ \delta_{jk}\sin j\theta\ \cos k\eta + \beta_{jk}\sin j\theta\ \sin k\eta) \Big\}
\end{aligned}
\tag{35}
$$

such that the marginals are of von Mises form, with constraints (24a) and (24b) both being satisfied.

This result is not surprising given that Bessel functions do not possess addition theorems in the strict sense of the term (Watson 1952, p.358). That is, $I_n(z_1+z_2)$ cannot be expressed as an algebraic function of $I_n(z_1)$ and $I_n(z_2)$ ($n \in \mathcal{N}$ and $z_1, z_2 \in \mathcal{R}_0^+$). However, the result does not imply that the marginals of θ and η cannot be close to a von Mises distribution. In fact, on the basis of (26) it is conjectured that any two von Mises distributions can be approximated arbitrarily closely by the marginals of (35), if p and q are sufficiently large.

CHAPTER 8

COMPARISON OF DIFFERENT MEASURES OF ASSOCIATION

In this chapter the different measures of angular association are compared numerically. In section 8.1 the case of independent and identically distributed bivariate data is considerd, while in section 8.2 the different measures are studied in the context of a wrapped autoregressive process. It is found that amongst all the coefficients considered the proposal σ_Q shows the best performance.

Given the aim of modelling the directional component (θ_t) of the residual wind series $(\mathbf{f}_t)$ as a wrapped autoregressive process, the task is to estimate the parameters of the underlying unwrapped process and thus to determine its autocovariance function C. However, each of the measures of association leads to a different definition of a circular autocovariance function c of (θ_t). In each case it will therefore be necessary to establish a relationship between the two autocovariance functions c and C. Based on the results of the simulation study performed in this chapter σ_Q is found to be the most appropriate measure of association for the analysis of directional time series.

8.1 Independent Bivariate Directional Data

In this section the different measures of angular association will be compared numerically given a sample of independent and identically distributed bivariate directional data. In the first Monte-Carlo study pairs (θ, η) of circular variables are generated from a bivariate uniform distribution so that they fall into the hatched area of Figure 8.1. Obviously, the smaller the width b of that area the larger the association between θ and η, with $b = 0$ and $b = 2\pi$ corresponding to $\theta = \eta$ and independence, respectively. Samples of 50 independent observations on (θ, η) are generated for each value of b. For the random number generator, see Ahrens, Dieter & Grube (1970).

Figure 8.1. Distribution of (θ, η) on the torus.

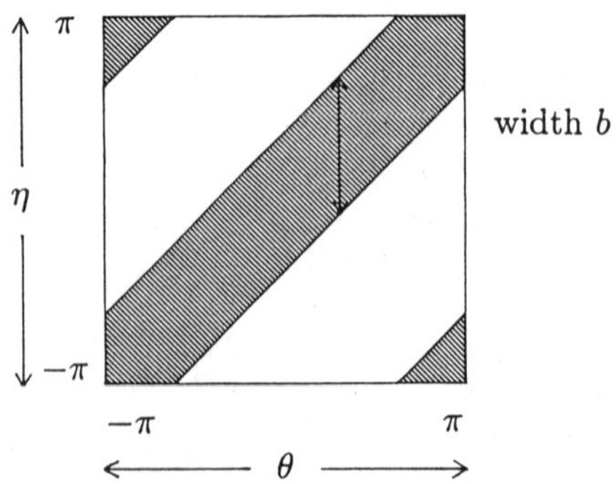

Various empirical correlation coefficients, which are defined as the sample analogues of the coefficients discussed in chapter 7, are plotted in Figure 8.2 as a function of b. In the case that θ and η are identical, r_{JM} would normally be equal to $\sqrt{2}$, as explained in section 7.3 (e). Therefore, in order to compare this measure of association with the others, the values of r_{JM} are divided by $\sqrt{2}$. Observe that the correlation coefficients depicted in Figure 8.2 all decrease monotonically as b moves from 0 to 2π. However, neither r_{JM} nor r_{JW} gets close to 0 if $b = 2\pi$. That is, the values of these measures do not suggest that θ and η are independent when indeed they are.

Given an unbiased estimator of the first and second order circular moments, r_Q and r_{WB} are obviously unbiased estimators of ρ_Q and ρ_{WB}, respectively. The fact that r_Q in Figure 8.2 does not reach 0 as b approaches 2π is therefore a reflection of the sampling error. However, it can be shown that all the other empirical correlation coefficients are biased estimators of the corresponding theoretical coefficients, even though they are *asymptotically* unbiased. In fact, this is the premium to be paid in order to adjust for the asymmetry between $\cos\theta \sin\eta$ and $\sin\theta \cos\eta$, as explained in the last chapter, and to ensure that property (vi) listed in section 7.1 is satisfied. Although this may be relevant if θ and η follow different distributions, it also means that none of these coefficients reaches 0 for $b = 2\pi$.

Figure 8.2. Angular correlation as a function of the width b of a bivariate uniform variable.

(a) Stephens' r_{St}.

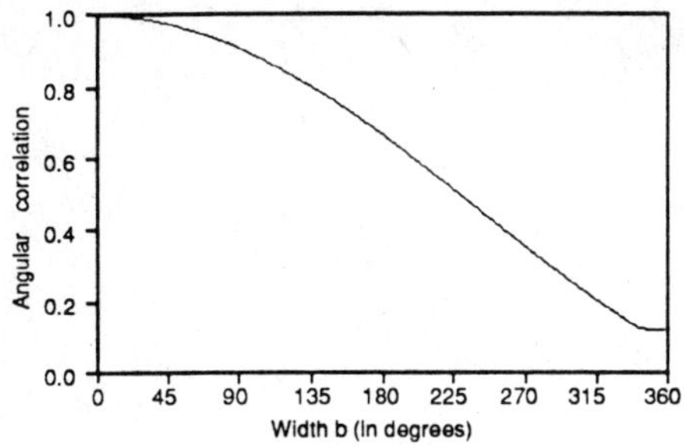

(b) Johnson & Wehrly's r_{JW}.

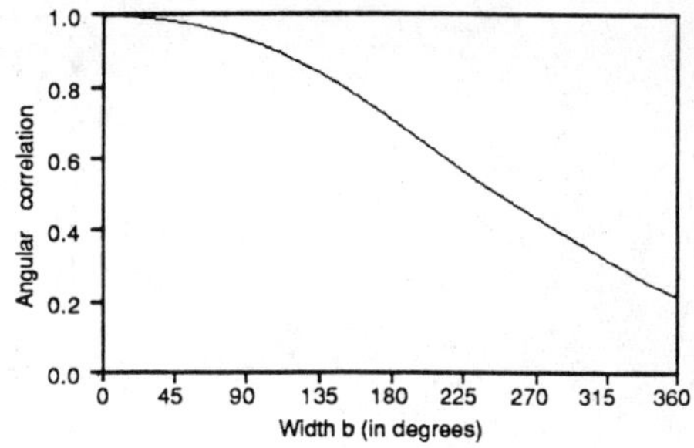

(c) Jupp & Mardia's r_{JM}.

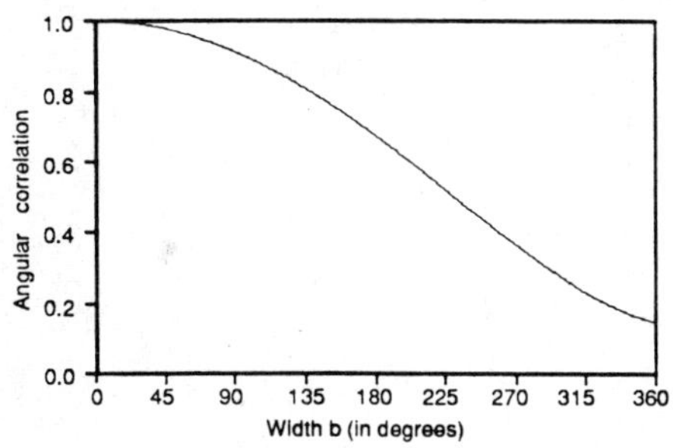

(d) Proposal r_Q.

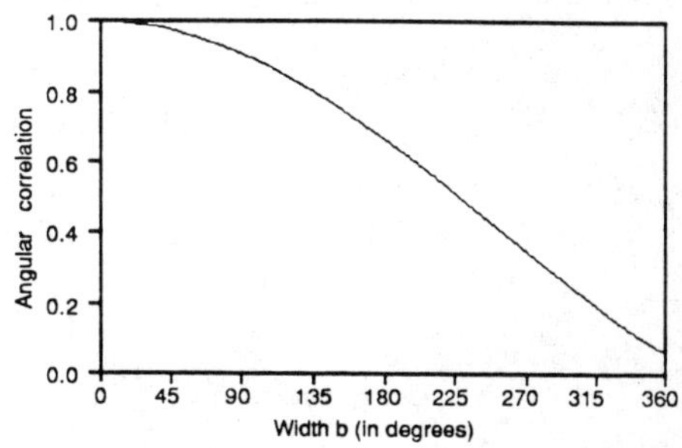

Nevertheless, for a bivariate uniform variable the different coefficients show a similar behaviour, as is evident from the scatterplots in Figure 8.3. A straight line through the origin in these diagrams would indicate that the two coefficients considered behave identically. Any difference between the coefficients will therefore be reflected in a departure of the plotted curve from such a straight line. Obviously and due to the bias, neither r_{JW} nor r_{JM} tend to 0 as quickly as r_Q.

Putting $\mathbf{X} = (\cos\theta, \sin\theta)'$ and $\mathbf{Y} = (\cos\eta, \sin\eta)'$ it follows that

$$\mathbf{\Sigma}_{XY} = \mathbf{\Sigma}_{YX} = \mathbf{\Pi}_{XY} = \mathbf{\Pi}_{YX} = \tfrac{1}{2}\,\frac{\sin(b/2)}{b/2}\,\mathbf{I}_2$$

and

$$\mathbf{\Sigma}_{XX} = \mathbf{\Sigma}_{YY} = \mathbf{\Pi}_{XX} = \mathbf{\Pi}_{YY} = \tfrac{1}{2}\,\mathbf{I}_2 \quad .$$

Hence, $\rho_Q = \rho_{WB} = \rho_{St} = \rho_{DM} = \rho_{JW} = \rho_{JM}/\sqrt{2} = \sin(b/2)/(b/2)$, while $\rho_{FL} = [\sin(b/2)/(b/2)]^2$. Since in this particular example $\rho_{FL} = \rho_Q^2$, the coefficient ρ_{FL} approaches 0 faster than ρ_Q. However, this can be attributed to a peculiarity in the definition of ρ_{FL}. Whilst all other coefficients can be regarded as linear functions of the corresponding eigenvalues, ρ_{FL} behaves more like a quadratic function as indicated in Table 7.1.

Figure 8.3. Scatterplots of r_Q versus various angular correlation coefficients given a bivariate uniform distribution.

(a) r_Q versus r_{St}.

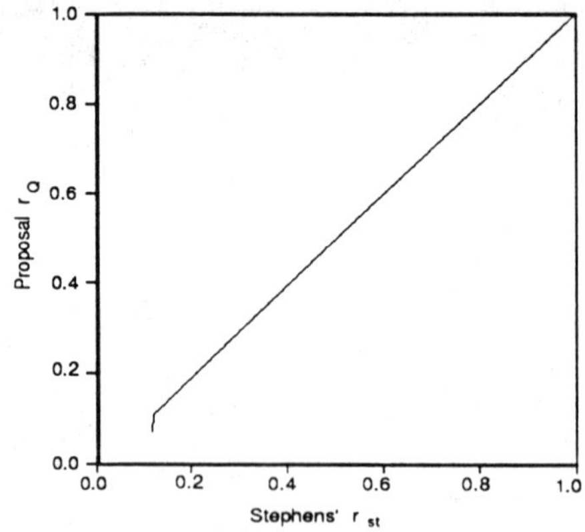

(b) r_Q versus r_{JW}.

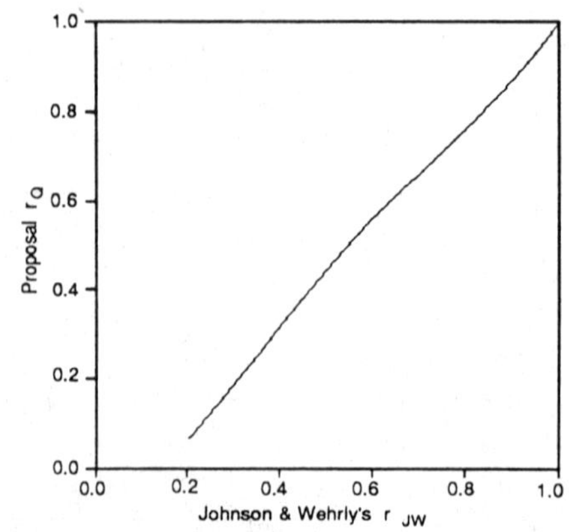

(c) r_Q versus r_{JM}.

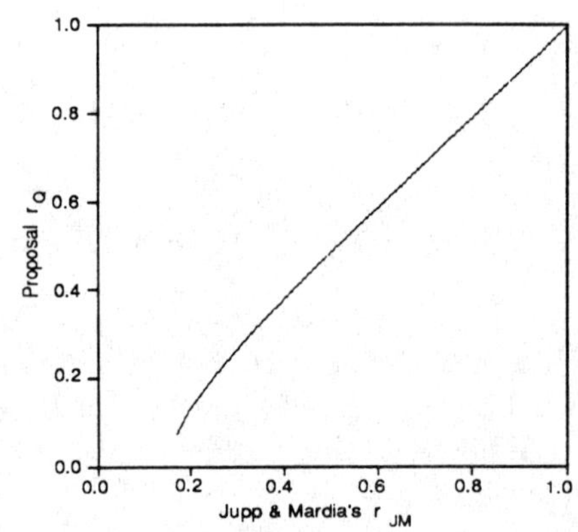

Figure 8.4. Angular correlation as a functions of the correlation ρ between the underlying unwrapped normal variables.

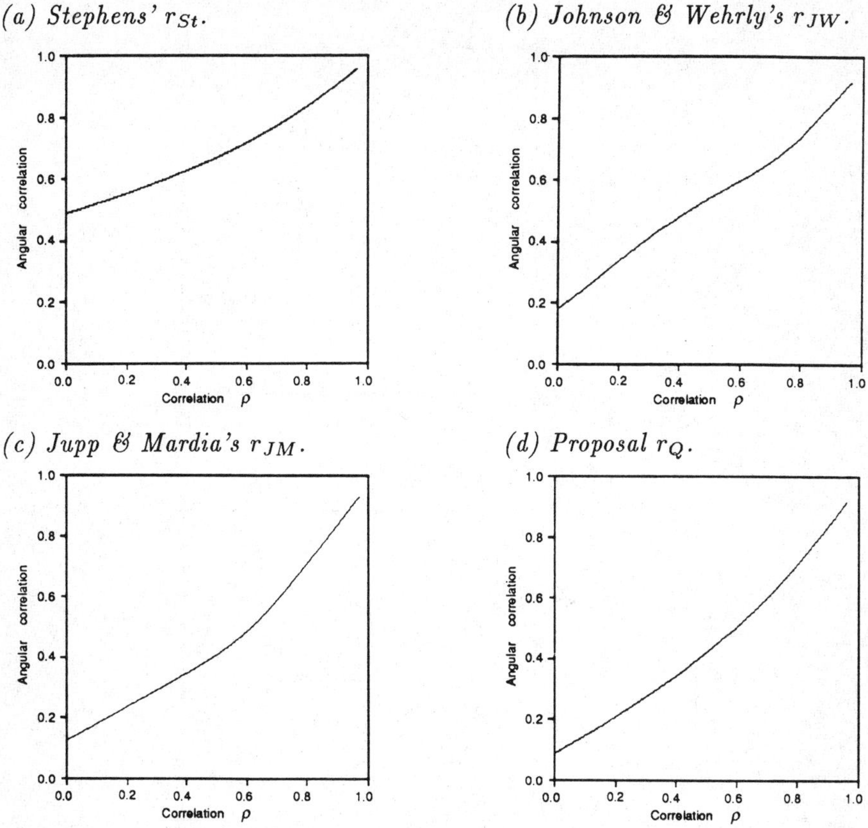

(a) Stephens' r_{St}. *(b) Johnson & Wehrly's r_{JW}.* *(c) Jupp & Mardia's r_{JM}.* *(d) Proposal r_Q.*

The second Monte-Carlo study is based on a wrapped bivariate normal distribution with correlation coefficient ρ. Figure 8.4 shows the behaviour of the same angular correlation coefficients, which were examined in Figure 8.2, as ρ varies between 0 and 1. For sampling from the normal distribution we implemented an extension of Forsythe's algorithm described in Ahrens & Dieter (1973).

All the correlation coefficients increase monotonically with ρ. However, whereas r_Q attains a value of 0.1 for $\rho = 0$, neither r_{JW} nor r_{JM} decrease below 0.2 as ρ tends to 0. Even more remarkable is the fact that r_{St}, with a value of about 0.5 for $\rho = 0$, provides no evidence that the two circular variables are actually independent. The tendency of r_{St}, r_{JW} and r_{JM} to attain non-zero values for $\rho = 0$ is also revealed by the scatterplots in Figure 8.5. As in the previous simulation the main difference between the correlation coefficients is in their tendency to go to 0 as θ and η become independent. Apart from this characteristic the most noticeable feature in these plots is the non-linearity of r_Q as a function of r_{JW}, which indicates that the nature of these two measures of association is quite distinct. Also note the change in the slope of r_{JM} for ρ greater than 0.6 in Figure 8.4 (c).

Figure 8.5. Scatterplots of r_Q versus various angular correlation coefficients given a wrapped bivariate normal distribution.

(a) r_Q versus r_{St}.

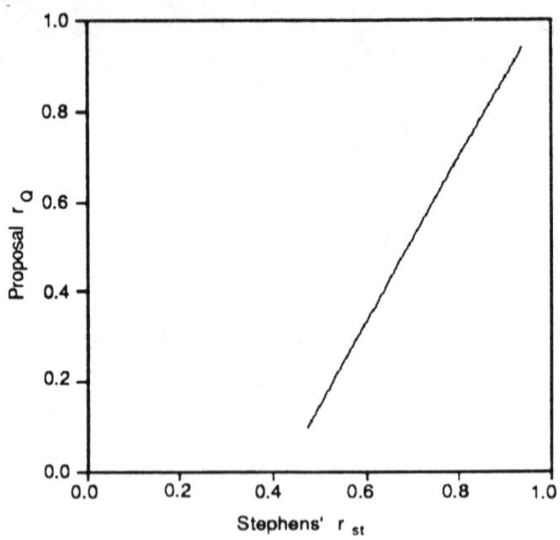

(b) r_Q versus r_{JW}.

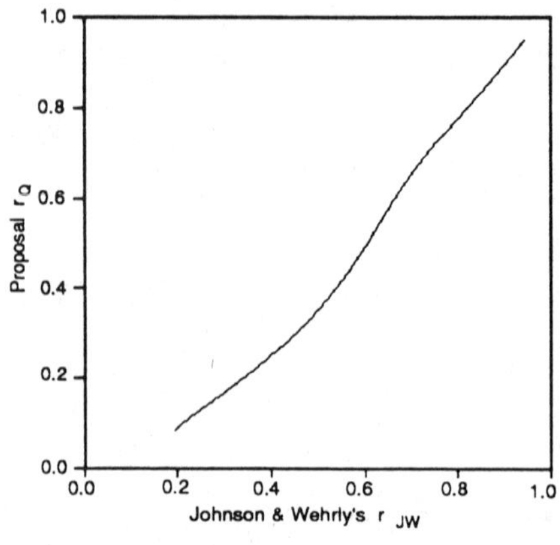

(c) r_Q versus r_{JM}.

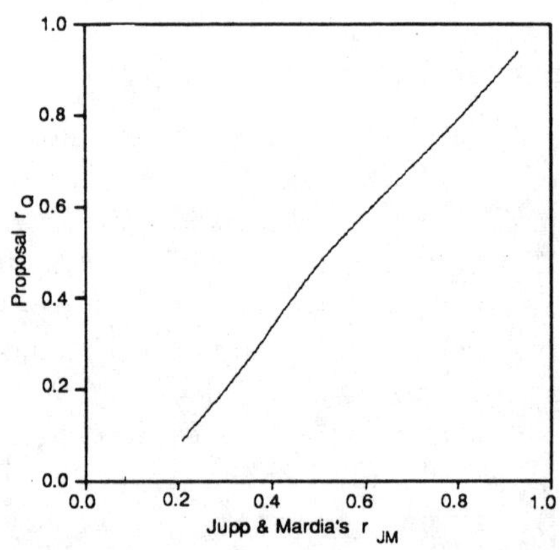

To calculate the theoretical correlation coefficients the characteristic function of a wrapped multivariate variable is required. Let $\mathbf{Y} = (Y_1, \ldots, Y_k)'$ $(k \geq 1)$ be a multivariate variable in $\mathcal{R}^k$ with characteristic function $\phi_{\mathbf{Y}}$, and suppose that $\theta_j = Y_j \pmod{2\pi}$ for $j = 1, \ldots, k$. If the characteristic function of $\theta = (\theta_1, \ldots, \theta_k)'$ is denoted by ϕ_θ, then it is easy to show that $\phi(\mathbf{m}) = \phi_\theta(\mathbf{m}) = \phi_{\mathbf{Y}}(\mathbf{m})$ for all $\mathbf{m} = (m_1, \ldots, m_k)' \in \mathcal{Z}^k$ (Mardia 1972, formula (3.4.25)). In particular, assume that $(X, Y)'$ has a bivariate normal distribution with mean $\mathbf{0}$ and covariance matrix $\mathbf{\Sigma} = (\sigma_{jk})_{j,k=1,2}$, and define $\theta = X \pmod{2\pi}$ and $\eta = Y \pmod{2\pi}$. Then,

$$\phi(s,t) = \mathcal{E}[\cos(s\theta + t\eta)] = \exp\{-\tfrac{1}{2}(s,t)\,\mathbf{\Sigma}\,(s,t)'\} \quad . \tag{1}$$

In view of the time series context it will generally be assumed that $\sigma_{11} = \sigma_{22}$. Hence,

$$\begin{aligned}
\mathcal{E}[\cos\theta] &= \mathcal{E}[\cos\eta] = \phi(1,0) = \exp\{-\tfrac{1}{2}\sigma_{11}\} \\
\mathcal{E}[\cos^2\theta] &= \mathcal{E}[\cos^2\eta] = \tfrac{1}{2} + \tfrac{1}{2}\,\phi(2,0) = \exp\{-\sigma_{11}\}\cosh(\sigma_{11}) \\
\mathcal{E}[\sin^2\theta] &= \mathcal{E}[\sin^2\eta] = \tfrac{1}{2} - \tfrac{1}{2}\,\phi(2,0) = \exp\{-\sigma_{11}\}\sinh(\sigma_{11}) \\
\mathcal{E}[\cos\theta\,\cos\eta] &= \tfrac{1}{2}\,\{\phi(1,-1) + \phi(1,1)\} = \exp\{-\sigma_{11}\}\cosh(\sigma_{12}) \\
\mathcal{E}[\sin\theta\,\sin\eta] &= \tfrac{1}{2}\,\{\phi(1,-1) - \phi(1,1)\} = \exp\{-\sigma_{11}\}\sinh(\sigma_{12})
\end{aligned}$$

and

$$\mathcal{E}[\sin\theta] = \mathcal{E}[\sin\eta] = \mathcal{E}[\cos\theta\,\sin\eta] = \mathcal{E}[\sin\theta\,\cos\eta] = 0\,.$$

It follows that

$$\mathbf{\Sigma}_{XY} = \mathbf{\Sigma}_{YX} = e^{-\sigma_{11}}\begin{pmatrix}\cosh(\sigma_{12}) - 1 & 0 \\ 0 & \sinh(\sigma_{12})\end{pmatrix} \tag{2a}$$

$$\mathbf{\Sigma}_{XX} = \mathbf{\Sigma}_{YY} = e^{-\sigma_{11}}\begin{pmatrix}\cosh(\sigma_{11}) - 1 & 0 \\ 0 & \sinh(\sigma_{11})\end{pmatrix} \tag{2b}$$

$$\mathbf{\Pi}_{XY} = \mathbf{\Pi}_{YX} = e^{-\sigma_{11}}\begin{pmatrix}\cosh(\sigma_{12}) & 0 \\ 0 & \sinh(\sigma_{12})\end{pmatrix} \tag{2c}$$

$$\mathbf{\Pi}_{XX} = \mathbf{\Pi}_{YY} = e^{-\sigma_{11}}\begin{pmatrix}\cosh(\sigma_{11}) & 0 \\ 0 & \sinh(\sigma_{11})\end{pmatrix} \tag{2d}$$

and

$$\mathbf{\Sigma}_{XX}^{-1}\mathbf{\Sigma}_{XY}\mathbf{\Sigma}_{YY}^{-1}\mathbf{\Sigma}_{YX} = \begin{pmatrix}\left(\frac{\cosh(\sigma_{12})-1}{\cosh(\sigma_{11})-1}\right)^2 & 0 \\ 0 & \left(\frac{\sinh(\sigma_{12})}{\sinh(\sigma_{11})}\right)^2\end{pmatrix}\,. \tag{2e}$$

Now suppose that $\sigma_{11} = \sigma_{22} = 1$ and $\sigma_{12} = \sigma_{21} = \rho$. This yields

$$\rho_{WB}(\theta,\eta) = e^{\rho-1} \tag{3a}$$

$$\rho_{St}(\theta,\eta) = e^{\rho-1} \tag{3b}$$

$$\rho_{DM}(\theta,\eta) = [\,e^\rho - 1\,]\,/\,[\,e - 1\,] \tag{3c}$$

$$\rho_{JW}(\theta,\eta) = \sinh(\rho)\,/\,\sinh(1) \tag{3d}$$

(3e) $$\rho^2_{JM}(\theta,\eta) = \left(\frac{\cosh(\rho)-1}{\cosh(1)-1}\right)^2 + \left(\frac{\sinh(\rho)}{\sinh(1)}\right)^2$$

(3f) $$\rho_{FL}(\theta,\eta) = \sinh(2\rho)\,/\,\sinh(2)$$

and

(3g) $$\rho_Q(\theta,\eta) = [\,e^\rho - 1\,]\,/\,[\,e-1\,] \quad .$$

It follows that $\rho_{FL} \le \rho_Q \le \rho_{JW}$ and hence, that ρ_{FL} tends to 0 slightly faster than ρ_Q. In fact, for $\rho \to 0$ we have $\rho_{FL} \to 2\rho/\sinh(2) \approx 0.551\rho$ and $\rho_Q \to \rho/[\,e-1\,] \approx 0.582\rho$. This means that for small ρ the coefficients ρ_{FL} and ρ_Q behave similarly. However, comparing the theoretical expressions (3d) and (3f), it can be seen that in this example ρ_{FL} has similar *overall* characteristics to ρ_{JW}.

In summary, the most pronounced difference between the various measures of angular association seems to be in their behaviour when θ and η are independent. It is this feature which will allow us in the following section to discriminate more clearly between the different correlation coefficients.

8.2 Time Series of Directional Data

In the last section the proposed measure of association ρ_Q was compared with others in the literature for samples of independent and identically distributed bivariate data. In this section the behaviour of the different correlation coefficients will be studied in the context of a directional time series generated from a first order WAR process.

The results derived in the last section can be used here to relate the autocovariance function $c_Q(\tau)$ based on σ_Q to the autocovariance function $C(\tau)$ of the underlying unwrapped process. As before, assume that $(X,Y)'$ has a bivariate normal distribution with mean $\mathbf{0}$ and covariance matrix $\mathbf{\Sigma} = (\sigma_{jk})_{j,k=1,2}$, let (θ,η) be the wrapped bivariate variable, and put $\mathbf{X} = (\cos\theta, \sin\theta)'$ and $\mathbf{Y} = (\cos\eta, \sin\eta)'$. From equation (1) it follows that

$$\sigma_Q(\theta,\eta) = \exp\left\{-\tfrac{1}{2}(\sigma_{11}+\sigma_{22})\right\}\,\left(\exp\{\sigma_{12}\}-1\right) .$$

Now, assume $\sigma_{11} = \sigma_{22}$, and identify σ_{11} and σ_{12}, respectively, with the variance $C(0)$ and covariance $C(\tau)$ for any particular $\tau \in \mathcal{N}$ of a stationary normal process (Y_t). The autocovariance function of the wrapped process (θ_t) is then obtained as

(4) $$c_Q(\tau) = \exp\{-C(0)\}\,\left(\exp\{C(\tau)\}-1\right) \quad , \quad \tau = 0,1,\dots \; .$$

Equivalently, one has

(5a) $$C(0) = -\ln(1 - c_Q(0))$$

and

(5b) $$C(\tau) = \ln\left[c_Q(\tau)\,\exp\{C(0)\}+1\right] \quad , \quad \tau = 1,2,\dots \; .$$

Once the circular autocovariance function $c_Q(\tau)$ has been estimated from the data, equations (5a) and (5b) can be used to get an estimate of $C(\tau)$ $(\tau = 0, 1, \ldots)$. A relationship similar to (5b) will now be derived for the other measures of association described in section 7.3.

Due to the standardisation, not all of these measures yield a value for the circular variance that is different from 1. In this section we will therefore always estimate $C(0)$ by use of (5a), and concentrate on the estimation of $C(\tau)$ for $\tau = 1, 2, \ldots$ The estimation of $C(0)$ will be discussed in detail in chapter 9.

Referring to equation (2), we are now in the position to express $C(\tau)$ $(\tau = 1, 2, \ldots)$ in terms of the circular autocovariance functions based on the different measures of association. These equations will be used in the subsequent Monte-Carlo study to estimate $C(\tau)$.

(a) Watson & Beran.
In this case the autocovariance function is obtained as

$$c_{WB}(\tau) = \exp\{-C(0) + C(\tau)\} \quad , \quad \tau = 0, 1, \ldots ,$$

and therefore

$$C(\tau) = C(0) + \ln\, c_{WB}(\tau) \quad , \quad \tau = 0, 1, \ldots . \tag{6}$$

(b) Stephens.
Since $c_{St}(\tau) = c_{WB}(\tau)$ the relationship between the circular autocovariance function and $C(\tau)$ is the same as in (6).

(c) Downs and Mardia.
Their measure of association implies

$$c_{DM}(\tau) = \frac{\exp\{C(\tau)\} - 1}{\exp\{C(0)\} - 1} \quad , \quad \tau = 0, 1, \ldots ,$$

which is equivalent to

$$C(\tau) = \ln\Big[c_{DM}(\tau)\,(\exp\{C(0)\} - 1) + 1\Big] \quad , \quad \tau = 0, 1, \ldots . \tag{7}$$

(d) Johnson & Wehrly.
According to (2e) the two eigenvalues of $\mathbf{\Sigma}_{XX}^{-1}\mathbf{\Sigma}_{XY}\mathbf{\Sigma}_{YY}^{-1}\mathbf{\Sigma}_{YX}$ are $\lambda_1 = [\cosh C(\tau) - 1]^2 / [\cosh C(0) - 1]^2$ and $\lambda_2 = \sinh^2 C(\tau)/\sinh^2 C(0)$, respectively. It is not hard to show that $\lambda_1 \le \lambda_2$. Consequently, $c_{JW}(\tau) = \sqrt{\lambda_2}$ and hence,

$$C(\tau) = \ln\big(z + \sqrt{z^2 + 1}\,\big) \quad , \quad \tau = 0, 1, \ldots , \tag{8}$$

where $z = c_{JW}(\tau) \cdot \sinh C(0)$.

(e) Jupp & Mardia.
In this case one has the relationship

$$c_{JM}^2(\tau) = \left(\frac{\cosh C(\tau) - 1}{\cosh C(0) - 1}\right)^2 + \left(\frac{\sinh C(\tau)}{\sinh C(0)}\right)^2 \quad , \quad \tau = 0, 1, \ldots ,$$

which yields

$$\text{(9)} \qquad C(\tau) = \ln\left(z + \sqrt{z^2 - 1}\,\right) \quad , \quad \tau = 0, 1, \ldots ,$$

where $z = \left[b + \sqrt{a^2 + (a + b)\, a\, b\, c_{JM}^2(\tau)}\;\right] \,/\, [a + b]$, $a = [\cosh C(0) - 1]^2$, and $b = \sinh^2 C(0)$.

(f) Fisher & Lee.
The autocorrelation function based on their measure of association is obtained as

$$c_{FL}(\tau) = \sinh(2C(\tau)) \;/\; \sinh(2C(0)) \quad , \quad \tau = 0, 1, \ldots ,$$

which, apart from the factor 2 in the argument of the sinh-function, is identical with $c_{JW}(\tau)$.

The different measures of association will now be compared in terms of the corresponding autocorrelation functions in the context of a first order WAR process. First, we generate $T = 50$ observations from the first order autoregressive process

$$\text{(10)} \qquad Y_t = \tfrac{1}{2}\, Y_{t-1} + U_t$$

where the U_t are independently distributed as $N(0, 0.6)$. Then we define $\eta_t = Y_t \pmod{2\pi}$ and calculate the circular autocorrelation functions based on the different measures of association. The autocovariance function $C(\tau)$ of the unwrapped process is estimated using formulae (6), (7), (8), (9) and (5b), respectively. Finally, we estimate the partial autocorrelation function and the parameters of the underlying unwrapped process using the *Yule-Walker equations*. Of course, the parameter estimates should be identical with the values in (10).

Table 8.1 shows the estimated autocorrelation functions based on the different measures of association, with the theoretical values presented in brackets. The two coefficients $\hat{c}_{St}(\tau)$ and $\hat{c}_{FL}(\tau)$ are omitted from that list because of their similarity with $\hat{c}_{WB}(\tau)$ and $\hat{c}_{JW}(\tau)$, respectively. The circular autocorrelation functions mainly differ in the way they converge as τ increases. This merely reflects the fact established in section 8.1, that the main difference between the corresponding measures of association is in their behaviour for *independent* variables. Only $\hat{c}_Q(\tau)$ decreases to 0 as τ increases, while all the others due to their inherent bias tend to fluctuate around non-zero values. Also note that all the measures listed in Table 8.1 overestimate the lag 1 covariance but not the lag 2 covariance. However, this can be largely attributed to sampling errors. As in the last section, $\hat{c}_{WB}(\tau)$ gives no indication that η_t and $\eta_{t+\tau}$ are virtually independent for τ greater than 3.

Table 8.1. Estimated autocorrelation functions of the wrapped process (η_t).

Time Lag τ	Watson Beran $\hat{c}_{WB}(\tau)$	Downs Mardia $\hat{c}_{DM}(\tau)$	Johnson Wehrly $\hat{c}_{JW}(\tau)$	Jupp Mardia $\hat{c}_{JM}(\tau)$	Proposal $\hat{c}_Q(\tau)$
1	0.80 (.67)	0.45 (.40)	0.55 (.46)	0.58 (.52)	0.45 (.40)
2	0.68 (.55)	0.15 (.18)	0.24 (.23)	0.24 (.23)	0.12 (.18)
3	0.64 (.50)	0.12 (.09)	0.23 (.11)	0.24 (.11)	0.02 (.09)
4	0.61 (.47)	0.13 (.04)	0.23 (.06)	0.23 (.06)	−0.07 (.04)
5	0.61 (.46)	0.13 (.02)	0.19 (.03)	0.21 (.03)	−0.05 (.02)

The theoretical values are given in brackets.

The negative values for $\hat{c}_Q(\tau)$ in contrast with the other autocorrelation functions are explained by the departure of $\widehat{\Sigma}_{XY}$ from symmetry. Whereas $\hat{c}_Q(\tau)$ is simply defined as the trace of $\widehat{\Sigma}_{XY}$, the other autocorrelation functions such as $\hat{c}_{DM}(\tau)$, $\hat{c}_{JW}(\tau)$ and $\hat{c}_{JM}(\tau)$ are based on symmetric matrices involving both $\widehat{\Sigma}_{XY}$ and $\widehat{\Sigma}_{YX}$. This causes the bias mentioned in the last section and is the major reason why these autocorrelation functions tend to fluctuate around *non-zero* values.

The findings presented in this section were supported by similar Monte-Carlo experiments involving a variety of first and second order WAR models. More specifically, in $Y_t = \alpha\, Y_{t-1} + U_t$ with $U_t \sim N(0, \sigma^2)$ the parameters $\alpha = 0.5, 0.75, 0.9$ were combined with $\sigma^2 = 0.2, 0.4, 0.6, 1.0, 3.0, 5.0$, while in $Y_t = \alpha_1\, Y_{t-1} + \alpha_2\, Y_{t-2} + U_t$ we chose $\alpha_1 = 0.75$, $\alpha_2 = 0.5$ and $\sigma^2 = 1.0, 3.0, 5.0$. These experiments suggested that the non-zero values mentioned above depend on the parameters of the AR process (Y_t).

Table 8.2. Estimated autocorrelation functions of the unwrapped process (Y_t).

Time Lag τ	Watson Beran $\hat{R}_{WB}(\tau)$	Downs Mardia $\hat{R}_{DM}(\tau)$	Johnson Wehrly $\hat{R}_{JW}(\tau)$	Jupp Mardia $\hat{R}_{JM}(\tau)$	Proposal $\hat{R}_Q(\tau)$	$R(\tau)$
1	0.51	0.51	0.56	0.53	0.51	0.50
2	0.14	0.18	0.24	0.24	0.15	0.25
3	0.00	0.15	0.24	0.24	0.02	0.13
4	−0.13	0.15	0.23	0.23	−0.08	0.06
5	−0.12	0.16	0.20	0.21	−0.07	0.03

Estimates of the autocorrelation function $R(\tau) = C(\tau)/C(0)$ of (Y_t), using formulae (6), (7), (8), (9) and (5b), are presented in Table 8.2. Depending on the formula used the estimates of these functions are denoted as $\hat{R}_{WB}$, $\hat{R}_{DM}$, $\hat{R}_{JW}$, $\hat{R}_{JM}$ and $\hat{R}_Q$, respectively. Except for $\hat{R}_{WB}$, they are all similar to the corresponding circular autocorrelation function. In fact, for these measures of association it is easy to show that the circular autocorrelation function is approximately equal to $R(\tau)$ if $C(0)$ is sufficiently small. Further, $\hat{R}_{WB}(\tau)$ and $\hat{R}_Q(\tau)$ tend to underestimate the true autocorrelation $R(\tau) = 0.5^\tau$ given in the last column of Table 8.2, whilst the other three start to fluctuate around constant non-zero values after only two lags, which implies that the higher order correlations are overestimated.

The value associated with independence is obviously attained much faster in case of a wrapped process compared with the corresponding unwrapped process. That is, by wrapping an AR process (Y_t) around the circle the correlation between the variables is diminished. The larger the variance σ^2 of the innovation process (U_t) the more pronounced this effect is. In fact, for $\sigma^2 > \pi$ this effect becomes so dominant that any dependence between the wrapped variables vanishes.

Table 8.3. Estimated partial autocorrelation functions of the unwrapped process (Y_t).

Lag	Watson Beran C^p_{WB}	Downs Mardia C^p_{DM}	Johnson Wehrly C^p_{JW}	Jupp Mardia C^p_{JM}	Proposal C^p_Q	$C^p(\tau)$
1	0.51	0.51	0.56	0.53	0.51	0.5
2	−0.16	−0.07	−0.11	−0.05	−0.13	0.0
3	0.00	0.17	0.22	0.18	0.01	0.0
4	−0.15	0.05	0.02	0.05	−0.10	0.0
order p	1	3	3	3	1	1

Estimates of the partial autocorrelation function $C^p(\tau)$ and of the order p of the AR process are given in Table 8.3. For lag 1 and 2 the various partial autocorrelation functions attain similar values but for lags greater than 2 their behaviour begins to differ. Whereas $C^p_{WB}(3)$ and $C^p_Q(3)$ are virtually 0, all the other lag 3 autocorrelations are about 0.2, which is roughly identical with the non-zero value referred to earlier. However, this distinction implies quite different results when using the ϕ-criterion of Hannan & Quinn (1979) to determine the order p of the underlying AR process $Y_t = \sum_{j=1}^p \alpha_j Y_{t-j} + U_t$ with $U_t \sim N(0, \sigma^2)$. In case of C^p_{WB} and C^p_Q we obtain $p = 1$, which in fact is the true order, while in case of C^p_{DM}, C^p_{JW} and C^p_{JM} we have $p = 3$, although this is marginal in the case of C^p_{DM}. Hence, the first and third order AR models can be regarded as close competitors in this example. To verify this result the reader is reminded that the time series length is $T = 50$. Parameter estimates using the Yule-Walker equations for both these models are presented in Table 8.4.

Table 8.4. Parameter estimates of fitted autoregressive model.

Order	Parameter	Watson Beran	Downs Mardia	Johnson Wehrly	Jupp Mardia	Proposal	True value
$p = 1$	α	0.51	0.51	0.56	0.53	0.51	0.5
	σ^2	0.59	0.59	0.55	0.58	0.59	0.6
$p = 3$	α_1	0.59	0.58	0.64	0.57	0.58	0.5
	α_2	−0.16	−0.19	−0.25	−0.16	−0.15	0.0
	α_3	0.00	0.14	0.22	0.19	0.01	0.0
	σ^2	0.58	0.57	0.52	0.55	0.58	0.6

When fitting a first order AR model the structure of the original process (10) is essentially recaptured in all cases, with the results based on c_{JW} being slightly inferior. On the other hand, the parameters of the third order AR model, as suggested by Hannan & Quinn's ϕ-statistic for *DM*, *JW* and *JM*, indicate in exactly these cases that the process (Y_t) has a strong high order correlation. Hence, the parameter estimates reflect the fact that for independent variables the corresponding circular correlation coefficients do not always decrease sufficiently fast to 0. As a result of the bias inherent in some of the empirical correlation coefficients the autocorrelation function of the underlying unwrapped process tends to be overestimated in these cases.

In summarising, it can be said that of all the methods applied in this study the one based on c_Q shows the best performance in terms of recapturing the parameters of the original process; that is $p = 1$, $\alpha = 0.5$ and $\sigma^2 = 0.6$. Even though the results presented in this section are based on a single WAR process, it is noted that they did not change fundamentally, when other first and second order WAR models were considered. In fact, given that most of the angular correlation coefficients discussed in section 7.3 fail to give sufficient indication when the underlying variables are independent, one would certainly not expect that other replicates could lead to different conclusions.

On the basis of the results presented in this chapter, it is therefore decided to use σ_Q as a measure of angular association in the analysis of directional time series. The coefficient ρ_{FL} may be regarded as a strong competitor. However, the choice of ρ_Q is supported by the fact that this measure is linear, related to the von Mises process in a natural way and easiest to calculate. The index Q in $c_Q(\tau)$ will subsequently be dropped and the autocovariance function simply be written as $c(\tau)$.

CHAPTER 9

INFERENCE FROM THE WRAPPED AUTOREGRESSIVE PROCESS

In the last chapter it was established that the circular autocovariance function c of a wrapped autoregressive process is best defined in terms of σ_Q. Given this definition, an asymptotically unbiased estimator of c is introduced in section 9.1. In sections 9.2 to 9.6 five different techniques of estimating the autocovariance function C of the underlying unwrapped process are discussed.

The first approach (EQ) presented in section 9.2 is simply based on equating empirical and theoretical covariances. Correction of the bias of this estimator, using a second order Taylor approximation, leads to EC, a technique described in section 9.3. In sections 9.4 to 9.6 we discuss a Bayesian approach (BA), maximum likelihood estimation using the EM-algorithm (ML) and a method based on characteristic functions estimation (CF), respectively.

In section 9.7 the five different techniques are compared in a number of Monte-Carlo studies. Both, EC and BA are expensive to implement and do not yield satisfactory results. For time series with a length of $T \leq 20$ the clearly best performance is shown by ML. However, as T gets larger the computational cost of the EM-algorithm also increases such that its implementation eventually becomes impractical. On the other hand, the results obtained by EQ and CF for $T \geq 50$ are almost as good as those obtained by ML. If only the variance $C(0)$ of the unwrapped autoregressive process is to be estimated then EQ and CF are identical, while EQ gives slightly better results if $C(0)$ as well as $C(1)$ are to be estimated. Since the given series (θ_t) of directional fluctuations is rather long, it is decided to use EQ in the estimation of C.

9.1 Introduction

Given a time series $(\eta_t)_{t=1,\dots,T}$ of directional data from a WAR process five techniques of estimating the autocovariance function $C(\tau)$ of the underlying unwrapped process are explored and compared in a Monte-Carlo study. First, the different techniques are briefly summarised assuming that the variables η_t are independent and identically distributed as $WN(0, C)$. Then the techniques are generalised to the case of an underlying first order autoregressive process. The independent and the dependent data case are both covered in the simulation study presented in section 9.7.

Suppose the underlying unwrapped process has the form

$$Y_t = \alpha Y_{t-1} + U_t$$

with $0 \le \alpha < 1$ and U_t independent and identically distributed as $N(0, \sigma^2)$. If $(y_t)_{t=1,\dots,T}$ was observed then it is easy to show that the estimator

$$\hat{C}(\tau) = \tag{1}$$
$$\frac{1}{T-\tau} \sum_{t=\tau+1}^{T} \left(y_{t-\tau} - \frac{1}{T-\tau} \sum_{s=\tau+1}^{T} y_{s-\tau} \right) \left(y_t - \frac{1}{T-\tau} \sum_{s=\tau+1}^{T} y_s \right)$$

has a bias which is approximately equal to

$$b_\alpha(\tau) = -\frac{1}{\alpha^\tau \, (T-\tau)} \frac{1+\alpha}{1-\alpha} C(\tau)\,.$$

The details of the derivation are omitted. That is, $\hat{C}(\tau)$ is an asymptotically unbiased estimator of $C(\tau)$. In a similar way it can be shown that this result also holds if the order of the autoregressive process (Y_t) is greater than 1. However, instead of (y_t) only (η_t) is observed so that $C(\tau)$ has to be estimated indirectly through the circular autocovariance function $c(\tau) = \sigma_Q(\eta_{t-\tau}, \eta_t)$. From (1) it follows that

$$\hat{c}(\tau) = s_c(\tau) + s_s(\tau) \tag{2}$$

with

$$s_c(\tau) = \frac{1}{T-\tau} \sum_{t=\tau+1}^{T} \left(\cos\eta_{t-\tau} - \sum_{s=\tau+1}^{T} \frac{\cos\eta_{s-\tau}}{T-\tau} \right) \left(\cos\eta_t - \sum_{s=\tau+1}^{T} \frac{\cos\eta_s}{T-\tau} \right)$$

and

$$s_s(\tau) = \frac{1}{T-\tau} \sum_{t=\tau+1}^{T} \left(\sin\eta_{t-\tau} - \sum_{s=\tau+1}^{T} \frac{\sin\eta_{s-\tau}}{T-\tau} \right) \left(\sin\eta_t - \sum_{s=\tau+1}^{T} \frac{\sin\eta_s}{T-\tau} \right)$$

is an asymptotically unbiased estimator of $c(\tau)$. Throughout this chapter, when concentrating on the estimation of $C(\tau)$, it will therefore be assumed that $\hat{c}(\tau)$ is asymptotically unbiased.

9.2 Equating Theoretical and Empirical Circular Variance (EQ)

The relationship between $C = C(0)$ and the circular variance $c = c(0)$ is given by

$$C = -\ln(1-c)\,,$$

which is sketched in Figure 9.1.

Figure 9.1. C as a function of c.

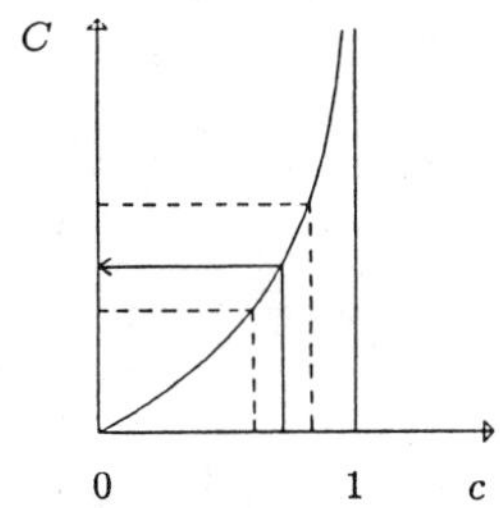

As a first approach to estimating C one could therefore use

$$C^{EQ} = -\ln(1-\hat{c})\,. \tag{3}$$

However, given an unbiased estimator $\hat{c}$ of c this will usually yield a biased estimate of C, since

$$\mathcal{E}[\,C^{EQ}\,] \geq -\ln(1-\mathcal{E}[\,\hat{c}\,]) = C\,,$$

as is evident from Figure 9.1. Indeed, if $\hat{c}$ is close to 1, that is if the wrapped variables are almost uniformly distributed, then the estimate C^{EQ} becomes extremely unstable and thus unreliable. In order to ensure numerical stability and at the same time to correct for the bias we therefore introduce a damping constant d and replace $\hat{c}$ by $\hat{c}\exp\{-d\hat{c}\}$ when estimating C. The effect of this damping is shown in Figure 9.2 by the departure of the curve $\hat{c}\exp\{-d\hat{c}\}$ from the diagonal.

Figure 9.2. Effect of damping on the estimate of c.

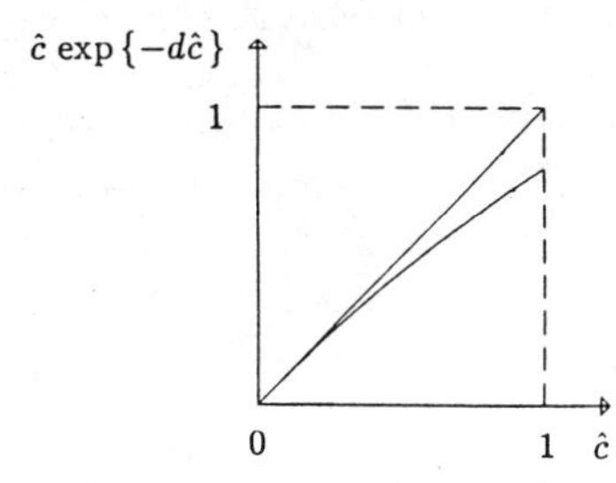

Given this exponential damping the bias of the estimate of C reduces to

$$\mathcal{E}[C_d^{EQ}] - C = \mathcal{E}[-\ln(1 - \hat{c}\exp\{-d\hat{c}\})] - C$$

$$\approx -\ln(1 - c\exp\{-dc\})$$

$$- \frac{(\exp\{dc\} - c)\, d + (d\, \exp\{dc\} - 1)\,(1 - dc)}{2\,(\exp\{dc\} - c)^2}\, \mathrm{var}\,(\hat{c}) - C$$

$$\approx -\frac{dc^2}{1-c} + \frac{1 - c - 2d}{2\,(1-c)^3}\,\mathrm{var}\,(\hat{c})$$

$$(4) \qquad = \frac{\mathrm{var}\,(\hat{c})}{2\,(1-c)^2} - d\,\frac{(1-c)^2c^2 + \mathrm{var}\,(\hat{c})}{(1-c)^3}\,.$$

The last term in (4) represents the major correction due to this damping. It follows that for any continuous distribution of $\hat{c}$ there exists a constant

$$(5) \qquad d \approx \tfrac{1}{2}\,[(1-c)\,\mathrm{var}\,(\hat{c})]\ /\ [(1-c)^2c^2 + \mathrm{var}\,(\hat{c})]$$

such that the modified estimator $C_d^{EQ} = -\ln(1-\hat{c}\exp\{-d\hat{c}\})$ is approximately unbiased if $\mathcal{E}[\hat{c}] = c$. In order to determine the appropriate damping constant d one therefore needs to know the size of $\mathrm{var}\,(\hat{c})$. A way of approximating this term will be presented in the following section (cf. equation (13)). Substituting that expression into (5) yields

$$(6) \qquad d_0 = [T + 2/(1-c)]^{-1}$$

where T is the length of the time series under consideration.

A brief numerical study indicated that the correction term in (4) is fairly accurate. Altogether 50 time series $(\eta_t)_{t=1,\ldots,T}$ were generated with all the variables η_t being independent and identically distributed as $WN(0, C)$. The parameter C was then estimated for each of the 50 time series using C_d^{EQ} with $d = 0.00, 0.02, 0.06, 0.10$. Average values for the 50 runs are listed in Table 9.1 for the two cases $C = 1$ and $C = \pi^2/4$. In the absence of any damping the bias amounts to 13 per cent and 17 per cent, respectively. Whilst $d = 0.06$ virtually eliminated the bias if $C = 1$, the best result for $C = \pi^2/4$ was obtained for $d = 0.02$. This smaller value for d is due to the fact that the length T of the time series is greater in the second case, and that the correction term in (4) becomes more sensitive as C increases and c tends towards 1. The last two columns in Table 9.1 list the variance $\mathrm{var}\,(\hat{c})$ and the damping constant d_0 calculated according to formulae (13) and (6), respectively. Whereas d_0 coincides with the best numerically established damping constant if $C = 1$ there is a slight discrepancy if $C = \pi^2/4$ reflecting the increasing instability of the estimator C^{EQ}. An expression for the variance of C^{EQ} will be derived below in section 9.3.

Table 9.1. Mean of C_d^{EQ} over 50 runs for various damping constants d.

C	c	T	Mean $C_{.00}^{EQ}$	Mean $C_{.02}^{EQ}$	Mean $C_{.06}^{EQ}$	Mean $C_{.10}^{EQ}$	var $(\hat{c})$	d_0
1	0.632	10	1.131	1.080	1.005	0.965	0.029	0.065
$\pi^2/4 = 2.467$	0.915	50	2.899	2.482	2.109	1.869	0.003	0.014

Note that (3) can be extended without difficulty to an estimator of $C(\tau)$ with τ different from 0. More specifically, if $\hat{c}(\tau)$ ($\tau = 0, 1, 2, \ldots$) denotes an estimate of the circular autocovariance function of a WAR process then

$$C^{EQ}(\tau) = \ln\left(\frac{\hat{c}(\tau)}{1 - \hat{c}(0)} + 1\right)$$

provides an estimate of the autocovariance function $C(\tau)$ of the unwrapped process, which is derived in the same way as C^{EQ} (cf. equation (8.2)).

9.3 Corrected EQ-Estimation (EC)

The introduction of the damping constant d was somewhat *ad hoc*. In this section we will therefore use a Taylor series expansion for $-\ln(1 - \hat{c})$ about the true parameter c to improve the estimator C^{EQ}. That is

(7) $$-\ln(1 - \hat{c}) = -\ln(1 - c) + \sum_{k=1}^{\infty} \frac{1}{k}\left(\frac{\hat{c} - c}{1 - c}\right)^k$$

leading to the approximation

(8) $$C \approx \mathcal{E}\big[-\ln(1 - \hat{c})\big] - \tfrac{1}{2}\operatorname{var}(\hat{c}) \;/\; (1 - c)^2 \,.$$

Note that this is the same as formula (4) for $d = 0$. On the basis of (8) let us now define the estimator

(9) $$C^{EC} = -\ln(1 - \hat{c}) - \tfrac{1}{2}\widehat{\operatorname{var}}(\hat{c}) \;/\; (1 - \hat{c})^2$$

where $\widehat{\operatorname{var}}(\hat{c})$ is an estimate of var $(\hat{c})$.

Subsequently we will concentrate on the estimation of

$$\operatorname{var}(\hat{c}) = \operatorname{var}(s_c^2) + 2\operatorname{cov}(s_c^2, s_s^2) + \operatorname{var}(s_s^2)\,,$$

where $s_c^2 = s_c(0)$ and $s_s^2 = s_s(0)$. First, expressing the right hand side in terms of the cumulants κ_{jk} of $(\cos\theta_t, \sin\theta_t)$ yields

$$\text{var}(\hat{c}) = \frac{1}{T}(\kappa_{40} + 2\kappa_{22} + \kappa_{04}) + \frac{2}{T-1}(\kappa_{20}^2 + 2\kappa_{11}^2 + \kappa_{02}^2) \tag{10}$$

(cf. Kendall & Stuart 1963, formulae (13.45) and (13.47)). To estimate var $(\hat{c})$ one can therefore employ corresponding k-statistics which in turn can be expressed in terms of a-statistics and then in terms of s-statistics. The result is given below without derivation since only straightforward, although tedious, algebra is required.

Given

$$\begin{aligned} k_4(x_t, y_t) = \big[T(T-1)(T-2)(T-3)\big]^{-1}\Big[& -6\sum_{q,r,s,t} x_q x_r y_s y_t \\ & + 2T\Big(\sum_{r,s,t} x_r x_s y_t^2 + 4\sum_{r,s,t} x_r y_s x_t y_t + \sum_{r,s,t} x_r^2 y_s y_t\Big) \\ & - (T^2 - T)\Big(\sum_{s,t} x_s^2 y_t^2 + 2\sum_{s,t} x_s y_s x_t y_t\Big) \\ & - 2(T^2 + T)\Big(\sum_{s,t} x_s x_t y_t^2 + \sum_{s,t} y_s y_t x_t^2\Big) + (T^3 + T^2)\sum_t x_t^2 y_t^2\Big] \end{aligned} \tag{11}$$

and

$$\begin{aligned} k_2^2(x_t, y_t) = \big[T(T-1)(T-2)(T-3)\big]^{-1}\Big[& \sum_{q,r,s,t} x_q x_r y_s y_t \\ & - \sum_{r,s,t} x_r^2 y_s y_t - 2(T-1)\sum_{r,s,t} x_r y_r x_s y_t - \sum_{r,s,t} x_r x_s y_t^2 \\ & + (T^2 - 3T + 2)\sum_{s,t} x_s y_s x_t y_t + \sum_{s,t} x_s^2 y_t^2 \\ & + 2(T-1)\Big(\sum_{s,t} x_s x_t y_t^2 + \sum_{s,t} y_s y_t x_t^2\Big) - (T^2 - T)\sum_t x_t^2 y_t^2\Big] \end{aligned} \tag{12}$$

the cumulants in (10) can be estimated by the following k-statistics:

$$\begin{aligned} &\kappa_{40} \text{ by } k_4(\cos\eta_t, \cos\eta_t) \\ &\kappa_{22} \text{ by } k_4(\cos\eta_t, \sin\eta_t) \\ &\kappa_{04} \text{ by } k_4(\sin\eta_t, \sin\eta_t) \\ &\kappa_{20}^2 \text{ by } k_2^2(\cos\eta_t, \cos\eta_t) \\ &\kappa_{11}^2 \text{ by } k_2^2(\cos\eta_t, \sin\eta_t) \\ \text{and } &\kappa_{02}^2 \text{ by } k_2^2(\sin\eta_t, \sin\eta_t)\,. \end{aligned}$$

Although the estimator obtained by substituting these expressions into (10) and then back into (9) is still biased it works well as long as η_t is not too close to a uniform distribution; more specifically that is, if the corresponding value of c is less than about 0.9. Further, to ensure numerical stability mild damping along the lines of section 9.2 is suggested.

Let us return to the case where $\eta_t \sim WN(0, C)$ for $t = 1, \ldots, T$. To derive equation (13) below one needs the relationship between cumulants and moments (cf. Kendall & Stuart 1963, formulae (12.28) and (13.2)), and the fact that the characteristic function of a wrapped variable is the same as for the unwrapped variable, as explained in section 8.2. Formula (10) then takes the form

$$\begin{aligned} \operatorname{var}(\hat{c}) &= -\frac{1}{T}\,(1-\exp\{-C\})^4 \\ &\quad + \frac{1}{T-1}\,(1-\exp\{-C\})^2 \cdot (1+\exp\{-2C\}) \\ &= \frac{2}{T}\,\exp\{-C\}\,(1-\exp\{-C\})^2 + o(T^{-2})\,. \end{aligned} \tag{13}$$

It follows that

$$\operatorname{var}(\sqrt{T}\,\hat{c}) \;\rightarrow\; 2\,(1-c)\,c^2\,.$$

Using the second order Taylor approximation

$$\hat{c} = 1-\exp\{-C^{EQ}\} \approx 1-\exp\{-C\}\,\left[1-C^{EQ}+C+\tfrac{1}{2}(C^{EQ}-C)^2\right] \tag{14}$$

implies

$$\operatorname{var}(C^{EQ}) \approx \frac{\operatorname{var}(\hat{c})}{(1-c)^2}\left(1-\frac{\operatorname{var}(\hat{c})}{4\,(1-c)^2}\right)$$

and hence,

$$\operatorname{var}(\sqrt{T}\,C^{EQ}) \;\rightarrow\; \frac{c^2}{1-c}\left(2-\frac{c^2}{(1-c)\,T}\right). \tag{15}$$

Expression (13) attains a maximum of $\frac{8}{27}T^{-1}$ for $C = \ln 3$, ignoring the term $o(T^{-2})$. On the other hand, if $C \rightarrow 0$ or $C \rightarrow \infty$ then $\operatorname{var}(\hat{c})$ will tend towards 0. The latter of these two results may be surprising and therefore require a comment. As $C \rightarrow \infty$, η will tend towards a uniform distribution. Consequently, $\hat{c} \rightarrow 1$ and $\operatorname{var}(\hat{c}) \rightarrow 0$.

9.4 Bayes Estimation (BA)

Again, consider $Y \sim N(0, C)$ and $\eta = Y \pmod{2\pi}$. The density of the wrapped variable η can then be well approximated for small C by

$$f(\eta) = \tag{16}$$
$$[2\pi C]^{-1}\left[\exp\left\{-\frac{\eta^2}{2C}\right\} + \exp\left\{-\frac{(\eta+2\pi)^2}{2C}\right\} + \exp\left\{-\frac{(\eta-2\pi)^2}{2C}\right\}\right].$$

The first term corresponds to no wrapping, the second to one positive and the third to one negative wrapping, while all wrappings of an order greater than one are ignored.

Table 9.2. Probability of Y conditional on η ($C = \pi^2/4$).

η	$-180°$	$-120°$	$-60°$	$0°$	$60°$	$120°$	$180°$
$P(Y = \eta - 2\pi \mid \eta)$	.000	.000	.000	.000	.005	.065	.500
$P(Y = \eta \mid \eta)$	.500	.935	.995	.999	.995	.935	.500
$P(Y = \eta + 2\pi \mid \eta)$	.500	.065	.005	.000	.000	.000	.000

Table 9.2 lists the probability that the angle η is obtained by wrapping Y k-times around the circle for $k = -1, 0$ and 1. It also shows that any wrapping of an order greater than 1 is unlikely to occur if the standard deviation $C^{1/2}$ takes a value of less than $\pi/2$. Of course, the probability that the wrapping of Y has no effect, that is $Y = \eta$, is 0.95 for $C^{1/2} = \pi/2$.

Let $g(C)$ denote the *a priori* density of C, and let C^{BA} be the Bayesian estimator of C based on the sample $\{\eta_1, \ldots, \eta_T\}$ with η_t $(t = 1, \ldots, T)$ independently distributed as $WN(0, C)$. It is further assumed that the cost function is $(C^{BA} - C)^2$.

To derive a formula for the Bayesian estimator C^{BA}, the following notation is introduced. Let $P, Q \subseteq \{1, \ldots, T\}$ contain the indices of all the observations η_t which are obtained through positive and negative wrapping, respectively. Obviously, $P \cap Q = \emptyset$, with $t \notin P \cup Q$ indicating that η_t is identical with the unwrappped observation y_t. Given the approximation (16), putting

$$b(P, Q) = \sum_{t \notin P \cup Q} \eta_t^2 + \sum_{t \in P} (\eta_t + 2\pi)^2 + \sum_{t \in Q} (\eta_t - 2\pi)^2$$

yields

$$C^{BA} = \frac{\int g(C)\cdot(2C)^{-n/2+1}\sum_{(P,Q)}\exp\{-b(P,Q)/2C\}\,dC}{2\int g(C)\cdot(2C)^{-n/2}\sum_{(P,Q)}\exp\{-b(P,Q)/2C\}\,dC}$$

where both sums are taken over all subset configurations $P, Q \subseteq \{1,\dots,T\}$ with $P \cap Q = \emptyset$. Choosing a uniform *a priori* density, that is

$$g(C) = \begin{cases} a^{-1} & \text{if } 0 \le C \le a \\ 0 & \text{otherwise} \end{cases} \quad , \; a \in \mathcal{R}^+$$

and letting a tend towards ∞ leads to the following estimator

$$C^{BA} = \frac{\Gamma(\frac{T}{2}-2)\sum_{(P,Q)} b(P,Q)^{-T/2+2}}{2\Gamma(\frac{T}{2}-1)\sum_{(P,Q)} b(P,Q)^{-T/2+1}}$$

(17) $$= \frac{1}{T-4}\,\frac{\sum_{(P,Q)} b(P,Q)^{-T/2+2}}{\sum_{(P,Q)} b(P,Q)^{-T/2+1}}\,.$$

For an *a priori* γ_r-distribtuion ($r \in \mathcal{N}$), on the other hand, the Bayesian estimator turns out to be

(18) $$C^{BA} = \frac{1}{T+2r-2}\cdot\frac{\sum_{(P,Q)}(b(P,Q)+1)^{-T/2-r+1}}{\sum_{(P,Q)}(b(P,Q)+1)^{-T/2-r}}\,.$$

For larger values of the variance C approximation (16) becomes worse, and an approximation based on the characteristic function representation

$$f(\eta) = [2\pi]^{-1}\left[1 + 2\sum_{\nu=1}^{\infty}\exp\{-\nu^2 C/2\}\cdot\cos\nu\eta\right]$$

might be more appropriate. For $C > \pi^2/4$ the density $f(\eta)$ is almost perfectly approximated if the summation only runs up to $\nu = 2$.

The drawback of using the Bayesian approach is that the expressions (17) and (18) become fairly intractable as T increases since all subset configurations (P,Q) have to be considered. Numerical results based on a number of generated series $(\eta_t)_{t=1,\dots,T}$ suggest that there does not exist a satisfactory approximation of the term $\sum_{(P,Q)} b(P,Q)^{-T/2}$ by a sum over only those (P,Q) with $|P \cup Q| \le k$ for some small $k \in \mathcal{N}_0$.

9.5 Maximum Likelihood Estimation (ML)

Given that the observed series (η_t) is obtained by wrapping (y_t) around the circle, each y_t can be written as $y_t = 2\pi m + \eta_t$ $(m \in \mathcal{N}_0)$ with η_t being observed and $2\pi m$ missing. In order to estimate the autocovariances $C(\tau)$ of the underlying process (Y_t) one could therefore employ the *EM-algorithm* (Dempster *et al.* 1977). Through an iterative procedure it determines the maximum likelihood estimate from incomplete data with each iteration consisting of an expectation step followed by a maximisation step.

Suppose the complete data specification is of the form

$$f(y \mid \mathbf{C}) = \frac{1}{a(\mathbf{C})} \exp\{\mathbf{C}' \, \mathbf{t}(y)\}$$

where $\mathbf{C}$, $a(\mathbf{C})$ and $\mathbf{t}(y)$ are a parameter vector, an integration constant depending on $\mathbf{C}$ and a complete data sufficient statistic, respectively. Suppose also that the random variable Y is observed only indirectly through a variable η, and that $f(y \mid \mathbf{C})$ and the incomplete data specification $g(\eta \mid \mathbf{C})$ are related through

$$g(\eta \mid \mathbf{C}) = \sum_{A(\eta)} f(y \mid \mathbf{C})$$

where $A(\eta) = \{y \mid y = \eta \pmod{2\pi}\}$.

Now, given an initial estimate $\mathbf{C}^{(0)}$ of $\mathbf{C}$ each iteration $(k = 1, 2, \ldots)$ of the EM-algorithm comprises the following two steps:

E-step: Estimate $\mathbf{t}(y)$ by

(19a) $$\mathbf{t}^{(k)} = \mathcal{E}[\,\mathbf{t}(y) \mid \eta, \mathbf{C}^{(k-1)}\,]\,.$$

M-step: Determine $\mathbf{C}^{(k)}$ as the solution of

(19b) $$\mathcal{E}[\,\mathbf{t}(y) \mid \mathbf{C}\,] = \mathbf{t}^{(k)}\,.$$

Note that the familiar maximum likelihood equations are of the form (19b), if the data are from a distribution which belongs to the regular exponential family. Only mild conditions have to be satisfied to ensure that a repeated application of the E- and M-step ultimately leads to the maximum likelihood estimate $\mathbf{C}^{ML}$. For a detailed discussion of the EM-algorithm the reader is referred to Dempster *et al.* (1977).

Let $Y \sim N(0, C)$ and $\eta = Y \pmod{2\pi}$. Given the sample $\{\eta_1, \ldots, \eta_T\}$ of independent observations from a wrapped normal distribution, the application of the

EM-algorithm involves the following steps:

$$E-step: \quad t^{(k)} = \sum_{t=1}^{T} \sum_{m=-\infty}^{\infty} \mathcal{P}(y_t = \eta_t + 2\pi m \mid \eta_t,\ C^{(k-1)}) \cdot (\eta_t + 2\pi m)^2$$

$$M-step: \quad C^{(k)} = t^{(k)}/T\ .$$

The probabilities required in the E-step are listed in Table 9.2 for a variance of $\pi^2/4$.

Now, consider a first order normal autoregressive process $(Y_t)_{t=0,\dots,T}$ and let $\eta_t = Y_t \pmod{2\pi}$. Then

$$f(y_t, y_{t-1}) = [2\pi \det^{1/2} \mathbf{C}]^{-1} \exp\left\{-\tfrac{1}{2}\,(y_t, y_{t-1})\,\mathbf{C}^{-1}\begin{pmatrix} y_t \\ y_{t-1}\end{pmatrix}\right\}$$

where $\mathbf{C} = \begin{pmatrix} C(0) & C(1) \\ C(1) & C(0)\end{pmatrix}$, and hence,

E-step:

$$\text{(20a)} \qquad \mathbf{t}^{(k)} = \sum_{t=1}^{T} \sum_{m,n=-\infty}^{\infty} \mathcal{P}_t(m,n) \begin{pmatrix} z_t^2(m) & z_t(m)\, z_{t-1}(n) \\ z_t(m)\, z_{t-1}(n) & z_{t-1}^2(n)\end{pmatrix}$$

with $\mathcal{P}_t(m,n) = \mathcal{P}(y_t = \eta_t + 2\pi m,\ y_{t-1} = \eta_{t-1} + 2\pi n \mid \mathbf{C}^{(k-1)}, \eta_t, \eta_{t-1})$ and $z_t(m) = \eta_t + 2\pi m$,

M-step:

$$\text{(20b)} \qquad \mathbf{C}^{(k)} = \mathbf{t}^{(k)}/T\ .$$

9.6 Characteristic Function Estimation (CF)

This method was proposed by Heathcote (1977) and may be preferred to standard methods of inference if the likelihood is more complicated than the characteristic function or if the characteristic function is smoother with respect to nuisance parameters. To motivate the extension of this technique to the dependent data case, the principle underlying the characteristic function estimation will be explained in slightly more detail.

Denote the characteristic function of a circular variable η, depending on a parameter C, by $\phi(\nu; C)$ and let

$$\phi_T(\nu) = T^{-1} \sum_{t=1}^{T} \exp\{i\nu\eta_t\}$$

be the empirical characteristic function based on the sample $\{\eta_1, \ldots, \eta_T\}$ of independent and identically distributed variables. The estimator C^{CF} which minimises the function

$$I(C) = \int_{-\infty}^{\infty} \|\phi_T(\nu) - \phi(\nu; C)\|^2 \, G(d\nu) ,$$

where G is a suitable monotonically increasing weight function with total variation 1, is then called the *integrated squared error estimator of* C (Heathcote 1977). The key statistic to this approach is given by

$$K(C) = \frac{d}{dC} \int_{-\infty}^{\infty} \| \exp\{i\nu\eta\} - \phi(\nu; C)\|^2 \, G(d\nu) \tag{21}$$

which plays a role analogous to that of $d/dC \ln f(\eta; C)$ in the context of maximum likelihood estimation. Indeed, the two are correlated in the following way:

$$\lambda(C) = \operatorname{cov}\left(K(C), \frac{d}{dC} \ln f(\eta; C)\right) = -2 \int_{-\infty}^{\infty} \|\phi'(\nu; C)\|^2 \, G(d\nu) .$$

Under mild regularity conditions it follows that

$$\sqrt{T}\,(C^{CF} - C) \xrightarrow{\text{d}} N(0, \lambda^{-2}(C) \cdot \operatorname{var}(K(C)))$$

(Heathcote 1977). If $\mathcal{F}(C) = \operatorname{var}\left(d/dC \ln f(\eta; C)\right)$ denotes the Fisher information contained in η then the efficiency of C^{CF} relative to the maximum likelihood estimator is defined as

$$e(C^{CF}) = \lambda^2(C) \,/\, \left[\operatorname{var}\left(K(C)\right) \cdot \mathcal{F}(C)\right] . \tag{22}$$

Given the right choice of G it is possible to bound $e(C^{CF})$ away from 0. In particular, if for all ν with $G(d\nu) \neq 0$, $\operatorname{corr}^2\left(\cos \nu\eta, \frac{d}{dC} \ln f(\eta; C)\right)$ and $\operatorname{corr}^2\left(\sin \nu\eta, \frac{d}{dC} \ln f(\eta; C)\right)$ are greater than $b \in \mathcal{R}_0^+$ then $b \leq e(C^{CF}) \leq 1$ (Heathcote 1977). This inequality provides some guidance concerning the support of G. A reasonable choice of G thus requires that its increase is confined to those points for which the correlations above attain their maxima.

As in the previous sections consider the case where all η_t, $(t = 1, \ldots, T)$ are independent and identically distributed as $WN(0, C)$. Following the discussion above concerning the choice of the weight function G, define

$$G(\nu) = \begin{cases} 0 & \text{if } \nu < 1 \\ 1 & \text{otherwise} . \end{cases}$$

For a discussion of the choice of the weight function G the reader is also referred to Heathcote (1977). By putting $I'(C^{CF}) = 0$ one obtains the characteristic function estimator

$$C^{CF} = -\ln r^2 \tag{23}$$

where $r = (1 - \hat{c})^{1/2}$ is an estimator of the resultant length $\rho = \exp\{-C/2\}$. Formula (23) shows that for this particular choice of G the characteristic function estimator C^{CF} coincides with C^{EQ}. Further, the statistics $K(C)$ and $\lambda(C)$ reduce to

$$K(C) = \cos\eta \ \exp\{-C/2\} - \exp\{-C\}$$

and

$$\lambda(C) = -\tfrac{1}{2}\exp\{-C\},$$

respectively. It follows that

$$\operatorname{var}\big(K(C)\big) = \tfrac{1}{2}\exp\{-C\}\ (1 - \exp\{-C\})^2$$

and hence,

$$\sqrt{T}\ (C^{CF} - C) \xrightarrow{\mathrm{d}} N\big(0, 2\exp\{C\}\ (1 - \exp\{-C\})^2\big)\,. \tag{24}$$

Observe that the variance in (24) is the same as the first term in (15) which can be derived by using a first order Taylor approximation instead of (14).

A lower bound for the relative efficiency of the characteristic function estimator C^{CF} is given by

$$\begin{aligned}\rho_c^2 &= \operatorname{corr}^2\Big(\cos\eta, \frac{d}{dC}\ln f(\eta; C)\Big)\\ &= \big[2\exp\{C\}\cdot(1 - \exp\{-C\})^2\cdot\mathcal{F}(C)\big]^{-1}\end{aligned}$$

which is identical with the asymptotic efficiency obtained by substituting the expression for the variance in (24) into (22).

Table 9.3 lists the values of $\lambda(C)$, $\operatorname{var}\big(K(C)\big)$, the Fisher information $\mathcal{F}(C)$ and the efficiency $e(C^{CF})$ for a range of values for C. From the last column it can be seen that C^{CF} is almost as efficient as the maximum likelihood estimator C^{ML}. Indeed, for $C^{1/2}$ smaller than 0.6 and $C^{1/2}$ greater than 2.0 the relative efficiency of characteristic function estimation is remarkably good. It is noted, however, that for larger values of

Table 9.3. Relative efficiency of the integrated squared error estimator C^{CF}.

St. dev. $C^{1/2}$	$\lambda(C)$	$\operatorname{var}\big(K(C)\big)$	$\mathcal{F}(C)$	Efficiency $e(C)$
0.5	$-3.9\cdot10^{-1}$	$1.9\cdot10^{-2}$	8.0	0.995
1.0	$-1.8\cdot10^{-1}$	$7.3\cdot10^{-2}$	$4.9\cdot10^{-1}$	0.937
1.5	$-5.3\cdot10^{-2}$	$4.2\cdot10^{-2}$	$7.0\cdot10^{-2}$	0.936
2.0	$-9.2\cdot10^{-3}$	$8.8\cdot10^{-3}$	$9.7\cdot10^{-3}$	0.983
2.5	$-9.7\cdot10^{-4}$	$9.6\cdot10^{-4}$	$9.7\cdot10^{-4}$	0.998
3.0	$-6.2\cdot10^{-5}$	$6.2\cdot10^{-5}$	$6.2\cdot10^{-5}$	0.999

C also maximum likelihood estimation becomes less satisfactory as is evident from the $\mathcal{F}(C)$-column in Table 9.3. Even the minimum relative efficiency of 0.921 attained at $C^{1/2} = 1.2$ is fairly high, suggesting that for *large* samples from a $WN(0, C)$ distribution, when maximum likelihood estimation becomes expensive to implement, C^{CF} and equivalently C^{EQ} provide appropriate alternatives to C^{ML} for the estimation of the variance C.

The method of characteristic function estimation will now be extended to the multivariate situation. First, assume that the variable (θ, η) has a density $f(\theta, \eta; \mathbf{C})$ and a characteristic function $\phi(\mu, \nu; \mathbf{C})$, both depending on the parameter vector $\mathbf{C}$. Then, just as in the univariate case, define the empirical characteristic function, based on the random sample $\{(\theta_t, \eta_t) \mid t = 1 \ldots, T\}$, as

$$\phi_T(\mu, \nu) = T^{-1} \sum_{t=1}^{T} \exp\{i\mu\theta_t + i\nu\eta_t\} .$$

As before, the estimator $\mathbf{C}^{CF}$ is obtained as the vector that minimises the integral

$$I(\mathbf{C}) = \iint_{-\infty}^{\infty} \|\phi_T(\mu, \nu) - \phi(\mu, \nu; \mathbf{C})\|^2 \, G(d\mu, d\nu) .$$

This integrated squared error estimator $\mathbf{C}^{CF}$ will be studied in the particular case of a first order autoregressive process $Y_t = \alpha Y_{t-1} + U_t$ with $U_t \sim N(0, \sigma^2)$. Hence, denoting the autocovariance function of (Y_t) by $C(\tau)$ $(\tau = 0, 1, \ldots)$,

$$\begin{aligned} \sigma^2 &= C(0) - C^2(1)/C(0) \\ \alpha &= C(1)/C(0) . \end{aligned}$$

In order to determine the parameter vector $\mathbf{C} = (C_1, C_2)' = (\sigma^2, \alpha)'$ it will therefore be sufficient to estimate $C(0)$ and $C(1)$. Similar to (21) we define

$$\mathbf{K}(\mathbf{C}) = \frac{d}{d\mathbf{C}} \iint_{-\infty}^{\infty} \|\exp\{i\mu\theta + i\nu\eta\} - \phi(\mu, \nu; \mathbf{C})\|^2 \, G(d\mu, d\nu) . \tag{25}$$

The estimating equations can then be written in the form

$$\frac{d}{d\mathbf{C}} I(\mathbf{C}^{CF}) = \frac{1}{T} \sum_{t=1}^{T} \mathbf{K}_t(\mathbf{C}^{CF}) = \mathbf{0} \tag{26}$$

where $\mathbf{K}_t(\mathbf{C}^{CF})$ is the expression in (25) with (θ, η) replaced by (θ_t, η_t) and $\mathbf{C}$ by $\mathbf{C}^{CF}$.

Just as in the univariate case one obtains the following relationship between $\mathbf{K}(\mathbf{C})$ and the maximum likelihood statistic $d/d\mathbf{C} \ln f(\theta, \eta; \mathbf{C})$:

$$\begin{aligned} \mathbf{\Lambda}(\mathbf{C}) &= \operatorname{cov}\left(\mathbf{K}(\mathbf{C}), \frac{d}{d\mathbf{C}} \ln f(\theta, \eta; \mathbf{C})\right) \\ &= -2\left(\iint_{-\infty}^{\infty} \left[\frac{\partial u}{\partial C_j} \frac{\partial u}{\partial C_k} + \frac{\partial v}{\partial C_j} \frac{\partial v}{\partial C_k} \right] G(d\mu, d\nu) \right)_{j,k=1,2} \end{aligned}$$

where u and v are the real and imaginary parts of the characteristic function $\phi(\mu,\nu;\mathbf{C}) = u + iv$. Define $\mathbf{\Sigma} = \mathrm{var}\,(\mathbf{K}(\mathbf{C}))$ and assume that all second partial derivatives of $I(\mathbf{C})$ are bounded by some G-integrable function. The strong law of large numbers for strictly stationary processes (Walters 1975, p.29) then yields at the true parameter vector $\mathbf{C}$

$$\left(\mathcal{E}\left[\frac{\partial^2}{\partial C_j \partial C_k}\, I(\mathbf{C})\right]\right)_{j,k=1,2} \overset{\mathrm{pr}}{\to} \mathbf{\Lambda} \tag{27}$$

and from the Central Limit Theorem for Markov processes (Doob 1953, p.232) one has

$$\sqrt{T}\,\frac{d}{d\mathbf{C}}\, I(\mathbf{C}) \overset{\mathrm{d}}{\to} N(\mathbf{0},\mathbf{\Sigma})\,. \tag{28}$$

Hence, on using (27), (28) and a first order Taylor approximation for $d/d\mathbf{C}\ I(\mathbf{C}^{CF})$,

$$\sqrt{T}\,(\mathbf{C}^{CF} - \mathbf{C}) \overset{\mathrm{d}}{\to} N(\mathbf{0},\ \mathbf{\Lambda}^{-1}\mathbf{\Sigma}\mathbf{\Lambda}^{-1})\,.$$

In order to determine a lower bound for the relative efficiency of the integrated squared error estimator $\mathbf{C}^{CF}$ assume that for all $(\mu,\nu) \in \{G(d\mu,d\nu) \neq 0\}$ the following conditions hold $(j = 1,2)$:

$$b_j \leq \mathrm{corr}^2\left(\cos(\mu\theta+\nu\eta)\,,\ \frac{\partial}{\partial C_j}\ln\, f(\theta,\eta;\mathbf{C})\right) \tag{29a}$$

$$b_j \leq \mathrm{corr}^2\left(\sin(\mu\theta+\nu\eta)\,,\ \frac{\partial}{\partial C_j}\ln\, f(\theta,\eta;\mathbf{C})\right)\,. \tag{29b}$$

Denoting the Fisher information contained in (θ,η) as $\mathcal{F}_j = \mathrm{var}\left(\frac{\partial}{\partial C_j}\ln f(\theta,\eta;\mathbf{C})\right)$ one can rewrite (29a) and (29b) as

$$b_j \leq \left[\frac{\partial}{\partial C_j}\ u(\mu,\nu;\mathbf{C})\right]^2 \Big/ \left[\mathcal{F}_j\ \mathrm{var}\left(\cos(\mu\theta+\nu\eta)\right)\right] \tag{30a}$$

$$b_j \leq \left[\frac{\partial}{\partial C_j}\ v(\mu,\nu;\mathbf{C})\right]^2 \Big/ \left[\mathcal{F}_j\ \mathrm{var}\left(\sin(\mu\theta+\nu\eta)\right)\right]\,. \tag{30b}$$

In the same way as in Heathcote (1977) it can be shown that $(j = 1,2)$

$$\begin{aligned}\mathrm{var}\left(K_j(\mathbf{C})\right) \leq 4\,\Bigg\{ \iint_{-\infty}^{\infty} \Bigg[&\frac{\partial}{\partial C_j}\ u(\mu,\nu;\mathbf{C})\ \mathrm{var}^{1/2}\left(\cos(\mu\theta+\nu\eta)\right) \\ &+ \frac{\partial}{\partial C_j}\ v(\mu,\nu;\mathbf{C})\ \mathrm{var}^{1/2}\left(\sin(\mu\theta+\nu\eta)\right)\Bigg]\ G(d\mu,d\nu)\Bigg\}^2\end{aligned} \tag{31}$$

where $K_j(\mathbf{C})$ is the j-th component of $\mathbf{K}(\mathbf{C})$. Further, let $\lambda_{j,k}$ $(j,k = 1,2)$ be the components of $\mathbf{\Lambda}$. Combining (30) and (31) then leads to

$$\mathrm{var}\left(K_j(\mathbf{C})\right) \leq \lambda_{jj}^2\ /\ [\mathcal{F}_j\ b_j] \qquad j = 1,2\,.$$

Defining the efficiency of the integrated squared error estimator $\mathbf{C}^{CF} = (C_1^{CF}, C_2^{CF})$ relative to the maximum likelihood estimator as

$$e(C_j^{CF}) = \lambda_{jj}^2 \ / \ [\mathcal{F}_j \ \text{var}\,(K_j(\mathbf{C}))] \qquad , \ j = 1,2 \ ,$$

finally yields

$$b_j \leq e(C_j^{CF}) \leq 1 \qquad j = 1,2 \ .$$

Let us now concentrate on the wrapped process $(\eta_t)_{t=1,\ldots,T}$ where

$$\eta_t = Y_t \ (\text{mod } 2\pi)$$

with $Y_t = \alpha Y_{t-1} + U_t$ and $U_t \sim N(0, \sigma^2)$. The characteristic function of (η_t, η_{t-1}) then reduces to

$$\phi(\mu, \nu) = \exp\{-\tfrac{1}{2} \ C(0) \ (\mu^2 + 2\alpha\mu\nu + \nu^2)\} \ .$$

Just as in the univariate situation the weight function $G(d\mu, d\nu)$ is defined such that its support is restricted to an area close to unity $(0 < \delta < 1)$:

$$G(d\mu, d\nu) = \begin{cases} \delta & \text{if } (\mu, \nu) = (-1, 1) \\ 1 - \delta & \text{if } (\mu, \nu) = (1, 1) \\ 0 & \text{otherwise} \ . \end{cases}$$

Hence, by defining

$$A_{\pm} = \sum_{t=1}^{T} \left[\cos(\eta_t \pm \eta_{t-1}) - \exp\{-C(0) \ (1 \pm \alpha)\}\right]$$

equation (26) reduces to

$$\begin{aligned} 0 = \delta \ (1 - \alpha^{CF}) \ &\exp\{C^{CF}(0)\alpha^{CF}\} \ A_- \\ + (1 - \delta) \ (1 + \alpha^{CF}) \ &\exp\{-C^{CF}(0) \ \alpha^{CF}\} \ A_+ \end{aligned}$$

and

$$\begin{aligned} 0 = -\ \delta \ &\exp\{C^{CF}(0) \ \alpha^{CF}\} \ A_- \\ + (1 - \delta) \ &\exp\{-C^{CF}(0) \ \alpha^{CF}\} A_+ \ , \end{aligned}$$

where α^{CF} is the integrated squared error estimator of α. However, this implies $A_{\pm} = 0$. Putting

$$L_{\pm} = -\ln\left(T^{-1} \sum_{t=1}^{T} \cos(\eta_t \pm \eta_{t-1})\right)$$

gives the final result

(32a) $$C^{CF}(0) = (L_+ + L_-) \ / \ 2$$

and

(32b) $$C^{CF}(1) = (L_+ - L_-) \ / \ 2$$

for estimation of the covariances $C(0)$ and $C(1)$, respectively.

9.7 Numerical Comparison of Estimators

In order to compare the different methods of estimating the variance $C(0)$ and covariance $C(1)$ of the underlying unwrapped process (Y_t), a Monte-Carlo study was carried out. Altogether three sets of simulations, which are described in Table 9.4, were performed. The purpose of the first comparatively small simulation was to identify those estimators that were worth further investigation, while the second and third study concentrated on the more successful estimators, first for independent and identically distributed variables and then for a first order wrapped autoregressive process.

Table 9.4. Design of Monte-Carlo study ($\theta_t = Y_t \pmod{2\pi}$).

Simulation	Time series length T	No. of runs	Model	$C(0)$	$C(1)$
1	10	50	i.i.d. $Y_t \sim N(0,1)$	1	0
2	10,20,50	1000	i.i.d. $Y_t \sim N(0,\pi^2/4)$	2.47	0
3	10,20,50	1000	$Y_t = 0.5Y_{t-1} + U_t$ with $U_t \sim N(0,1)$	4/3	2/3

In the first simulation study the five different techniques, that were discussed in sections 9.2 to 9.6, were compared on the basis of 50 independent samples. Each of these contained $T = 10$ observations which were independently generated from a $WN(0,1)$ distribution, implying $c = 0.632$ and $\text{var}\,(\hat{c}) = 0.029$ for the true values. First, c and $\text{var}\,(\hat{c})$ were estimated for each of the 50 samples using (2), and (10) in conjunction with (11) and (12), respectively. On average, an estimate of 0.631 was obtained for $\hat{c}$ and of 0.036 for $\widehat{\text{var}}\,(\hat{c})$, compared with a value of 0.031 for the variance of $\hat{c}$ over the 50 runs. Then the different estimators based on (3), (9), (20), (17) and (18) were applied in an attempt to recapture the variance $C = 1$ of the underlying unwrapped variable as well as possible. Average estimates for the 50 runs and corresponding standard deviations are listed in Table 9.5. Note that C^{EQ} and C^{CF} are identical in case of independent observations.

As could be expected from the discussion in section 9.2, C^{EQ} tends to overestimate the variance, in contrast with C^{EC} which generally yields a conservative estimate. That is, the additional term in C^{EC} tends to overcorrect, implying that the first two terms in the Taylor expansion (7) do not provide a satisfactory approximation. It comes as no surprise that the maximum likelihood estimator C^{ML} is most accurate with the EM-algorithm converging after an average of 3.3 iterations. However, as the sample size increases the number of iterations required also increases, which implies that the EM-algorithm becomes expensive to implement.

Table 9.5. Mean estimates over 50 runs in Monte-Carlo Study No.1 ($C = 1$).

Estimator	Mean estimate
EQ (=CF)	1.13 (0.53)
EC	0.86 (0.34)
ML	1.01 (0.46)
BA (uniform)	1.29 (0.63)
BA (γ_2)	0.96 (0.41)

The numbers in brackets are standard deviations over the 50 runs.

Of the two Bayesian estimators with uniform and γ_2 *a priori* distributions, the former certainly is unsatisfactory while the latter seems to produce reasonable results. However, for $\text{var}(Y_t) = \pi^2/4 = 2.47$ ($t = 1, \ldots, 10$) the situation is exactly reversed, with the results for the *a priori* uniform and γ_2-distribution being 2.443 and 1.950, respectively. Apart from this unreliability C^{BA} is extremely expensive to implement, even if the sample size is only $T = 10$, and was therefore dropped from subsequent Monte-Carlo experiments.

Hence, the second set of simulations concentrated on C^{EQ}, C^{EC} and C^{ML} with $\text{var}(Y_t) = \pi^2/4 = 2.47$ ($t = 1, \ldots, T$). We also introduced the damping constant $d = 0.02$ in connection with C^{EQ} and C^{EC}, and repeated the experiment 1000 times. The results of this study are summarised in Table 9.6.

Table 9.6. Mean estimates over 1000 runs in Monte-Carlo study No.2 ($C = 2.47$).

Estimator	$T = 10$	$T = 20$	$T = 50$
EQ (=CF)	2.28 (0.79)	2.73 (0.67)	2.47 (0.34)
EC	2.21 (0.84)	2.61 (0.70)	2.42 (0.34)
ML	2.59 (1.41)	2.55 (0.68)	2.50 (0.37)

The numbers in brackets are standard deviations over the 1000 runs.

Figure 9.3. Histograms of variance estimates in study No.2 ($T = 10$).

(a) Frequencies of C^{EQ}.

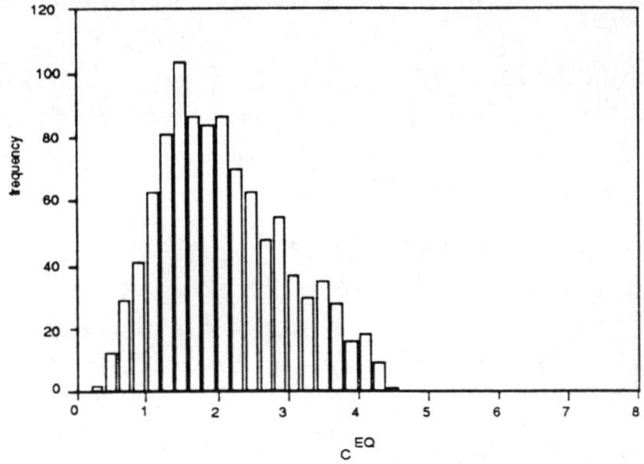

(b) Frequencies of C^{EC}.

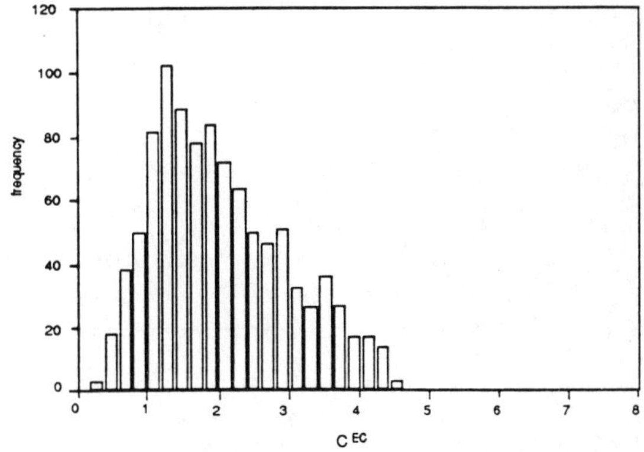

(c) Frequencies of C^{ML}.

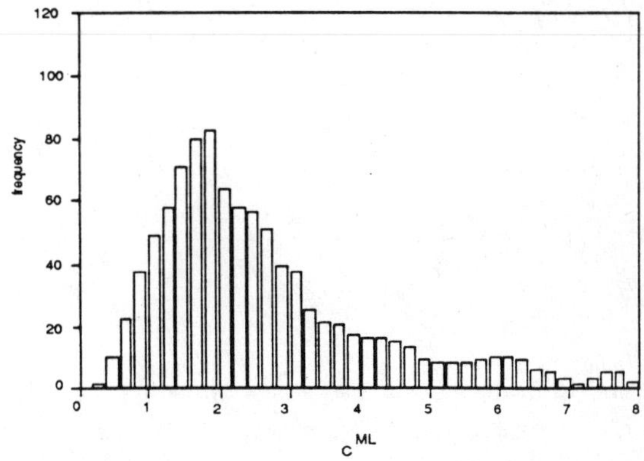

For small sample sizes C^{ML} lies much closer to the true value of $C = 2.47$ than the other two estimators. On the other hand, the variance of C^{ML} for $T = 10$ is distinctly larger due to a heavy upper tail of its distribution. Figure 9.3 shows the histograms of C^{EQ}, C^{EC} and C^{ML} based on simulation No. 2 for $T = 10$. All three estimators exhibit a mode well below 2.47 and a positive bias towards larger values. This bias is strongest for maximum likelihood estimation which, in comparison with the other two techniques, yields a significant proportion of values greater than 4.0 and hence, has a relatively large standard deviation.

In Figure 9.4 the values of C^{EQ} are plotted against C^{ML} for each of the 1000 runs. About 90 per cent of the points lie close to the diagonal with the remaining points scattered below that line. These points, in particular, account for the heavy tail of C^{ML} and represent those samples for which maximum likelihood gave larger values than C^{EQ}. On the other hand, there is no sample with C^{EQ} being significantly larger than C^{ML}. This asymmetry between maximum likelihood and the other two techniques seems to contradict the results presented in Table 9.5 where the standard deviation of C^{ML} was within the range of that for C^{EQ} and C^{EC}. However, by introducing a damping constant for the latter two estimators, their bias is partly corrected and their upper tail virtually cut. Given the fairly large value of $\pi^2/4$ for C, the damping starts to have a pronounced effect on the estimate. Since it is absent from C^{ML}, on the other hand, this estimator turns out to be significantly greater than C^{EC} in some cases, even though it yields a similar value for most of the time.

Figure 9.4. Scatterplot of variance estimates C^{EQ} and C^{ML} in study No.2 ($T = 10$).

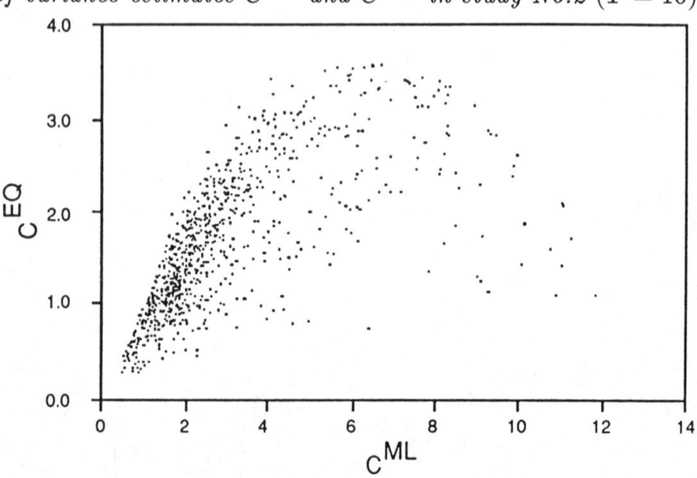

As the sample size T increases the effect of the damping vanishes. Indeed, for $T = 50$ all three estimators give comparable results with C^{EQ} being slightly superior. Furthermore, the two estimators C^{EQ} and C^{EC} show a very similar behaviour independent of T. On the grounds that C^{EC} is a lot more expensive to implement it was decided to drop it from the final Monte-Carlo experiment.

Table 9.7. Mean estimates over 1000 runs in Monte-Carlo study No.3 ($C(0) = 1.33$, $C(1) = 0.67$).

Estimator	T=10 Lag 0	T=10 Lag 1	T=20 Lag 0	T=20 Lag 1	T=50 Lag 0	T=50 Lag 1
EQ	1.24 (0.65)	0.39 (0.59)	1.31 (0.42)	0.57 (0.37)	1.30 (0.17)	0.59 (0.15)
CF	0.84 (0.27)	0.26 (0.22)	1.05 (0.23)	0.43 (0.21)	1.28 (0.21)	0.61 (0.20)
ML	1.40 (0.60)	0.57 (0.52)	1.32 (0.32)	0.59 (0.27)	1.30 (0.17)	0.62 (0.15)

The numbers in brackets are standard deviations over the 1000 runs.

The third set of simulations was based on the first order autoregressive process $Y_t = \frac{1}{2}Y_{t-1} + U_t$ with $U_t \sim N(0,1)$ which implied $C(0) = 1.33$ and $C(1) = 0.67$ (cf. Table 9.4). The results are presented in Table 9.7. Note that no damping was applied in this simulation.

Clearly, for small sample sizes maximum likelihood shows the best behaviour, while the integrated squared error estimator C^{CF} gives no indication of the true parameter values. Even though C^{EQ} tends to underestimate $C(0)$ and $C(1)$ for $T = 10$ it becomes more comparable to C^{ML} as T increases. Further, the covariance $C(1)$ seems to be always underestimated, especially for smaller sample sizes. Figure 9.5 shows the histograms of the three variance and covariance estimators C^{EQ}, C^{CF} and C^{ML}, respectively, obtained in simulation No. 3 for $T = 10$. As in simulation No. 2 there is a positive bias which is slightly more pronounced for C^{ML}. Further, all modes are smaller than the true parameter values, with standard deviations for C^{EQ} and C^{ML}, that are relatively big.

In Figure 9.6(a) the values of $C^{EQ}(1)$ are plotted against $C^{EQ}(0)$ for each of the 1000 runs, while Figures 9.6(b) and (c) show pairs of corresponding estimates, namely $(C^{CF}(0), C^{EQ}(0))$ and $(C^{ML}(0), C^{EQ}(0))$, respectively. The following three comments emerge from these graphs. First, there does not seem to exist a strong relationship between the variance and corresponding covariance estimates as exemplified by C^{EQ} in Figure 9.6(a). Second, the plot of $C^{EQ}(0)$ against $C^{CF}(0)$ indicates that these two estimators behave similarly apart from the fact that $C^{EQ}(0)$ tends to be slightly greater than $C^{CF}(0)$ (cf. Figure 9.6(b)). And third, Figure 9.6(c) shows that the results obtained for $C^{EQ}(0)$ and $C^{ML}(0)$ can be very different for the individual run even though their average characteristics are somewhat similar, judged on the basis of Table 9.7 and Figure 9.5.

In summarising these results, it can be said that C^{EQ} and C^{CF} gave comparable results with C^{EQ} being superior. Whereas for T less than 20 C^{ML} clearly demonstrated the best performance, for $T = 50$ the three estimators showed a similar behaviour overall. These conclusions were confirmed by a number of small simulation studies based on a variety of first and second order WAR models. On the grounds that the time series (θ_t) of residual wind directions consists of fairly long segments and the EM-algorithm becomes uneconomical as T increases, C^{EQ} seems to be the appropriate choice for fitting a first order WAR process to (θ_t).

Figure 9.5. Histograms of covariance estimates in study No.3 ($T = 10$).

(a) $C^{EQ}(0)$.

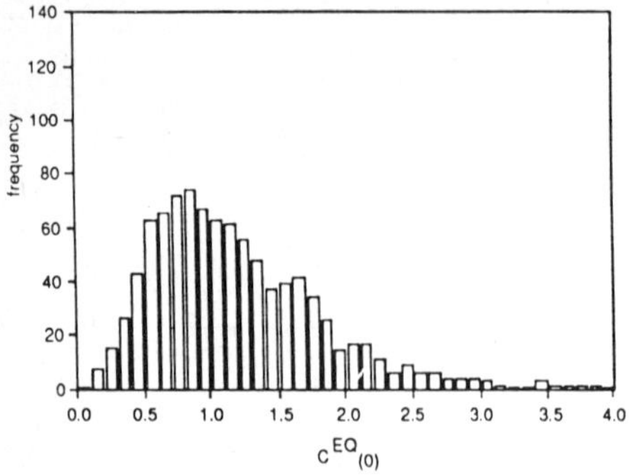

(b) $C^{EQ}(1)$.

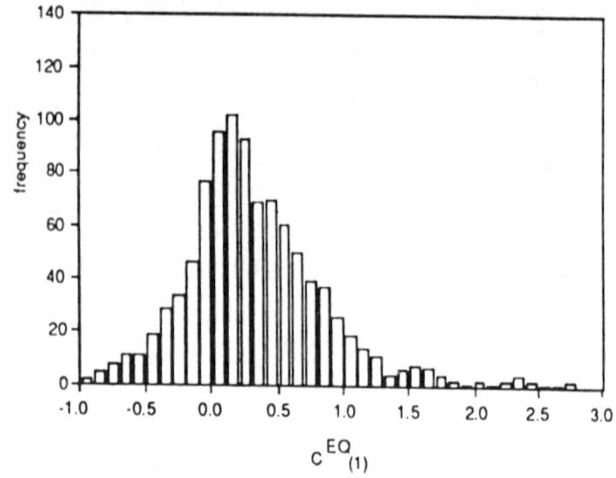

(c) $C^{CF}(0)$.

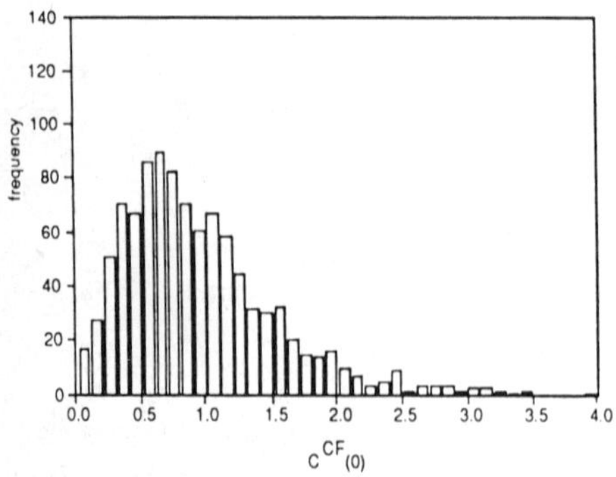

(d) $C^{CF}(1)$.

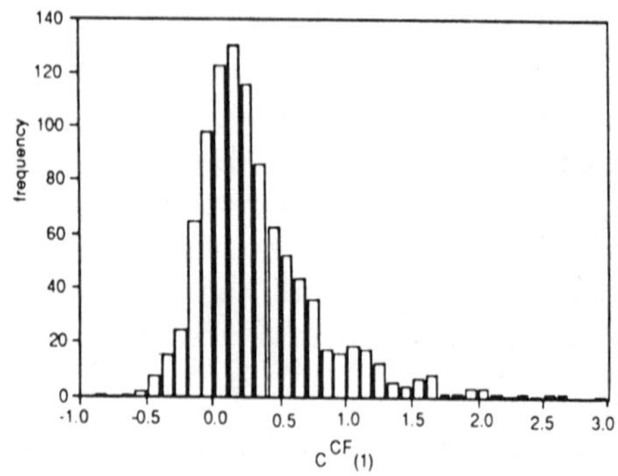

(e) $C^{ML}(0)$.

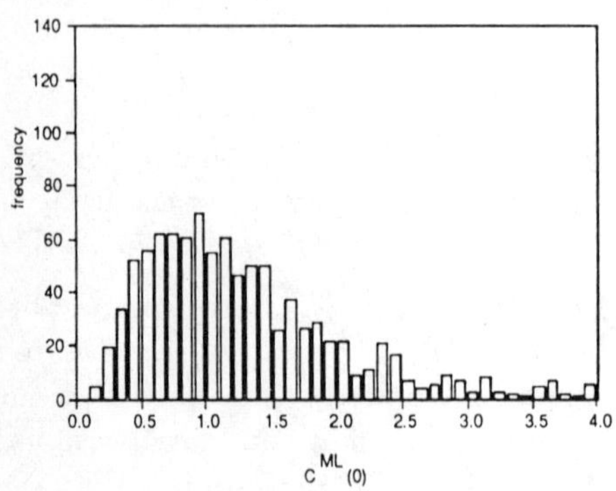

(f) $C^{ML}(1)$.

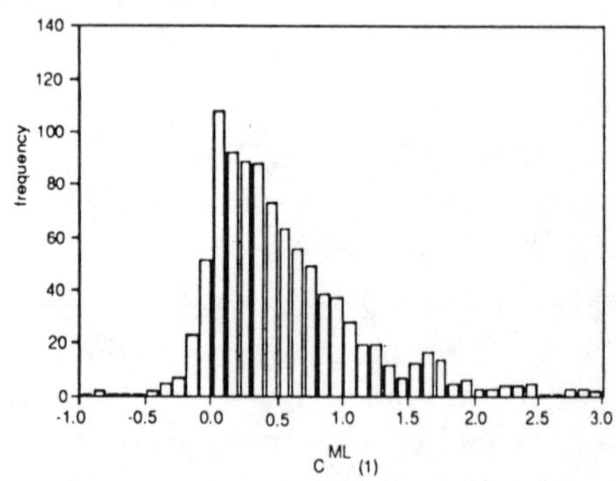

Figure 9.6. Scatterplots of covariance estimates in study No.3 ($T = 10$).

(a) $C^{EQ}(1)$ *against* $C^{EQ}(0)$.

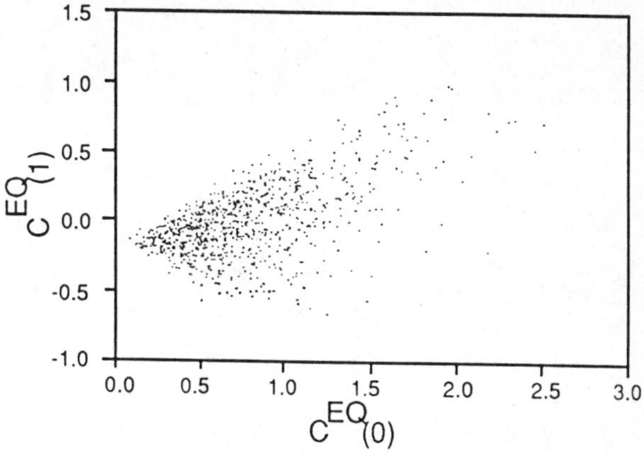

(b) $C^{EQ}(0)$ *against* $C^{CF}(0)$.

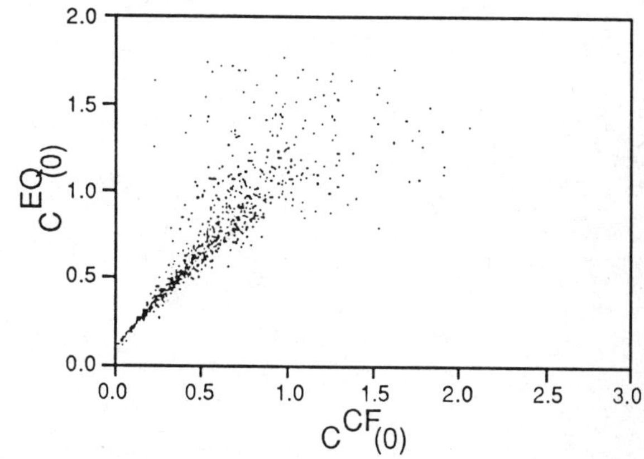

(c) $C^{EQ}(0)$ *against* $C^{ML}(0)$.

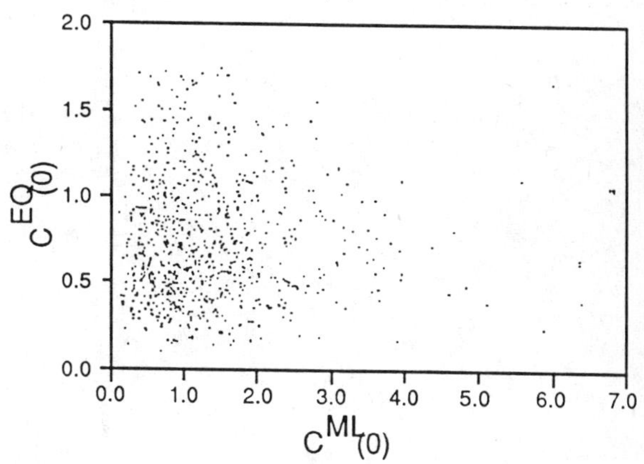

CHAPTER 10

APPLICATION TO SERIES OF RESIDUAL WIND DIRECTIONS

Part II of this monograph will now be concluded with an application of the concepts presented in chapters 6 to 9. In this chapter we shall therefore return to the analysis of the wind series $(\mathbf{w}_t)$ which was discussed in detail in Part I. Both, a first order wrapped autoregressive and a first order von Mises model are fitted to the directional component (θ_t) of the residual series $(\mathbf{f}_t)$. It is also shown that the wrapped normal approximation described in section 6.3 works well in this example. Using this approximation it is found that the von Mises model is slightly better suited for describing time series in which periods with a prevailing wind direction alternate with periods when the direction changes more rapidly.

It is seen that the series (θ_t) of residual wind directions can be well described by a first order von Mises process with a small value for κ_0, the parameter associated with a distinct mean direction. Since the wrapped autoregressive model fitted to (θ_t) is close to the non-stationary boundary, the series (θ_t) will be differenced yielding a time series (ξ_t) of wind direction changes. When a first order wrapped autoregressive model is fitted to this series the value of the dependence parameter α is found to be negligible. This suggests that successive wind direction differences are independent and hence, that the wind directions in $(\mathbf{f}_t)$ can be regarded as random fluctuations about the previous record.

Furthermore, it is noted that the first order wrapped autoregressive model with $\alpha = 0$ fitted to (ξ_t) is basically identical with the first order von Mises model with $\kappa_0 = 0$ fitted to (θ_t). It is shown that this constrained model achieves a fit almost as good as the unconstrained wrapped autoregressive model; however, that the unconstrained von Mises model performs marginally better. That is, relaxing the constraint on κ_0 slightly improves the fit, whereas relaxing the constraint on α makes virtually no difference.

The objective in this chapter is to model the series of residual wind directions, and to fit a first order WAR as well as a first order von Mises process to the directional component (θ_t) of the residual series $(\mathbf{f}_t)$. Referring to chapter 6 the WAR model can be written as

$$\text{(1a)} \qquad \theta_t = Y_t \pmod{2\pi}$$

where the density of Y_t conditional on Y_{t-1} is

$$\text{(1b)} \qquad f(y_t \mid y_{t-1}) = [2\pi\sigma^2]^{-1/2} \exp\Big\{-\frac{1}{2\sigma^2}\Big(y_t - \alpha y_{t-1}\Big)^2\Big\}.$$

The von Mises model, on the other hand, is defined via the density of θ_t conditional on θ_{t-1}, that is

$$\text{(2a)} \qquad f(\theta_t \mid \theta_{t-1}) = [2\pi I_0(\kappa_t)]^{-1} \exp\{\kappa_t \cos(\theta_t - \mu_t)\}$$

with κ_t and μ_t determined as follows

$$\text{(2b)} \qquad \kappa_t \begin{pmatrix} \cos\mu_t \\ \sin\mu_t \end{pmatrix} = \kappa_0 \begin{pmatrix} 1 \\ 0 \end{pmatrix} + \kappa_1 \begin{pmatrix} \cos\theta_{t-1} \\ \sin\theta_{t-1} \end{pmatrix}.$$

Assuming mean direction 0° the parameters of the two models are α and σ^2, and κ_0 and κ_1, respectively. Whereas the technique *EQ* described in section 9.2 is employed for the estimation of α and σ^2, the parameters κ_0 and κ_1 are estimated using maximum likelihood (cf. equation (6.14)).

Suppose that a stationary time series (Y_t) in $\mathcal{R}$ with variance C_Y can be decomposed into a model component ($\mathcal{E}[Y_t \mid Y_{t-1}, Y_{t-2}, \ldots]$) and an error process (U_t) with variance C_U. This usually implies that the values for (U_t) are *smaller* than those for (Y_t). Further, the term $1 - C_U/C_Y$ is generally conceived as the amount explained by the model component. The decomposition is regarded as complete if the error process can be interpreted as white noise; that is, if the U_t are independent and identically distributed with a variance C_U as small as possible.

In this form, however, the argument cannot be extended to directional time series. Since an angle can always be identified with a unit vector, the *size* of a value is a meaningless concept for directional data. Hence, the decomposition of a directional process cannot lead to a reduction in size, corresponding to the step from (Y_t) to (U_t) above. This means that the uniform distribution can generally be regarded as the only error distribution for directional time series, in contrast with the *family* of error distributions which exist for time series in $\mathcal{R}$ and are characterized by the parameter C_U.

However, part of the statement above can still be formulated for directional processes. By analogy with a time series in $\mathcal{R}$ we will regard the decomposition of a directional process (θ_t) as complete if the error process (ζ_t) constitutes white noise; that is, if the ζ_t have a uniform marginal distribution and are independent. This implies $c_\zeta(1) = 0$ and $c_\zeta(0) = 1$ corresponding to independence and uniform marginals, respectively. The circular variance can therefore be expected to *increase* as the regular components are removed from a directional time series, which is the reverse of the situation for time series in $\mathcal{R}$. This result is an immediate consequence of the scaling that is associated with directional data. Further, since $c_\zeta(0) = 1 - \rho_\zeta^2$, where ρ_ζ is the resultant length of ζ_t, $c_\zeta(0) = 1$ is equivalent to $\rho_\zeta = 0$. Similar to a time series in $\mathcal{R}$ one could therefore define

$$R^2 = 1 - \frac{\rho_\zeta^2}{\rho_\theta^2} = \frac{c_\zeta(0) - c_\theta(0)}{1 - c_\theta(0)} \tag{3}$$

and interpret the size of R^2 as the amount that is explained by the model component.

Given 24 observations per day and about 90 days per season, a time series corresponding to a particular season consists of roughly 2160 values. Since only two parameters are to be estimated and since at the same time it may be interesting to know the spread of these parameters, the record for each particular season is divided into 45 subseries such that each subseries comprises data for two consecutive days. Separate WAR and von Mises models are then fitted to each of the 45 series in winter 1971 and each of the 45 series in summer 1971/72.

Table 10.1. Parameter estimates for the WAR model (1).

	Mean $\hat{C}(0)$	Mean $\hat{C}(1)$	Mean $\hat{\alpha}$	Mean $\hat{\sigma}^2$
Winter 1971	4.38 (2.13)	4.10 (2.13)	0.89 (0.10)	0.53 (0.23)
Summer 1971/72	4.04 (1.87)	3.65 (1.88)	0.86 (0.08)	0.71 (0.22)

The numbers in brackets are standard deviations over the 45 series.

Average estimates of the variance $C(0)$, the covariance $C(1)$, and the parameters α and σ^2 of the underlying unwrapped process (1) are presented in Table 10.1. Both means and standard deviations refer to the 45 estimates obtained in each season. The large mean value for $\hat{\alpha}$ shows that consecutive wind directions in $(\mathbf{f}_t)$ are highly dependent. Consequently, the variance $\hat{C}_U = \hat{\sigma}^2$ of the error process (U_t) is much smaller than $\hat{C}_Y = \hat{C}(0)$, which implies that most of the variation in (Y_t) can be explained by a model of the form (1). The term $1 - \hat{C}_U/\hat{C}_Y$ amounts to an average of 0.81 in winter 1971 and 0.75 in summer 1971/72. It is noted that appropriate damping is applied in the estimation of $C(0)$ and $C(1)$, as described in section 9.2.

Fitting the model (2) to the same 45 subseries of the directional component (θ_t) yields the results presented in the first half of Table 10.2. The large values for $\hat{\kappa}_1$ compared with the relatively small values for $\hat{\kappa}_0$ show that for any 2-day period the wind θ_t is largely determined by θ_{t-1} and only to a small extent by a constant, that is

Table 10.2. Parameter estimates for the von Mises model (2).

	Mean $\hat{\kappa}_0$	Mean $\hat{\kappa}_1$	Mean $\hat{\alpha}_{\mathrm{vM}}$	Mean $\hat{\sigma}^2_{\mathrm{vM}}$
Winter 1971	0.31 (0.17)	3.53 (1.41)	0.77 (0.12)	0.44 (0.16)
Summer 1971/72	0.30 (0.12)	2.30 (0.60)	0.72 (0.10)	0.59 (0.15)

The numbers in brackets are standard deviations over the 45 series.

prevailing wind direction. This observation is consistent with the parameter estimates obtained for the WAR model above.

In a von Mises process there is no *explicit* error component as in a WAR process. However, it is *implicit* in the uncertainty of the von Mises distribution. Just as for an autoregressive process, let us define this error component as (ζ_t) where

$$r_t \begin{pmatrix} \cos\zeta_t \\ \sin\zeta_t \end{pmatrix} = \begin{pmatrix} \cos\theta_t \\ \sin\theta_t \end{pmatrix} - A_1(\kappa_t) \begin{pmatrix} \cos\mu_t \\ \sin\mu_t \end{pmatrix} \tag{4}$$

with $r_t \in \mathcal{R}_0^+$, and μ_t and κ_t as in (2). The fact that (ζ_t) does not involve r_t reflects the scaling process referred to earlier. Note that the second term on the right hand side is equal to $\mathcal{E}[(\cos\theta_t, \sin\theta_t)' \mid \theta_{t-1}]$, and therefore that (4) represents a decomposition into a model and an error component as mentioned above. The reader is reminded that the maximum likelihood estimates of κ_0 and κ_1 minimise the expected norm square of the right hand side of (4). Now, if $\mathcal{E}[(\cos\zeta_t, \sin\zeta_t)' \mid \theta_{t-1}] = \mathbf{0}$, then the resultant length of ζ_t conditional on θ_{t-1} would be zero and the marginal distribution of ζ_t would be uniform. Similarly, it would follow that ζ_t and ζ_{t-1} are uncorrelated, which means that (ζ_t) could be interpreted as white noise. In general this will not be the case. However, using a second order Taylor approximation for the denominator of

$$\cos\zeta_t = \frac{\cos\theta_t - A_1(\kappa_t)\cos\mu_t}{[1 - 2A_1(\kappa_t)\cos(\theta_t - \mu_t) + A_1^2(\kappa_t)]^{1/2}}$$

and

$$\sin\zeta_t = \frac{\sin\theta_t - A_1(\kappa_t)\sin\mu_t}{[1 - 2A_1(\kappa_t)\cos(\theta_t - \mu_t) + A_1^2(\kappa_t)]^{1/2}}$$

it can be shown that $\| \mathcal{E}[\, (\cos\zeta_t, \sin\zeta_t)' \mid \theta_{t-1}] \|^2$ is small and hence, that (ζ_t) is close to being white noise. In particular,

$$R_\zeta^2(\kappa_t) = \left(\mathcal{E}[\cos \zeta_t \mid \theta_{t-1}] \right)^2 + \left(\mathcal{E}[\sin \zeta_t \mid \theta_{t-1}] \right)^2$$

$$\approx \frac{[4A_1 - A_1^3 - 8A_1^5 + 4A_1A_2 - 2A_1^3A_2 + 3A_1^2A_3]^2}{64\,[1 + A_1^2]^5}$$

where $A_\nu = A_\nu(\kappa_t)$ ($\nu = 1, 2, 3$). Using the recursion formula $A_{\nu+1}(\kappa) = A_{\nu-1}(\kappa) - 2\nu A_\nu(\kappa)/\kappa$ for $\nu \geq 2$ (cf. Abramowitz & Stegun, formula (9.6.26)) then yields the approximation

$$R_\zeta(\kappa_t) \approx \frac{1}{[1 + 1/A_1^2]^{1/2}} \cdot \frac{1 - A_1^2}{1 + A_1^2} - \frac{(5 - A_1^2)\, A_1^2/\kappa_t - 6A_1^3/\kappa_t^2}{2\,[1 + A_1^2]^{5/2}} .$$

The maximum of this expression is 0.14 and is attained for $\kappa_t = 0.98$. Given that the average value for κ_t is about 3.78 in winter and 2.54 in summer, the conditional mean resultant length of the error component (ζ_t) will be roughly 0.03 in winter and 0.06 in summer. Given a value of 0.87 for $\hat{c}_\theta(0)$ in winter and of 0.90 in summer, this implies, according to (3), that more than 99 per cent of the circular variation in $(\mathbf{f}_t)$ in winter and 96 per cent in summer are explained by the von Mises model (2).

Let us now comment on the wrapped normal approximation discussed in section 6.3. Average values for $\hat{\alpha}_{\text{vM}}$ and $\hat{\sigma}^2_{\text{vM}}$, derived by using that approximation for each of the 45 von Mises models fitted to (θ_t), are presented in the second half of Table 10.2. Note that the values for $\hat{\sigma}^2_{\text{vM}}$ are within one standard deviation of the estimates $\hat{\sigma}^2$ listed in Table 10.1, which were obtained by fitting a WAR model directly to the data series. In fact, the correlation between $\hat{\sigma}^2_{\text{vM}}$ and $\hat{\sigma}^2$ is about 0.96 in both seasons. This illustrates that the wrapped normal approximation works well in this example. Observe, however, that the mean value for $\hat{\sigma}^2_{\text{vM}}$ is slightly smaller than that for $\hat{\sigma}^2$. Even though these differences are only marginal this has been a consistent pattern across different data series and years. The standard deviations for $\hat{\sigma}^2_{\text{vM}}$ are also smaller than those for $\hat{\sigma}^2$. This smaller variation of $\hat{\sigma}^2_{\text{vM}}$ suggests that the von Mises process is slightly better suited to describe a series, where periods with a prevailing wind direction alternate with periods when the direction changes more rapidly. This result is due to the fact that the concentration κ_t in the von Mises process (2) varies with time, which seems to be more appropriate if the underlying process exhibits some long-term non-stationarity.

It was mentioned earlier that a prevailing wind direction and thus the term involving κ_0 is only marginally important for most 2-day periods. Of course, the longer the subseries the less well-defined this mean wind direction is. In fact, there does not seem to exist any dominant direction in $(\mathbf{f}_t)$ even though the data are still highly correlated. Figure 5.7 suggests indeed, that the directions are uniformly distributed. However, fitting the model (1) with $0 \leq \alpha < 1$ to any data series would result in a well defined mean direction at 0°. A uniform marginal distribution, on the other hand, implies $\kappa_0 = 0$ and thus $\kappa_t = \kappa_1$ for all t in the von Mises process (2), and $\alpha = 1$ in the WAR process (1). In particular the latter means that the underlying unwrapped process is non-stationary. In fact, both the high value for $\hat{\alpha}$ in Table 10.1 as well as the small value for $\hat{\kappa}_0$ in Table 10.2 indicate that for most 2-day periods we are close to the non-stationary boundary. This suggests that we take first differences

$$\xi_t = \theta_t - \theta_{t-1} \pmod{2\pi} \tag{5}$$

and fit a WAR model to the series (ξ_t). Referring to this differencing such a model will be called an *integrated* WAR model.

Table 10.3. Distribution of wind direction differences (in %).

Winter 1971	−120°	−90°	−60°	−30°	0°	30°	60°	90°	120°	Total	Mean direction
$(\mathbf{w}_t)$	0	1	2	16	70	8	2	1	0	100	357°
$(\mathbf{d}_t)$	2	2	5	21	55	9	3	1	2	100	353°
$(\mathbf{r}_t)$	2	3	6	21	47	13	4	2	2	100	355°
$(\mathbf{f}_t)$	2	3	7	21	45	14	4	2	2	100	355°
Summer 1971/72	**−120°**	**−90°**	**−60°**	**−30°**	**0°**	**30°**	**60°**	**90°**	**120°**	**Total**	**Mean direction**
$(\mathbf{w}_t)$	0	0	1	9	81	6	1	1	1	100	357°
$(\mathbf{d}_t)$	2	2	5	24	56	7	2	1	1	100	350°
$(\mathbf{r}_t)$	4	4	6	22	38	15	5	3	3	100	355°
$(\mathbf{f}_t)$	4	4	6	22	37	15	5	3	4	100	356°

Table 10.3 lists the frequencies of wind direction differences for the four series $(\mathbf{w}_t)$, $(\mathbf{d}_t)$, $(\mathbf{r}_t)$ and $(\mathbf{f}_t)$ for winter 1971 and for summer 1971/72. Recall that these denote the raw data series, the daily component, and the residual components before and after removing the storms, respectively. Notice that the directional differences become larger as the geostrophic, sea breeze and storm components are removed from $(\mathbf{w}_t)$. Whereas 70 per cent of all differences are close to 0° in the case of $(\mathbf{w}_t)$ in winter, it is only 45 per cent in the case of $(\mathbf{f}_t)$, an effect that is even more pronounced in summer. The increasing spread of the directional differences in Table 10.3 reflects the amount of variation in $(\mathbf{w}_t)$ which is explained by the different components subtracted from $(\mathbf{w}_t)$. In contrast with the histograms of wind direction, the wind direction differences exhibit a distinct mean direction around 0°. It thus seems reasonable to model the series (ξ_t) of wind direction differences as a first order WAR process.

It is also found that the mean direction for each of the eight data series lies between 350° and 357°, which can be attributed to a predominantly anticlockwise rotating wind. For $(\mathbf{w}_t)$, for example, it is −3° in both seasons, which amounts to a full rotation of the wind direction within five days. It is interesting to note that this is roughly in the order of one synoptic cycle as described in section 3.1. Constant trends corresponding to these mean directions are removed from the data series before the WAR and von Mises models are fitted.

Table 10.4. Parameter estimates for the integrated WAR model.

Winter 1971	Mean $\hat{c}_{\mathrm{d}}(0)$	Mean $\hat{c}_{\mathrm{d}}(1)$	Mean $\hat{C}_{\mathrm{d}}(0)$	Mean $\hat{C}_{\mathrm{d}}(1)$
$(\mathbf{w}_t)$	0.165 (.076)	0.011 (.018)	0.191 (.140)	0.012 (.028)
$(\mathbf{d}_t)$	0.291 (.084)	0.012 (.028)	0.362 (.146)	0.015 (.047)
$(\mathbf{r}_t)$	0.382 (.128)	0.018 (.030)	0.535 (.166)	0.028 (.054)
$(\mathbf{f}_t)$	0.394 (.121)	0.013 (.031)	0.551 (.157)	0.019 (.056)
Summer 1971/72	**Mean $\hat{c}_{\mathrm{d}}(0)$**	**Mean $\hat{c}_{\mathrm{d}}(1)$**	**Mean $\hat{C}_{\mathrm{d}}(0)$**	**Mean $\hat{C}_{\mathrm{d}}(1)$**
$(\mathbf{w}_t)$	0.125 (.060)	0.017 (.012)	0.139 (.063)	0.021 (.016)
$(\mathbf{d}_t)$	0.256 (.070)	0.017 (.024)	0.305 (.088)	0.023 (.035)
$(\mathbf{r}_t)$	0.507 (.097)	0.027 (.048)	0.754 (.246)	0.041 (.190)
$(\mathbf{f}_t)$	0.512 (.094)	0.023 (.050)	0.765 (.241)	0.034 (.191)

The numbers in brackets are standard deviations over the 45 series.

Average estimates of the circular variance $c_{\mathrm{d}}(0)$ and covariance $c_{\mathrm{d}}(1)$ as well as of the variance $C_{\mathrm{d}}(0)$ and the covariance $C_{\mathrm{d}}(1)$ of the underlying unwrapped process are presented in Table 10.4. The index d is to denote reference to the model (5). As stated above the marginal distributions of the directional components of the different wind series are almost all uniform. However, the observations are highly autocorrelated. By forming the process (5) of directional differences we would expect to have accounted for most of the dependence in (θ_t). Further, as the different components are removed from the original series $(\mathbf{w}_t)$ we would expect that the estimates of the circular variance $c_{\mathrm{d}}(0)$ increase. Indeed, as evident from Table 10.4, $\hat{c}_{\mathrm{d}}(0)$ rises from 0.17 for $(\mathbf{w}_t)$ to 0.39 for $(\mathbf{f}_t)$ in winter, and from 0.13 to 0.51 in summer, whilst the covariance estimates are virtually all zero.

In particular, using (3) as a measure for the amount that is explained by the model component, we get that in winter 15 per cent of the circular variation can be attributed to the geostrophic component, 11 per cent to the land and sea breeze circulation, and 1 per cent to the storm component, while the corresponding numbers for summer are 15, 28 and 1 per cent. However, these numbers are based on the differenced *directional* components of the four data series. Therefore, the numbers should not be interpreted as a measure of importance of the various components subtracted from $(\mathbf{w}_t)$, even though they give an indication of their *relative* importance.

At the same time, as the different components are removed from the original series $(\mathbf{w}_t)$ the variance estimates $\hat{C}_d(0)$ increase. This effect is more pronounced in summer, and basically reflects the greater importance of the sea breeze circulation during this time of the year, as is evident from a comparison of the figures for $(\mathbf{d}_t)$ and $(\mathbf{r}_t)$. Further, the standard deviation of $\hat{C}_d(0)$ increases together with the mean of $\hat{C}_d(0)$, except for the change from $(\mathbf{r}_t)$ to $(\mathbf{f}_t)$. The latter is a result of simply smoothing the series during storm periods according to formula (5.1) without changing the series in any other way. None of the estimates $\hat{C}_d(1)$ for the lag 1 covariances is significantly different from zero. Nevertheless, the ratio $\hat{\alpha}_d = \hat{C}_d(1)/\hat{C}_d(0)$ decreases as the different components are removed from $(\mathbf{w}_t)$, suggesting that successive differences become less and less dependent.

As for the variance $\hat{\sigma}_d^2 = \hat{C}_d(0) - \hat{C}_d^2(1)/\hat{C}_d(0)$ of the integrated WAR model of $(\mathbf{f}_t)$ we have 0.55 in winter and 0.76 in summer. Comparing these values with the variance $\hat{\sigma}^2$ presented in Table 10.1 shows that the data series are almost as well described as by the non-integrated model (1). The small values for the estimates $\hat{\alpha}_d$ suggest that successive wind direction changes are almost independent. This means that the differences virtually constitute white noise and that the winds in $(\mathbf{f}_t)$ can be regarded as fluctuations about the previous record. In particular, a value of 0.55 for $\hat{\sigma}_d^2$ implies that for 95 per cent of the time one can expect the wind direction in $(\mathbf{f}_t)$ to be within 83° of the previous record. Compare this number with the winter-$(\mathbf{f}_t)$-row in Table 10.3.

Table 10.5. Parameter estimates for von Mises model with $\kappa_0 = 0$.

	Mean ($\hat{\kappa}$)	Mean ($\hat{\sigma}_c^2$)
Winter 1971	3.28 (1.20)	0.54 (0.23)
Summer 1971/72	2.21 (0.54)	0.74 (0.23)

The numbers in brackets are standard deviations over the 45 series.

Fitting a first order von Mises model with $\kappa_0 = 0$ and $\kappa = \kappa_1$ to the 45 subseries yields the estimates given in the first half of Table 10.5. Note that this model is the one considered in example 6.1. On the other hand, when using the wrapped normal approximation we get the values listed in the second half of Table 10.5. Notice that the variance in this case is identical with that obtained for an integrated WAR model with $\alpha = 0$. This confirms that the integrated first order WAR model with $\alpha = 0$ and the first order von Mises model with $\kappa_0 = 0$ are in principle identical. As the concentration κ_t does no longer vary with time, the difference between the two models reduces to that between the von Mises and the wrapped normal distribution. Both models will therefore be simply referred to as the *constrained* model.

From Tables 10.1 and 10.5 we see that the values for the variance $\hat{\sigma}_{\mathrm{c}}^2$ of the constrained model are almost identical with those for $\hat{\sigma}^2$ of model (1). Since the parameter α in model (1) is virtually equal to 1, one can explicitly set $\alpha = 1$ without greatly reducing its predictability. The constrained model, however, is superior in that it involves one parameter less. The von Mises model (2), on the other hand, performs slightly better than both these models, judging by the smaller values for the associated variance $\hat{\sigma}_{\mathrm{vM}}^2$ presented in Table 10.2. This result is due to the fact that the parameter κ_0 in model (2) is sufficiently different from 0 to improve the predictability of the constrained model.

CHAPTER 11

CONCLUSIONS AND SUMMARY OF RESULTS

The aim of this study has been to analyse the structure of a particular time series of wind speeds and directions. Special techniques were developed to divide the series into the following four components, as mentioned in the introduction:

(i) *a geostrophic component,*
(ii) *a land and sea breeze cycle,*
(iii) *a storm component, and*
(iv) *a residual series of short-term fluctuations.*

All these components were analysed in detail and were related to certain meteorological phenomena. The interaction between the geostrophic component, the land and sea breeze cycle, and the storm component was also investigated.

The knowledge of general wind patterns is relevant whenever one is concerned with long-term prediction and average weather conditions, as for example, in the context of wind energy or air pollution control. To make use of wind energy, it is important to know how the distribution of wind speed changes during the year, and how susceptible the wind is to directional fluctuations. The results presented in this monograph may also be helpful for the design of industrial areas. Knowing the intensity and geostrophic dependence of the sea breeze pattern, for instance, can provide some guidance as to when to release certain pollutants into the air, to ensure that major residential areas are not effected.

Synoptic scale winds as captured by the geostrophic component were found to dominate the wind pattern in the Fremantle area throughout the year. A similar degree of importance could be attributed to the land and sea breeze circulation in summer but even in winter this cycle was found to be a significant factor in the local wind pattern. The storm component effectively accounted for the outlier behaviour in the residual series after the removal of the geostrophic and sea breeze components from the given wind series. As such the storm component did not play a major role in the decomposition but contained valuable information of a different kind. The residual series of short-term fluctuations still exhibited a high amount of autocorrelation and, albeit of small magnitude, could not be ignored.

In chapter 2 we separated the geostrophic from the daily component by using a filter which smoothed the wind record uniformly over 24-hour periods. Various data representations and filters, which are robust against storms, were considered. As far as the particular wind series was concerned, however, the mean filter gave comparable results to robust filtering methods for most of the time. It was therefore decided to use the mean filter for the initial decomposition, since it was less expensive to implement. This led to a distortion in the daily component particularly when storms and gusty winds prevailed. However, this effect was subsequently corrected by applying special robust techniques in the analysis of the daily component.

In chapter 3 the geostrophic component was related to the synoptic pressure field. Using a simplistic model to predict the geostrophic wind from the atmospheric pressure field, it was shown that the geostrophic component could be physically interpreted as the synoptic-scale wind depicted on weather charts. In order to determine wind shifts and sequences of weather patterns, a number of synoptic states was defined in terms of wind speed and direction. The classification of the geostrophic component according to these states roughly reflected the movement of large-scale weather systems.

It was established that strong southerlies in summer are followed by moderate easterlies, due to the high pressure system over the Indian Ocean extending into the Bight. The trough, which at this stage begins to develop from the heat low over the north-west splitting an eastern cell away from the high pressure ridge, is usually accompanied by light north-easterlies. As the trough moves across the region the wind strengthens and swings back to a southerly direction. The situation in winter is as follows: the low pressure systems approaching the continent from the south-west are associated with strong north-westerlies; as the depressions move to the east the wind turns south-easterly and begins to moderate; at this stage Fremantle may come under the influence of the high pressure belt, giving rise to light winds from an easterly to northerly direction; then, as the next front approaches strong north-westerlies prevail again.

In summer the weather patterns have a prolonged duration due to the relative stability of the atmospheric pressure field, whereas the extratropical cyclones in winter tend to pass by more rapidly and cause the synotic states to change more frequently. Usually, the wind swings in an anticlockwise direction, indicating a predominantly southerly influence.

We showed how embedding the geostrophic component in a sequence of only a few synoptic states helped us to find regularities in the record of wind speed and direction, and to detect a pattern in the seemingly random flow of high and low pressure systems. Our results suggest that a refinement of this analysis may substantially improve the understanding of synoptic weather patterns.

In chapter 4, studying all the effects which have a 24-hour pattern, we found that the major constituent of the daily component is the land and sea breeze circulation. Its basic feature is the complete anticlockwise circle described by the wind direction each day, with wind speed peaking in the morning in form of a land breeze and in the afternoon in form of a sea breeze. Robust techniques enabled us to select those days which exhibited a pronounced land and sea breeze cycle, and by averaging over these days we established a typical circulation pattern. The similarity between the summer and winter patterns is surprising as they differed only in intensity, that is the magnitude of wind speed. Whilst the land and sea breeze cycle was the dominant feature in summer, we found two major patterns in winter: first extratropical cyclones approaching from the south-west, and second the sea breeze cycle, which occured on about one third of the days and was associated with the high pressure belt lying across the Fremantle region. Another interesting result was the consistency of this circulation pattern over the years.

The sea breeze circulation was almost negligible when the geostrophic component indicated a westerly wind, but was particularly pronounced when south-easterlies prevailed. The effect of using a mean filter, instead of more robust techniques for the initial decomposition, was that the ideal land and sea breeze cycle registered on fewer days.

Having removed the land and sea breeze cycle from the daily component, the residual series was then investigated in chapter 5 in order to detect and study particular short-term events such as calms, storms, and oscillating winds. Whereas calms indicated that the geostrophic and sea breeze components virtually constituted the actual wind, storms were interpreted as abrupt deviations from the smooth flow as defined by these two components. Consequently, storms tended to occur when the synoptic pressure field was highly asymmetric or changing rapidly. In winter they were often associated with cold fronts approaching from the southwest, while in summer they were usually generated by the trough developing from the north and splitting the high pressure belt into a western and an eastern cell. Calms, on the other hand, generally reflected stable atmospheric conditions and thus occured more often in winter when the south-west came under the influence of the high pressure belt, at a time when also continental heating was less significant.

The number of storms and calms changed dramatically over the years, although their main characteristics, such as their distribution over the year and their time of onset, were similar. It was found that the times of day could be partitioned corresponding to the onset of these events on the one hand, and the peak of the land and sea breeze circulation on the other hand. Whereas calms tended to begin just before the onset of either breeze, the storms usually began when the breezes started to weaken. The geostrophic and sea breeze components were therefore a better predictor of the actual wind during the developing stages rather than during the ebbing stages of the land and sea breeze circulation. In this study, only elementary events were considered, but a more sophisticated analysis may be performed along similar lines.

The residual series obtained after the removal of the storms could in general not be related to any physical phenomenon. However, the data were still highly autocorrelated. In Part II of this monograph we concentrated on the directional aspect, and showed how the series of short-term fluctuations could be described by an autoregressive model. It was found that successive wind direction differences were almost independent, suggesting that the residual wind direction can be regarded as resulting from a fluctuation about the previous record.

Although this Part was motivated by an analysis of the residual series, it was mainly concerned with mathematical aspects and as such was self-contained. Given that the von Mises and wrapped normal distributions play a similarly central role for circular variables as the normal distribution for real random variables, two models for directional time series were introduced which are related to the von Mises and wrapped normal distribution, respectively. The dilemma was that the latter of the two distributions does not belong to the exponential family, while the former cannot be extended to a bivariate distribution with von Mises marginals. It the same way it turned out that each of the two time series models had some but not all of the desirable properties. However, just like the two distributions, it was shown that the two time series models can be approximated by one another so that whichever model is more appropriate under the given circumstances may be chosen. The main difference between the two models is that the von Mises process will generally involve a certain kind of non-stationarity, while the wrapped autoregressive process will usually be assumed to be stationary. This implies that the von Mises process is slightly better suited to describe a time series in which periods with a prevailing wind direction alternate with periods when the direction changes more rapidly.

Referring to the maximum entropy characterization, the von Mises process can be associated with a particular measure of angular correlation. When this correlation coefficient was compared with other measures in the literature, it clearly exhibited the best performance in a number of Monte-Carlo studies, apart from possessing such favourable properties as simplicity and linearity. The definition of an autocovariance function of a directional process was therefore based on this measure of association.

Having introduced the two processes of circular variables, the main problem was to fit these models to time series of directional data. In fact, much of Part II was devoted to the estimation of the parameters of a wrapped autoregressive process and hence, to the estimation of the autocovariance function of the underlying unwrapped process. Although the EM-algorithm yielded the best estimates, it became increasingly expensive to implement for large data sets. At the same time, however, the method based on equating empirical and theoretical covariances became more accurate, provided the data were not too close to functional dependence. It involved only a few percentage points loss in efficiency and thus provided an appropriate alternative when estimating the covariances of the unwrapped process underlying the directional component of the residual wind series.

As the different components were removed from the original wind series, a wrapped autoregressive model was fitted to each of the residual series, more precisely to each of the differenced directional components. As the circular variance increased it gave an indication of the importance of the component that had just been removed. It was seen that the land and sea breeze cycle dominated the weather pattern for most of the year. Even in winter it was an important factor beside the geostrophic wind, whose influence did not seem to vary much between summer and winter.

Apart from the particular application presented in this study it should be stressed that the techniques developed for the analysis of the given wind series are generally applicable in the following situations:

(i) robust filtering of multivariate time series,
(ii) classification of the atmospheric pressure field,
(iii) derivation of land and sea breeze patterns, and
(iv) detection of particular events.

Further, the von Mises and wrapped autoregressive processes provide two models for time series of angular data and may be useful in a wide range of applications. We would like to emphasize that it is the wind direction and not the speed which was the dominant aspect of our analysis. Using a polar representation for the wind records enabled us to interpret the results immediately and to separate the angular aspect from wind speed.

Finally, we suggest that a similar analysis be performed for different geographical regions, for example along the west coast of Australia, both to study the variation of the above patterns with respect to location and to examine their typicality for those areas.

BIBLIOGRAPHY

Abramowitz, M. and **Stegun, I.A.** (1966): *Handbook of Mathematical Functions*, U.S. Government Printing Office, Washington.

Ahrens, J.H. and **Dieter, U.** (1971): An Exact Determination of Serial Correlation of Pseudo-Random Numbers, *Numerische Mathematik 17*, 101-123.

Ahrens, J.H. and **Dieter, U.** (1973): Extensions of Forsythe's Method for Random Sampling from the Normal Distribution, *Mathematics of Computation 27*, 927-937.

Ahrens, J.H., Dieter, U. and **Grube, A.** (1970): Pseudo-Random Numbers - A new Proposal for the Choice of Multiplicators, *Computing 6*, 121-138.

Anderson, T.W. (1971): *The Statistical Analysis of Time Series*, John Wiley & Sons, New York.

Anderson, T.W. (1984): *An Introduction to Multivariate Statistical Analysis*, John Wiley & Sons, New York.

Andrews, D.F., Bickel, P.J., Hampel, F.R., Huber, P.J., Rogers, W.H. and **Tukey, J.W.** (1972): *Robust Estimates of Location - Survey and Advances*, Princeton University Press, New Jersey.

Barndorff-Nielsen, O. (1979): *Information and Exponential Families in Statistical Theory*, John Wiley & Sons, New York.

Barndorff-Nielsen, O.E., Blæsild, P., Jensen, J.L. and **Sørenson, M.** (1985): The Fascination of Sand,
in Atkinson, A.C. and Fienberg, S.E. (eds.): *A Celebration of Statistics - The ISI Centenary Volume*, Springer-Verlag, New York, 57-87.

Barnett, V. (1981) (ed.): *Interpreting Multivariate Data*, John Wiley & Sons, New York.

Barnett, V. and **Lewis, T.** (1978): *Outliers in Statistical Data*, John Wiley & Sons, New York.

Bauer, H. (1981): *Probability Theory and Elements of Measure Theory*, Academic Press, London.

Bellman, R. (1960): *Introduction to Matrix Analysis*, McGraw-Hill Book Company, New York.

Best, D.J. and **Fisher, N.I.** (1979): Efficient Simulation of the von Mises Distribution, *Applied Statistics 28*, 152-157.

Billingsley, P. (1965): *Ergodic Theory and Information*, John Wiley & Sons, New York.

Box, G.E.P. and **Jenkins, G.M.** (1976): *Time Series Analysis*, Holden-Day, San Francisco.

Breckling, J. and **Chambers, R.** (1988): M-Quantiles, *Biometrika 75*, 761-771.

Brillinger, D. (1975): *Time Series, Data Analysis and Theory*, Holt, Reinhart & Winston, New York.

Brown, B.M. (1983): Statistical Uses of the Spatial Median, *Journal of the Royal Statistical Society 45B*, 25-30.

Brown, B.M. (1985): Multiparameter Linearization Theorems, *Journal of the Royal Statistical Society 47B*, 323-331.

Bureau of Meteorology (1973): The Climate and Meteorology of Western Australia, *Western Australian Yearbook 12*, 36-54.

Chanda, K.C. (1976): Some Comments on Sample Quantiles for Dependent Observations, *Communications in Statistics - Theory and Methods 5A*, 1385-1392.

Collins, J.R. (1976): Robust Estimation of a Location Parameter in the Presence of Asymmetry, *The Annals of Statistics 4*, 68-85.

Cox, D.R. and **Isham, V.** (1980): *Point Processes*, Chapman & Hall, London.

Cramer, H. and **Leadbetter, M.R.** (1967): *Stationary and Related Stochastic Processes*, John Wiley & Sons, New York.

Dempster, A.P., Laird, N.M. and **Rubin, D.B.** (1977): Maximum Likelihood from Incomplete Data via the EM - Algorithm, *Journal of the Royal Statistical Society 39B*, 1-22.

Denby, L. and **Martin, R.D.** (1979): Robust Estimation of the First-Order Autoregressive Parameter, *Journal of the American Statistical Association 74*, 140-146.

Doob, J.L. (1953): *Stochastic Processes*, John Wiley & Sons, New York.

Downs, T.D. (1966): Some Relationships among the von Mises Distributions of Different Dimensions, *Biometrika 53*, 269-272.

Downs, T.D. (1972): Orientation Statistics, *Biometrika 59*, 665-676.

Downs, T.D. (1974): Rotational Angular Correlations,
in: Ferin, Halberg, Richart, van der Wiele (eds.): *Biorhythms and Human Reproduction*, John Wiley & Sons, New York, 97-104.

Epp, R.J., Tukey, J.W. and **Watson, G.S.** (1971): Testing Unit Vectors for Correlation, *Journal of Geophysical Research 76*, 8480-8483.

Essenwanger, O. (1976): *Applied Statistics in Atmospheric Science*, Elsevier Scientific Publishing Company, Amsterdam.

Estoque, M.A. (1962): The Sea Breeze as a Function of the Prevailing Synoptic Situation, *Journal of the Atmospheric Sciences 19*, 244-250.

Feller, W. (1960): *An Introduction to Probability Theory and its Applications, vol.1*, John Wiley & Sons, New York.

Feller, W. (1966): *An Introduction to Probability Theory and its Applications, vol.2*, John Wiley & Sons, New York.

Fisher, E.L. (1961): A Theoretical Study of the Sea Breeze, *Journal of Meteorology 18*, 216-233.

Fisher, N.I. (1985): Spherical Medians, *Journal of the Royal Statistical Society 47B*, 342-348.

Fisher, N.I. and **Lee, A.J.** (1982): Nonparametic Measures of Angular-Angular Association, *Biometrika 69*, 315-321.

Fisher, N.I. and **Lee, A.J.** (1983): A Correlation Coefficient for Circular Data, *Biometrika 70*, 327-332.

Fisher,N.I. and **Willcox, M.E.** (1978): A useful Decomposition of the Resultant Length for Samples from von Mises-Fisher Distributions, *Communications in Statistics - Simulation and Computation 7B*, 257-267.

Fisher, R.A. (1953): Dispersion on a Sphere, *Proc. of the Royal Society 217A*, 295-305.

Fisz, M. (1973): *Wahrscheinlichkeitsrechnung und Mathematische Statistik*, VEB Deutscher Verlag der Wissenschaften, Berlin.

Gastwirth, J.L. and **Rubin, H.** (1975): The Behavior of Robust Estimators on Dependent Data, *The Annals of Statistics 3*, 1070-1100.

Gebelein, H. (1941): Das statistische Problem der Korrelation als Variations- und Eigenwertproblem und sein Zusammenhang mit der Ausgleichsrechnung, *Zeitschrift für angewandte Mathematik und Mechanik 21*, 364-379.

Gnanadesikan, R. (1977): *Methods for Statistical Data Analysis of Multivariate Observations*, John Wiley & Sons, New York.

Gnedenko, B.W. (1968): *Lehrbuch der Wahrscheinlichkeitsrechnung*, Akademie-Verlag, Berlin.

Gordon, L. and **Hudson, M.** (1977): A Characterization of the von Mises Distribution, *The Annals of Statistics 5*, 813-814.

Gould, A.L. (1969): A Regression Technique for Angular Variates, *Biometrics 25*, 683-700.

Graybill, F.A. (1983): *Matrices with Applications in Statistics*, Wadsworth, Belmont, California.

Haltiner, G.J. (1971): *Numerical Weather Prediction*, John Wiley & Sons, New York.

Hampel, F.R. (1968): *Contributions to the Theory of Robust Estimation*, Ph.D. thesis, University of California, Berkeley.

Hampel, F.R. (1971): A General Qualitative Definition of Robustness, *The Annals of Mathematical Statistics 42*, 1887-1896.

Hampel, F.R. (1973): Robust Estimation: A Condensed Partial Survey, *Zeitschrift für Wahrscheinlichkeitstheorie und verwandte Gebiete 27*, 87-104.

Hampel, F.R. (1974): The Influence Curve and its Role in Robust Estimation, *Journal of the American Statistical Association 69*, 383-393.

Hampel, F.R. (1975): Beyond Location Parameters: Robust Concepts and Methods, *Proc. of the 40th Session, Warsaw, XLVI-1*, 375-382.

Hampel, F.R. (1978): Modern Trends in the Theory of Robustness, *Mathematische Operationsforschung und Statistik 9*, 425-442.

Hampel, F.R. (1980): Robuste Schätzungen: Ein anwendungsorientierter Überblick, *Biometrical Journal 22*, 3-21.

Hampel, F.R., Ronchetti, E.M., Rousseeuw, P.J. and **Stahel, W.A.** (1985): *Robust Statistics - The Approach Based on Influence Functions*, John Wiley & Sons, New York.

Hannan, E.J. (1960): *Time Series Analysis*, Methuen, London.

Hannan, E.J. (1970): *Multiple Time Series*, John Wiley & Sons, New York.

Hannan, E.J. and **Quinn, B.G.** (1979): The Determination of the Order of an Autoregression, *Journal of the Royal Statistical Society 41B*, 190-195.

Hardy, D.M. and **Walton, J.J.** (1978): Principal Components Analysis of Vector Wind Measurements, *Journal of Applied Meteorology 17*, 1153-1162.

Harris, B. (1967): *Spectral Analysis of Time Series*, John Wiley & Sons, New York.

Haslett, J. and **Raftery, A.E.** (1989): Space-Time Modelling with Long-Memory Dependence: Assessing Ireland's Wind Power Resource (with discussion), *Journal of the Royal Statistical Society 38C*, 1-50.

Heathcote, C.R. (1977): The Integrated Squared Error Estimation of Parameters, *Biometrika 64*, 255-264.

Holton, J.R. (1979): *An Introduction to Dynamic Meteorology*, International Geophysics Series , vol. 23, Academic Press, New York.

Houghton, J.T. (1977): *The Physics of Atmospheres*, Cambridge University Press, Cambridge.

Huber, P.J. (1964): Robust Estimation of a Location Parameter, *The Annals of Mathematical Statistics 35*, 73-101.

Huber, P.J. (1972): Robust Statistics: A Review, *The Annals of Mathematical Statistics 43*, 1041-1067.

Huber, P.J. (1973): Robust Regression: Asymptotics, Conjectures and Monte-Carlo, *The Annals of Statistics 1*, 799-821.

Huber, P.J. (1981): *Robust Statistics*, John Wiley & Sons, New York.

Jenkins, G.M. (1979): *Practical Experiences with Modelling and Forecasting Time Series*, Titus Wilson Ltd., Kendal.

Jenkins, G.M. and **Watts, D.G.** (1968): *Spectral Analysis and its Applications*, Holden-Day, San Francisco.

Jensen, J.L. (1980): *On the Hyperboloid Distribution*, Research Report No. 59, Dept. of Theoretical Statistics, University of Århus.

Johnson, R.A. and **Wehrly, T.E.** (1977): Measures and Models for Angular Correlation and Angular-Linear Correlation, *Journal of the Royal Statistical Society 39B*, 222-229.

Johnson, R.A. and **Wehrly, T.E.** (1978): Some Angular-Linear Distributions and Related Regression Models, *Journal of the American Statistical Association 73*, 602-606.

Jupp, P.E. and **Mardia, K.V.** (1980): A General Correlation Coefficient for Directional Data and Related Regression Problems, *Biometrika 67*, 163-173.

Justus, C.G., Hargraves, W.R., Mikhall, A. and **Graber, D.** (1978): Methods for Estimating Wind Speed Frequency Distributions, *Journal of Applied Meteorology 17*, 350-353.

Justusson, B.I. (1979): *Order Statistics on Stationary Random Processes, with Applications to Moving Medians*, Technical Report, Dept. of Mathematics, Royal Institute of Technology, Stockholm.

Justusson, B.I. (1981): Median Filtering - Statistical Properties, in Huang, T.S. (ed.): *Two-Dimensional Digital Signal Processing II - Transforms and Median Filters*, Springer-Verlag, Berlin, 161-196.

Kagan, A.M., Linnik, Yu.V. and **Rao, C.R.** (1973): *Characterization Problems in Mathematical Statistics*, John Wiley & Sons, New York.

Kailath, T. (1977) (ed.): *Linear Least-Squares Estimation*, Dowden, Hutchinson & Ross, Stroudsburg, Pennsylvania.

Kawata, T. (1966): On the Fourier Series of a Stationary Stochastic Process, *Zeitschrift für Wahrscheinlichkeitsrechnung und verwandte Gebiete 6*, 224-245.

Kawata, T. (1972): *Fourier Analysis in Probability Theory*, Academic Press, New York.

Kendall, M.G. (1973): *Time Series*, Griffin, London.

Kendall, M.G. and **Stuart, A.** (1963): *The Advanced Theory of Statistics, vol.1*, Griffin, London.

Kent, J.T. (1977): The Infinite Divisibility of the von Mises-Fisher Distribution for all Values of the Parameter in all Dimensions, *Proc. of the London Mathematical Society 35 (third series)*, 359-384.

Kent, J.T. (1978a): Some Probabilistic Properties of Bessel Functions, *The Annals of Probability 6*, 760-770.

Kent, J.T. (1978b): Limiting Behaviour of the von Mises-Fisher Distribution, *Mathematical Proc. of the Cambridge Philosophical Society 84*, 531-536.

Kent, J.T. (1982): The Fisher-Bingham Distribution on the Sphere, *Journal of the Royal Statistical Society 44B*, 71-80.

Khatri, C.G. and **Mardia, K.V.** (1977): The von Mises-Fisher Matrix Distribution in Orientation Statistics, *Journal of the Royal Statistical Society 39B*, 95-106.

Kleiner, B., Martin, R.D. and **Thomson, D.J.** (1979): Robust Estimation of Power Spectra (with discussion), *Journal of the Royal Statistical Society 41B*, 313-351.

Koopmans, L.H. (1974): *The Spectral Analysis of Time Series*, Academic Press, New York.

Kreyszig, E. (1965): *Statistische Methoden und ihre Anwendungen*, Vandenhoeck & Ruprecht, Göttingen.

Krickeberg, K. (1963): *Wahrscheinlichkeitstheorie*, B.G. Teubner, Stuttgart.

Krishnaiah, P.R. (1977) (ed.): *Multivariate Analysis - IV, Proc. of the Fourth International Symposium on Multivariate Analysis*, North-Holland Publishing Company, Amsterdam.

Kullback, S. (1959): *Information Theory and Statistics*, John Wiley & Sons, New York.

Lorenz, E.N. (1967): *The Nature and Theory of the General Circulation of the Atmosphere*, World Meteorological Organization, Geneva.

Mackenzie, J.K. (1957): The Estimation of an Orientation Relationship, *Acta Crystallographica 10*, 61-62.

Mahrer, Y. and **Pielke, R.A.** (1977): The Effects of Typography on Sea and Land Breezes in a Two-Dimensional Model, *Monthly Weather Review 105*, 1151-1162.

Mak, M.K. and **Walsh, J.E.** (1976): On the Relative Intensities of Land and Sea Breezes, *Journal of the Atmospheric Sciences 33*, 242-251.

Mardia, K.V. (1972): *Statistics of Directional Data*, Academic Press, London.

Mardia, K.V. (1975a): Distribution Theory for the von Mises-Fisher Distribution and its Application, *Statistical Distributions in Scientific Work 1*, 113-130.

Mardia, K.V. (1975b): Statistics of Directional Data (with discussion), *Journal of the Royal Statistical Society 37B*, 349-393.

Mardia, K.V. (1976): Linear-Circular Correlation Coefficients and Rhythmometry, *Biometrika 63*, 403-405.

Mardia, K.V. and **Puri, M.L.** (1978): A Spherical Correlation Coefficient Robust Against Scale, *Biometrika 65*, 391-395.

Mardia, K.V. and **Sutton, T.W.** (1978): A Model for Cylindrical Variables with Applications, *Journal of the Royal Statistical Society 40B*, 229-233.

Maronna, R.A. (1976): Robust M-Estimators of Multivariate Location and Scatter, *The Annals of Statistics 4*, 51-67.

Martin, R.D. (1978): Robust Estimation of Autoregressive Models,
in Brillinger, D.R. and Tiao, G.C. (eds.): *Directions in Time Series, Proc. of the IMS Special Topics Meeting on Time Series Analysis*, Iowa State University, Ames, 228-254.

Miller, R.G. (1962): *Statistical Prediction by Discriminant Analysis*, American Meteorological Society, Meteorological Monographs 4, No. 25.

Neumann, J. and **Mahrer, Y.** (1971): A Theoretical Study of the Land and Sea Breeze Circulation, *Journal of the Atmosperic Sciences 28*, 532-542.

Olkin, I. and **Press, S.J.** (1969): Testing and Estimation for a Circular Statistical Model, *Annals of Mathematical Statistics 40*, 1368-1378.

Otness, R. and **Enochsen, L.** (1972): *Digital Time Series Analysis*, John Wiley & Sons, New York.

Palmen, E. and **Newton, C.W.** (1969): *Atmospheric Circulation Systems*, Academic Press, New York.

Physick, W. (1966): A Numerical Model of the Sea Breeze Phenomenon over a Lake or Gulf, *Journal of the Atmospheric Sciences 33*, 2107-2135.

Pielke, R.A. (1974): A Three-Dimensional Numerical Model of the Sea Breezes over South Florida, *Monthly Weather Review 102*, 115-139.

Portnoy, S.L. (1977): Robust Estimation in Dependent Situations, *The Annals of Statistics 5*, 22-43.

Portnoy, S.L. (1979): Further Remarks on Robust Estimation in Dependent Situations, *The Annals of Statistics 7*, 224-231.

Puri, M.L. and **Rao, J.S.** (1977): Problems of Association for Bivariate Circular Data and a new Test of Independency,
in Krishnaiah, P.R. (ed.): *Multivariate Analysis - IV, Proc. of the Fourth International Symposium on Multivariate Analysis*, North-Holland Publishing Company, Amsterdam, 513-522.

Rao, C.R. (1973): *Linear Statistical Inference and its Applications*, John Wiley & Sons, New York.

Rényi, A. (1959): On Measures of Dependence, *Acta Mathematica Academiae Scientarium Hungaricae 10*, 441-451.

Riordan, J. (1958): *An Introduction to Combinatorial Analysis*, John Wiley & Sons, New York.

Riordan, J. (1968): *Combinatorial Identities*, John Wiley & Sons, New York.

Rivest, L.P. (1982): Some Statistical Methods for Bivariate Circular Data, *Journal of the Royal Statistical Society 44B*, 81-90.

Rosenblatt, M. (1974): *Random Processes*, Springer-Verlag, New York.

Rothman, E.D. (1971): Tests of Coordinate Independence for a Bivariate Sample on a Torus, *The Annals of Mathematical Statistics 42*, 1962-1969.

Rozanov, J.A. (1967): *Stationary Random Processes*, Holden-Day, San Francisco.

Rye, P.J. (1980): Environmental Applications of a Numerical Sea Breeze Model, *Proc. of the Second Conference on Coastal Meteorology of the American Meteorological Society*, Los Angeles, 10-15.

Schmetterer, L. (1966): *Mathematische Statistik*, Springer-Verlag, Wien.

Serfling, R.J. (1980): *Approximation Theorems of Mathematical Statistics*, John Wiley & Sons, New York.

Smart, W.M. (1962): *Textbook on Spherical Astronomy*, Cambridge University Press, Cambridge.

Steedman, R.K. and **Craig, P.D.** (1979): *Numerical Model Study of Circulation and other Oceanographic Aspects of Cockburn Sound*, Report No. 64, R.K. Steedman & Associates, Perth.

Stephens, M.A (1962a): Exact and Approximate Tests for Directions I, *Biometrika 49*, 463-477.

Stephens, M.A (1962b): Exact and Approximate Tests for Directions II, *Biometrika 49*, 547-552.

Stephens, M.A. (1964): The Testing of Unit Vectors for Randomness, *Journal of the American Statistical Association 59*, 160-167.

Stephens, M.A (1969): Multi-Sample Tests for the Fisher Distribution for Directions, *Biometrika 56*, 169-181.

Stephens, M.A (1979): Vector Correlation, *Biometrika 66*, 41-48.

Steutel, F.W. (1979): Infinite Divisibility in Theory and Practice (with discussion), *Scandinavian Journal of Statistics 6*, 57-64.

Tennekes, H. and **Lumley, J.L.** (1972): *A First Course in Turbulence*, MIT Press, Cambridge, Massachusetts.

Tukey, J.W. (1960): A Survey of Sampling from Contaminated Distributions, in Olkin, I. (ed.): *Contributions to Probability and Statistics*, Standford University Press, Palo Alto, 448-485.

Tukey, J.W. (1977): *Exploratory Data Analysis*, Addison-Wesley, Reading, Massachusetts.

Walters, P. (1975): *Ergodic Theory*, Lecture Notes in Mathematics No. 458, Springer-Verlag, Berlin.

Watson, G.N. (1952): *A Treatise on the Theory of Bessel Functions*, Cambridge University Press, Cambridge.

Watson, G.S. (1960): More Significance Tests on the Sphere, *Biometrika 47*, 87-91.

Watson, G.S. (1966): The Statistics of Orientation Data, *The Journal of Geology 74*, 786-797.

Watson, G.S. (1967): Another Test for the Uniformity of a Circular Distribution, *Biometrika 54*, 675-676.

Watson, G.S. (1982): Distributions on the Circle and Sphere,
in Gani, J.M. and Hannan, E.J. (eds.): *Essays in Statistical Science*, Journal of Applied Probability 19A (special volume), 265-280.

Watson, G.S. and **Beran, R.J.** (1967): Testing a Sequence of Unit Vectors for Serial Correlation, *Journal of Geophysical Research 72*, 5655-5659.

Wehrly, T.E. and **Johnson, R.A.** (1979): Bivariate Models for Dependence of Angular Observations and a Related Markov Process, *Biometrika 66*, 255-256.

Zurmühl, R. (1965): *Praktische Mathematik*, Springer-Verlag, Berlin.

LIST OF SYMBOLS

$\mathcal{N}$ set of all positive intergers.
$\mathcal{N}_0$ set of all non-negative intergers.
$\mathcal{Z}$ set of all intergers.
$\mathcal{R}$ set of all real numbers.
$\mathcal{R}^+$ set of all positive real numbers.
$\mathcal{R}_0^+$ set of all non-negative real numbers.
$\Re(\cdot)$ real part of a complex number.
$(\mathbf{w}_t)$ time series of observed wind speed and direction.
$(\mathbf{g}_t)$ geostrophic component obtained through filtering.
$(\mathbf{d}_t)$ daily component.
$(\mathbf{b}_t)$ land and sea breeze component.
$(\mathbf{r}_t)$ residual wind component before removal of storms.
$(\mathbf{e}_t)$ storm component.
$(\mathbf{f}_t)$ residual series of short-term fluctuations.
$(\mathbf{p}_t)$ time series containing central pressure and location of low and high pressure systems.
$(\mathbf{h}_t)$ time series of geostrophic winds estimated from atmospheric pressure configuration.
0900 h 9am on the 24-hour clock.
$\mathbf{r}$ position vector.
$\mathbf{k}$ vertical unit vector.
$\mathbf{\Omega}$ angular velocity of the earth.
ψ longitude.
ϕ latitude (negative in the Southern Hemisphere).
p atmospheric pressure.
p_0 = 1013.25 mb standard sea level pressure.
ϱ = 1.225 kg m^{-3} atmospheric density.
R_0 = 6378 km radius of the earth.
$f = 2\,\|\mathbf{\Omega}\|\,\sin\phi$ Coriolis parameter.
ψ_0 = 116° longitude of Fremantle.
ϕ_0 = −32° latitude of Fremantle.
$c_L = 3\,(200\text{ km})^2$ low pressure system parameter.
$c_H = 3\,(400\text{ km})^2$ high pressure system parameter.
$\mathcal{P}(A)$ probability of the event A.
$\mathbf{U}_\theta$ random unit vector in the direction of θ.

$\mathcal{E}[\cdot]$ expectation of a random variable.
$\mathrm{var}(\cdot)$ variance of a random variable.
$\mathrm{cov}(\cdot,\cdot)$ covariance of two random variables.
$c(\cdot)$ circular autocorrelation function based on ρ_Q, cf. equation (8.4).
$C(\cdot)$ autocovariance function of a stationary autoregressive process.
$\phi(\cdot;C)$ characteristic function depending on the parameter C.
$\mathcal{F}(C)$ Fisher information of a random variable depending on the parameter C.
$N(\mu,\sigma^2)$ normal distribution with mean μ and variance σ^2.
$vM(\mu,\kappa)$ von Mises distribution with mean direction μ and concentration κ.
$WN(\mu,\sigma^2)$ wrapped normal distribution with mean μ and variance σ^2.
$\sim$ distributed as.
$\overset{\text{approx.}}{\sim}$ approximately distributed as.
$\overset{\text{d}}{\to}$ convergence in distribution.
$\overset{\text{pr}}{\to}$ convergence in probability.
$\overset{\text{asympt.}}{\approx}$ asymptotically approximately equal to.
$\arg(\mathbf{v})$ direction of the vector $\mathbf{v} \in \mathcal{R}^2$.
$\mathbf{I}_k$ identity matrix of order k ($k \in \mathcal{N}$).
$\mathrm{tr}\,\mathbf{A}$ trace of the square matrix $\mathbf{A} \in \mathcal{R}^{k\times k}$ ($k \in \mathcal{N}$).
$\mathbf{A}'$ transpose of the matrix $\mathbf{A} \in \mathcal{R}^{k\times l}$ ($k,l \in \mathcal{N}$).
$\Gamma(\cdot)$ gamma function.
$I_n(\cdot)$ modified Bessel function of the first kind and order n, $n \in \mathcal{N}_0$.
$A_n(\kappa)$ $= I_n(\kappa)/I_0(\kappa)$, $n \in \mathcal{N}_0$ and $\kappa \in \mathcal{R}_0^+$.
$J_0(\cdot)$ Bessel function of the first kind and order 0.
$\square$ end of example or proof.
$\delta(\cdot,\cdot)$ measure of angular association.
ρ_{WB} Watson & Beran's measure of association, cf. equation (7.7).
ρ_{St} Stephens' measure of association, cf. equation (7.8).
ρ_{DM} Downs and Mardia's measure of association, cf. equation (7.9).
ρ_{JW} Johnson & Wehrly's measure of association, cf. equation (7.10).
ρ_{JM} Jupp & Mardia's measure of association, cf. equation (7.12).
ρ_{FL} Fisher & Lee's measure of association, cf. equation (7.13).
σ_P proposed measure of association, cf. equation (7.14).
ρ_P corresponding correlation coefficient, cf. equation (7.15).
σ_Q measure of association proposed for directional time series, cf. (7.16).
ρ_Q corresponding correlation coefficient.
c_{WB} circular autocorrelation function based on ρ_{WB}.
c_{DM} circular autocorrelation function based on ρ_{DM}.
c_{JW} circular autocorrelation function based on ρ_{JW}.
c_{JM} circular autocorrelation function based on ρ_{JM}.
c_{FL} circular autocorrelation function based on ρ_{FL}.
C^{EQ} estimator of C using *EQ*, cf. equation (9.3).
C^{EC} estimator of C using *EC*, cf. equation (9.9).
C^{BA} estimator of C using *BA*, cf. equation (9.17) and (9.18).
C^{ML} estimator of C using *ML*, cf. equation (9.20).
C^{CF} estimator of C using *CF*, cf. equation (9.32).

SUBJECT INDEX